Mapping Modernities

When the communist governments of Eastern Europe and the Soviet Union collapsed between 1989 and 1991, there was a revived interest in a region that had been largely neglected by western geographers. *Mapping Modernities* draws on the resulting work and other original theoretical and empirical sources to describe, interpret and explain the place and spatial order of modernities in Central and Eastern Europe since 1920, to give a theoretically underpinned, regional geography of the area.

The book interprets the geography of Central and Eastern Europe from 1920 to 2000 in terms of spatial modernity. It details the individual and collective development of places produced within the three modernising projects of Nationalism, Communism and Neo-liberalism. These ideologies are seen as three geo-historical, time/space conjunctions of modernity competing for the hegemonic imposition of order onto the maelstrom of ideas, people and events in the central–eastern territories of Europe. Within each geo-historical period, localities, regions, states, Europe and the globe are systematically debated as constructed and contested places.

Spatial modernity theorises an experience of time and space mediated through place. Place is conceptualised as bounded, multi-scalar, meaningful, relational space. Place expresses individual and environmental development in ideas and practices. Spatial modernity directs attention to the multi-scalar properties of place, place particularities and place relations that are embedded in socio-cultural and locational attitudes and values.

Alan Dingsdale is Principal Lecturer in the Department of International Studies at Nottingham Trent University.

Routledge Studies in Human Geography

This series provides a forum for innovative, vibrant, and critical debate within human geography. Titles will reflect the wealth of research that is taking place in this diverse and ever-expanding field.

Contributions will be drawn from the main subdisciplines and from innovative areas of work which have no particular subdisciplinary allegiances.

Mapping Modernities
Geographies of Central and Eastern Europe, 1920–2000

Alan Dingsdale

London and New York

First published 2002
by Routledge
11 New Fetter Lane, London EC4P 4EE

Simultaneously published in the USA and Canada
by Routledge
29 West 35th Street, New York NY 10001

Routledge is an imprint of the Taylor & Francis Group

© 2002 Alan Dingsdale

Typeset in Galliard by
Florence Production Ltd, Stoodleigh, Devon

Printed and bound in Great Britain by
MPG Books Ltd, Bodmin

British Library Cataloguing in Publication Data
A catalogue record for this book is available from the British Library

Library of Congress Cataloging in Publication Data
Dingsdale, Alan.
  Mapping modernities: geographies of Central and Eastern
  Europe, 1920–2000/Alan Dingsdale.
    p. cm.
  Includes bibliographical references and index.
  1. Human geography – Europe, Eastern.   2. Spatial
  behaviour – Europe, Eastern.   3. Europe, Eastern – Historical
  geography.   4. Nationalism – Europe, Eastern.   5. Europe,
  Eastern – Boundaries.   I. Title.
GF645.E92 D56 2001
304.2′0947–dc21   2001034994

ISBN 0–415–21620–6

This book is dedicated to Zsuzsi, my wife, and Simon, our son

Contents

Figures

Tables

Acknowledgements

In May 1980, whilst preparing a new course on the Geography of Eastern Europe, I visited Hungary, supported by a grant from the British Council. Within two years I had married a Hungarian and become a two or three times a year visitor to Hungary. Academic contacts made on the first visit multiplied and I worked closely with several Hungarian geographers. I also made a wide range of acquaintances outside the academic sphere and discussed with them topics of current mutual concern. After 1989, when the European Commission promoted academic contacts by means of the Trans-European Programme for University Studies (TEMPUS), my contacts intensified and widened. I became the director of two TEMPUS projects in Hungary and soon had a similar position in a third in Poland. This last introduced me to many new academic colleagues.

In the early 1990s through the initiative of Dr Derek Hall, the Institute of British Geographers set up the Working Party on Post-Socialist Areas. After a year or so I became its secretary. As the forerunner of the present Post-Socialist Areas Research Group, the Working Party attracted an increasing number of geographers, mostly with a new interest in the region. My role brought me into contact with many of them.

Over the period since 1980 the ideas and practices of Geography in Western Europe and the United States of America have changed a great deal. I would say that geography, as an academic discipline, has fragmented into a great number of subdisciplines. These subdisciplines have rigorously developed and pursued research ever closer to other academic disciplines with which they have overlapped. Geography has hardly maintained a real sense of central unity. At the same time geography has moved from being a discipline whose debates and practices were relatively undertheorised to an arena of contest between many theoretical positions. These trends have been the source of much formal debate in the literature. In my case, the debate has been more informal, and conducted with colleagues drawn from a wide variety of academic disciplines within and outside geographical sciences.

This book owes a great deal to all those with whom I have spoken over the past twenty years in all of these contexts. As regards the practical task of shaping and organising the preparation of the book there are some particular individuals and institutions whose contribution I would like to particularly acknowledge. Much of the research for the book was conducted in a period of sabbatical leave awarded by the Faculty of Humanities, Nottingham Trent University, Sabbatical Leave Programme. During that time I spent three months in Hungary and was assisted by the Geography Division Research Fund. Whilst in Hungary, the Hungarian Geographical Research Institute provided me (not for the first time) with a room and a computer. Mrs Judit Simonfalvi, the Institute's Librarian, and her assistant were painstakingly helpful. Dr Zoltan Kovács was the cornerstone of the whole operation. I thank him warmly. Whilst in Hungary, conversations with Dr Attila Karampai of the Department of Economic Geography of the Budapest University of Economic Sciences, and Professor György Enyedi, Vice-President of the Hungarian Academy of Sciences, were especially fruitful. Nearer to home Linda Dawes of Belvoir Cartographics and Design did her, usual, excellent work in preparing the figures, as did Janet Elkington in preparing the tables. Professor Brian Chalkley of the University of Plymouth, Director of the Geography Subject Centre, and my colleague Professor John Tomlinson played crucial roles as sounding boards for ideas and in commenting on early drafts of the text. The two anonymous reviewers also made valuable comments. Andrew Mould and his editorial assistant Ann Michael at Routledge worked effectively in shepherding the book along.

Whilst I acknowledge the help of all these individuals and others, if there are any errors of fact or interpretation of views present in the book, they are entirely my own.

The author and publishers would like to thank the following for granting permission to reproduce material in this work: Akadémiai Kiadó, Budapest, for Figure 2.1 by Bennett, on page 37 of *Social Change and Urban Restructuring in Central Europe*, edited by G. Enyedi, 1998 (Figure 8.1). Blackwell Publishers, for the figure 'Emerging and Frontier Markets', by Sidaway and Pryke, from the *Transactions of the Institute of British Geographers*, Vol. 25 (2) pp. 192–3, 2000 (Figure 16.3). The Geographical Association for figures from 'Redefining Eastern Europe', by A. Dingsdale, in *Geography*, Vol. 84 (3), pp. 205–8, 1999 (Figures 7.1, 11.1, 13.5 and 15.4). Dr Zoltán Hajdú, Hungarian Academy of Sciences, for Figure 2 from the chapter 'Geography and Reforms of Administrative Areas in Hungary' in *Geographical Essays in Hungary*, edited by G. Enyedi and M. Pécsi, 1984 (Figure 9.3). HarperCollins Publishers, for Figure 1 on page 441, Figure 21 on page 475 and Table 14 on page 473 (from *Regional Statistical Year Book*), by A. Dingsdale in the chapter 'Socialist Industrialisation in Hungary', in *Societies, Choices and Economies*, edited by F. Slater, 1991 (Figures 9.1 and 9.2, and Table 9.1). Dr Karoly Kocsis, for an adaptation of Figure 4 in his book, *Jugoszlávia: Egy felrobbant etnikai mozaik esete*, 1993 (Figure 5.3). Kartográfiai Vállalat, Budapest, for a map from *Képes politikai és Gazdasági Világatlasz*, 1966 (Figure 6.3). Kluwer Academic

Publishers, The Netherlands, and Professor U. Sailer-Fliege, for Figure 1 by U. Sailer-Fliege, from *GeoJournal*, 49: 1, 1999 (Figure 8.2). Lock Haven University International Review for the figure 'Republic boundaries and ethnic settlement in Yugoslavia, *c.* 1990', by A. Dingsdale, in the *Lock Haven University International Review*, p. 19, 1999 (Figure 14.2). Macmillan Ltd, for the figure 'Marchlands of Civilisations', by Hupchick and Cox, in *A Concise Historical Atlas of Eastern Europe*, 1996 (Figure 2.2). Osiris Kiadó Publishers, Budapest, for material from pages 321, 607 and 665 of *Kösztes-Európa 1793–1993*, compiled by L. Pándi, 1997 (Figures 3.2, 5.1 and 13.4). Phaidon Press, for the figure 'Redesigning Bucharest under Systematazare', from *Razing Roumania's Past*, by Giurescu, 1990 (Figure 8.3). Püski Publishers, Budapest, for material from *The Atlas of Central Europe*, edited by A. Rónai, 1993 (Figures 3.2, 3.3, 4.1, 5.2, 5.4a and 5.4b). Dr Craig Young, Manchester Metropolitan University, for the figure 'Urban Governance and Local Economic Development in Lodz, Poland', from *Urban Regeneration and Development in Post-Socialist Towns and Cities*, edited by A. Dingsdale, Trent Geographical Papers, number 2, pp. 32–51, 1997 (Figure 12.3). The United Nations for material from the *World Development Report*, 1999 (Figure 16.1 and Table 16.1).

Every effort has been made to contact copyright holders for their permission to reprint material in this book. The publishers would be grateful to hear from any copyright holder who is not here acknowledged and will undertake to rectify any errors or omissions in future editions of this book.

Alan Dingsdale
February 2001

Marchland: land comprising the marches of a country; a border territory; border-land, frontier land.

> (*Oxford English Dictionary*, Second edition,
> Vol. IX 1989)

Marches: Boundary or debatable strip between countries.

> (*Oxford Pocket Dictionary* 1961)

There are two 'marchland' or border regions making patchworks of small states on the present map of Europe: . . . the eastern marchland region between Russia and Germany is the larger and more complicated of the two, and its origins and growth are worth studying.

> (H. Wanklyn, *The Eastern Marchlands of Europe* 1941:3)

The fall of the Berlin Wall threw Eastern Europe into social, political and cultural turmoil. As old alliances break down, East and West are jockeying for position across the fault line at the edge of Europe . . . Poland, Belarus and Ukraine – disputed lands . . . are emerging from Communism to find their new international identities.

> (Alan Little, Assignment – 'Disputed Borderlands',
> BBC 2, 5.11.1994)

The question of 'place', at various scales and in various guises is a topic of widespread concern. And so too are the questions of boundaries, borders and spatiality more generally. And this has raised issues of theoretical approach, and even theory itself; of the conceptualisation of places and their practical definition; and of what should be studied 'within' them. Today these questions are at the centre of the agenda of economic and social geography, as well as integral to developments in social and cultural studies.

> (J. Allen, D. Massey, A. Cochraine,
> *Rethinking the Region* 1998:1)

Modernity, an idea so all-embracing and deeply embedded in our cultural self-understanding as to provide the implicit context against which other analytical descriptions must apparently situate themselves: 'western modernity', 'capitalist modernity', 'post-modernity', and so 'global modernity'.

> (J. Tomlinson, *Globalisation and Culture* 1999:32)

To be modern is perceived as being essentially positive, it is about 'moving with the times', being up to date, following the latest fashion or using the newest gadgets.

> (P. Taylor, *Modernities* 1999:1)

Introduction

Geographies, modernity and transformations in Central and Eastern Europe

The collapse of communist regimes in Eastern Europe and the Soviet Union between 1989 and 1991 was hailed by Fukuyama (1992) as 'the end of History' and by Johnston (1994) as 'the ascendancy of Geography'. It looks as though neither of these claims is valid, but there is no doubt that the dramatic events of that short period had a radical effect. There was a burst of interest among western geographers in a part of Europe that had before attracted little attention. More importantly, during the 1990s the geographical realities and meanings for the people, almost 400 million of them, who live in the area have been subjected to dramatic changes.

This book explores three sets of geographies that have been constructed after 1920 in the area usually referred to as Central and Eastern Europe. Each set of geographies is distinctive. Each was constructed from new structuring of locational, societal and cultural dimensions, promoted by clearly formed projects, that challenged the attitudes and values which made up the antecedent ruling order. Each redrew the contours of geographical space and place. These are the Nationalist Project (1920–39), the Communist Project (1945–90) and the Neo-liberalist Project (1990–).

The three sets of geographies are, however, linked in several ways. First, the material artefacts of places are rarely, if ever, completely obliterated from the landscape. There are always remnants; reminders of how things used to be. Second, new principles, ideas and projects, however dynamic, are never wholly accepted by everyone. All antecedent world views have their adherents, either on the grounds of vested interests, principled support, or inertia. Radicals always encounter conservatives in their attempts to reconstruct spatial, societal, cultural and environmental relationships. All new world views also have contemporary alternatives. Competing and contesting ideas make up the vital processes of constructing and reconstructing geographies.

But, the most significant link between the three sets of geographies is the claim made by the Projecteers for the ideas, the actions and processes that they set in motion. Each new Project – Nationalist, Communist and Neoliberal – was motivated by 'modernity', looking to the future, destroying the old to develop new locational, societal and cultural relationships and so constructing new geographies. They all claimed to be modernisers.

Thus the theme of the book is how three distinctive, but linked conceptualisations of 'modernity' that were imposed on Central and Eastern Europe in the twentieth century, constructed and reconstructed three new sets of geographies, in three geo-historical periods, that I will call three spatial modernities.

These themes arise from discourses around two concepts that are of a very high order of generalisation and that are, also, highly contested as to their meaning – Geographical Space and Modernity. I will not attempt an exhaustive analysis of either in this book. I will try to open some avenues of connection between them that, through new ideas and new practices, produced new places within Central and Eastern Europe, and redefined and reconceptualised Central and Eastern Europe themselves in European and global arenas.

The approach adopted in the book

Whilst much has been written about post-socialist Europe, there has not been any attempt to present a text that explores the area from the geographical perspective of place and space. It is the aim of this book to make a modest contribution to meeting this need by elaborating the ideas of place and space in relation to the eastern areas of Europe during its 'modern' period, 1920–2000. The book is about the nature and impact of modernity in Central and Eastern Europe as it has been negotiated through and by geographical space. It aims to chart the interplay of geographical space and modernity.

The approach adopted here, therefore, differs, on the one hand from traditional geographies of Central and Eastern Europe, and on the other hand from more recent approaches to post-socialist Europe. Traditionally, geographers approached Central and Eastern Europe using either a country by country survey such as Wanklyn (1941), or by focusing on one or more topical branches, for example industry or planning, such as Dawson (1987). Sometimes both these approaches were used together as separate sections within a single book, for example Pounds (1966) and Turnock (1979).

In my approach neither a topical nor a country by country approach is adopted. Instead, the development of places at different scales is discussed, focusing in turn on localities, regions, states, Europe and the globe. This approach situates places within Central and Eastern Europe, and Central and Eastern Europe themselves, as places, in a multi-scalar setting stressing the main issues and debates about modern development relevant to each scalar category. It stresses place particularities and place relations.

A number of recently published geographical texts contain references to, and analyses of, aspects of the geography of the east of Europe in cultural and societal terms. Elementary texts, of a compendium type, include the third edition of T. Jordan's *The European Culture Area* (1996) and the seventh edition of W. Berentsen's (ed.) *Europe in the 1990s* newly titled

Contemporary Europe (1997). More imaginatively presented and more selective are G. Drake's *Issues in the New Europe* (1995) and Gowland, O'Neill and Reid's (eds) *The European Mosaic* (1995). More advanced texts are D. Pinder's (ed.) *The New Europe* (1998) and B. Graham's (ed.) *Modern Europe* (1998). The former is of a more traditional 'economy, society and environment' type, the later is more reflective of recent trends in geo-cultural analysis.

Four recent books of note have a more focused regional scope. M. Bradshaw's (ed.) (1997) *Geography and Transition in the Post-Soviet Republics*, F. Carter, P. Jordan and F. Rey's (eds) *Central Europe After the Fall of the Iron Curtain*, D. Hall, D. Danta's (eds) (1996) *Reconstructing the Balkans: A Geography of the New Southeast Europe* and D. Turnock's (ed.) (2001) *East Central Europe and the Former Soviet Union*. Whilst the editors have done a good job to marshal contributions into coherent themes, they lack the unison of voice or direction needed.

Staddon (1999) has pointed out that recent writing has generally been either strongly theoretical or specifically empirical. The theoretical work has been notably abstract in its approach, lacking the support of detailed empirical cases. Empirical work, on the other hand, has often lacked any theoretical base. Exceptions to this are, for example, Pickles and Smith's edited (1998) volume *Theorising Transition* and Pávlinek and Pickles' (2000) *Environmental Transitions*. Each in its way has developed theoretical propositions in empirical case material. Their theoretical position, however, has been firmly rooted in regulation theory. My approach attempts to chart a course that offers some new theoretical insight for the theoretician, whilst at the same time providing sufficient detail for the empiricist. It tries to overcome the problem of pure theory without facts, and empirical facts without theory.

There has also been a marked tendency in recent work to focus on the period since 1990. It is this period that has captured a new interest among geographers and social scientists. Within the debates that this has stimulated, some interest has been shown in the communist period (Pávlinek and Pickles 2000; Turnock 1997a), as the question of continuity and discontinuity before and after 1989–90 has attracted theoretical and empirical attention.

My approach embraces the period between the First and Second World Wars, as well as the communist period. This is because I argue that modernity had its full impact in Central and Eastern Europe after 1920. Thus, in terms of modernity, as an experience of practice and a way of thinking, the First World War imposes a key discontinuity on spatial development, forming a geo-historical period, in this area of Europe. In the period before the First World War, Central and Eastern Europe had only fragmented signs of modernity in the modernisation of societal and cultural relations, though it was constructed as a distinctive place in the European and global arena of western modernity. After 1920, first the Nationalist Project, then the Communist Project and finally the Neo-liberalist Project shaped development by a commitment to modernisation that was driven by the search for new futures, rather than the conservatism of past traditions. Each Project sought to clear out the past. However, each Project actually looked to the past as one important means of creating its vision of the future, and imagined the future as a new remaking of the past, mainly by clearing out most of what was there. Theorising spatial development around modernity in Central and Eastern Europe demands that attention is given to the whole period since 1920. It also implies that an understanding of the present is to be found in this geo-historical period. The approach brings Central and Eastern Europe into the geographical debate on modernity.

My approach aims to bring together theory and empirical, factual detail related to space, place and place relations. This is intended to complement and extend the above selection of texts by offering a theoretically informed regional geography. Throughout the writing of the book, however, I have felt a tension between the theoretical and regional aims. I have struggled with a balance of regional factual content and theoretical discussion. A secondary tension was associated with the time period and its triadic division. How much attention could be given to each distinct period since 1920? I had hoped originally to give the same multi-scalar framework to each period. In the end, however, editorial guidance led to about half the book being devoted to the post-1990 period. This meant that whilst in the neo-liberal period a chapter could be devoted to each scale of analysis, in the earlier periods this was not possible. Therefore in the nationalist and communists periods I have had to run together a discussion of regions and state, and also European and global scales. I hope that this has not produced a serious imbalance in my references to either the material practices or the discourses that produced these places.

The structure of the book

Mapping Modernities sets out the geographies of Central and Eastern Europe, from 1920 to 2000, as Marchlands of Modernities in three geo-historical periods, three spatial modernities. Marchlands are disputed border areas and I use the term in the sense of material practices and discourses to interpret the changing geography of Central and Eastern Europe. To achieve this the book is arranged in four parts. Part 1 sets out, first of all, the theory of spatial modernity that forms the theoretical grounding of my interpretation. It then seeks to situate Central and Eastern Europe, constructed as places, within several academic discourses, stressing the ambiguities of bounding and defining them. It provides six key features that have given identity and meaning to Central and Eastern Europe as geo-historical Marchlands places. Part 2 explores the Nationalist Project, Part 3, the Communist Project and Part 4, the Neo-liberal Project. Each Project is a specific geo-historical period, a time/space conjunction mediated through multi-scalar place and place relations – a spatial modernity. Each part follows a similar format. The first chapter in each part describes the main socio-cultural features of the particular modernity and establishes the boundaries of the Marchlands in that period. This is followed by discussions of the ideas and practices, the thoughts and actions, that have created new places at different scales – local, regional, state, European and global – by developing the particularities of places, and resituating them in place relations. Each chapter examines the production of places in two principal modes. Places as material and spatial practices, and places as perceived and imagined in discourses. The balance between these modes varies, but both modes of producing places are elaborated from two perspectives. First, the perspective of place development as a lattice of experienced, perceived and imagined socio-cultural spatiality that changes place particularities. Second, from the perspective of insiders and outsiders in a multi-scalar renegotiation of place relations. At all times attention is paid to contemporary commentaries as well as later academic representation. Together these modes and perspectives create new places, embedded in the new socio-cultural modernities, as they have transformed and modernised historical legacies and antecedent spatialities. Finally, a short conclusion brings together the threads and restates the essential features of the argument that I am making in the book.

Part 1

Geography, modernity and Central and Eastern Europe as Marchlands

1 Geographical space, modernity and spatial modernity

Geographical space and place

Geographers have always been interested in people, space and environment. From these three categories emerges an idea of two central pillars of geographical interest: the relationship between people and space, and the relationship between people and environment. This has been resolved in geography for many decades in the idea of place.

Place/space dilemmas and place/space tensions

Recent attention has been paid to the relationship between place and space. M. Crang (1998) talks about place or space, P. Taylor (1999) talks about space/place tensions.

Crang poses 'Place *or* Space?' (my emphasis) as an important question for cultural geography and reveals the underlying theory of cultural approaches as dilemmas over a sense of place. He contrasts the space of the nomothetic approach with the place of the idiographic approach. People do not just locate themselves in space, they define themselves by a sense of place. Places are not just locations on the globe, they stand for a set of cultural characteristics. Places emerge from spaces as they become 'time-thickened'. They have a past and a future that allow people to identify themselves, share experiences and form communities. Individuals are shaped by the places they know and they reproduce them through their cultural identity informed by a 'genius loci'. Place-centred knowledge leads to personal and group identity through the creation of local bounded territories to reinforce a sense of them and us, insiders and outsiders. Current trends in physical and electronic communications are homogenising places and creating global space, leading to a 'loss of place' and the creation of 'non-places'. An industry is growing up to manufacture places, to create pseudo-places. But perhaps the locational link between a single culture and a single place is no longer an appropriate idea for understanding human experience.

Taylor has a different perception, he recognises place/space tensions. He complained that in social science space and place were too often treated as the same and furthermore were regarded as 'dead' – merely containers for social and historical processes. He followed Yi-Fu Tuan in asserting that space and place are different. The essential contrast between the ideas of place and space for Taylor is related to abstraction. Space is general, place is particular; space is everywhere, place is somewhere. Space is meaningless, place is meaningful. He also contended that place was widely seen as local, leading to the idea that space is the stable framework that contains places. He rejected this and again cites Tuan, 'Places exist at different scales'. He acknowledges that space is generally thought of as occurring at different scales, but 'providing place with the same multi-scalar property means that relations between space and place can be explored beyond the local up to and including the geographical limit of the whole Earth as both place and space . . . the same location can be both a place and a space; everywhere has the potential for being both place and space. This can be historical when a space is transformed into a place or vice-versa, or contemporaneous when the same location is viewed from different perspectives' (Taylor 1999: 99). This he asserts creates a 'place/space tension' as there are producers of space and makers of place.

Space/place transpositions – the idea and usage of place and space in this book

As these two examples show, understandings of place and space in geography are clearly contested. Crang wrestles with the competition between space and place as global and local. Taylor unifies place and space through the idea of degrees of abstraction and gives them scalar properties. How are space and place related? How do places and spaces come into existence? How do they become endowed with attributes?

Space and place are different, intertwined, transpositional concepts. Places exist within spaces, and spaces exist within places. Space, that is everywhere, is transposed into place, that is somewhere, when it is differentiated by formal and functional bounding, scaling and naming, that define and delimit meaning, that is embedded in socio-cultural ideas and practices. Bounding, scaling and naming are imposed by individuals, organisations (such as political parties or commercial corporations), and institutions (such as states or the law), that have a particular perspective or a particular purpose. These human agents construct an objective spatial order of differentiated places, using a subjective mixture of experience, perception and imagination of locational, societal and cultural values and attributes. In these ways, the multi-scalar place ordering of space transposes continuous space into discontinuous places.

Places created by bounding, scaling and naming impose order on space by practices, on the ground, by representations on maps, and by thoughts, in the mind. Places are thus produced in material and discursive modes. They are physical attributes of social and spatial practices as they are experienced. They are attributes of perception, that is formed from observation, reflection and interpretation of experience. They are attributes, also, of imagination and conceptualisation that have no direct link with the material world. Each individual, organisation or institution has a self-made spatial order of places and spaces that forms a distinctive spatiality. These subjective spatialities differ. Places as objects in these subjective spatialities become contested. Places are produced in constructed, contested spatialities, here and now, in material and discursive modes.

The differentiation of places is relational. Places exist in relation to other places, from which they are differentiated. The fine art concept of simultaneous contrast adds precision to the general concept of competing spatialities by enabling us to think of a place as being, at once, many different places depending on the other places with which it is simultaneously contrasted. A place is simultaneously many places because it is constructed by contrasts with different other places at the same time, in competing spatialities. The identity and meaning of a place is a layered synthesis of many different places as place particularities.

Places exist not only as internal particularities, but also in external connectivities. The connectivity of places is also embedded in social and spatial practices. There is a complex connectivity between places in transport and communications technologies and the movement of people, commodities and ideas. Places are multi-scalar, multi-structured coherences. Structured coherence was conceptualised by Harvey as functional in the form of urban–regional labour market areas and is associated with a 'spatial fix' regulating capitalist 'creative destruction'. Duncan, Goodwin and Halford extended this idea to include the norms and values of socio-cultural relations (Pávlinek and Pickles 2000). This represents, in fact, a variety of competing functional and formal spatialities.

Terminology of place in geographical discourse

Spatiality gives form to a maelstrom of people, artefacts, events and ideas. Geographers have developed a complex terminology of place divisions. However, too often there is confusion in this terminology and terms are ill-defined and used for convenience in a personal intellectual enterprise. This is unfortunate for it is vital that terminology conveys some common, publicly understandable meaning. Geographical terminology has a surprisingly weak development of scalar conceptualisation.

There has been much debate about place terminology and it continues, but the scalar divisions used in this book are in the mainstream of geographical usage. **Localities** are the places in which people spend most of their lives, enact their daily routines and with which they are most familiar. **Regions** are generally in geographical practice statistical divisions of states. 'Region' is, however, geography's utility player and can be qualified by many adjectives such as formal, functional, dominant, backward, growth or continental and global implying modes of conceptualisation, formation and scalar contexts as theorised in spatiality. **States** are most commonly thought of as comprising a set of public powers and legal institutions of which 'sovereignty', the right to make its own laws, is a critical attribute. Territorial sovereignty is of particular importance in understanding the state as a place. **Continents** are major territorial units of the earth's surface, generally marked by clear physical boundaries. Europe is so marked in the north, south and west. There is no strong physical boundary in the east. The question of identifying an eastern boundary of Europe is one of the key questions permeating this book. **Global** space is the limiting category of geographical space; boundless global space as proposed by globalisation and globality theorists is a single, deterritorialised, place at the apex of the space/place transposition.

Modernity

Modernity as an organising and ordering category is so firmly embedded in western societal and cultural discourse that other categories are referred to it (Tomlinson 1999) – post-modernity, capitalist modernity, global modernity – and I will add spatial modernity.

Cultural, societal and locational modernity

Therborn (1995) defines cultural modernity as 'an epoch turned to the future'. As an epochal concept it had to have: an empirically recognised beginning and end; a definition relating its beginning to major social and cultural change, not used to define it; and relation to its etymological meaning.

He suggests that the second half of the eighteenth century saw the commencement of modernity in Western Europe, but recognised that there were different 'routes to and through modernity' in different locations, societies and cultures. He classified them as:

- European;
- New Worlds, represented primarily, but not exclusively, by America, South as well as North;
- the Colonial Zone stretching from northwestern Africa to Papua New Guinea and the South Pacific;
- countries of externally induced Modernisation, challenged and threatened by the new imperial powers of Europe and America.

These are 'actual existing historical trajectories' that, Therborn argues, can also be regarded as Weberian ideal types. He writes 'Russia from Peter the Great onwards, for instance, contains features of the fourth road, as well as of Europe' (Therborn 1995: 7). Thus he introduces a broad geographical component and hints at the possibility of further articulation and diversity of his routes. I will extend this idea to think in terms of distinctive trajectories of subcontinental, state, regional and local modernities.

It is from the ideas of M. Berman's book *All that is Solid Melts into Air* that my idea of modernity is most fully derived. With Berman I take three additional steps in forming my idea of modernity. First, modernity, he writes, is 'an experience – experience of space and time, of self and others, of life's possibilities and perils . . . to be modern is to find ourselves in an environment that promises us adventure, power, joy, growth, transformation of ourselves and the world – and at the same time threatens to destroy everything we have, everything we know, everything we are' (Berman 1983: 15). Modern life is exciting and uncertain.

Second, Berman's brilliant interpretation of Goethe's Faust shows us a profile of modernity in the unifying of human thought and action. Therborn's 'epoch looking to the future' is energised by dynamism and purpose, first by the idea of the Dreamer, then metamorphosed into the Lover, and ultimately into the Developer. The dynamism of development – self-development and environmental development – is a powerful image of individuals, societies and cultures. It is a development that must destroy all vestiges of the past in thought worlds and in material worlds. Development is motivated by a restless, optimistic, impatient arrogance and cowardice spurred on by a sense of insecurity to shape gigantic visions into symbols of material progress.

Third, Berman reveals that this experience is not a profile exclusive to capitalism as is often thought. It is:

> equally central to the collectivist mystique of twentieth century socialism
> . . . The aspirations are universally modern, regardless of the ideology
> under which modernisation takes place.
>
> (Berman 1983: 50)

I shall use this interpretation to unify the three ideologies of nationalism, communism, and neo-liberalism, in tracing the passage of modernities through Central and Eastern Europe.

Development is the centrepiece of Barman's idea of modernity: self-development and environmental development. Development unifies the experience of time and space; is the motivation of self- and institutional change; and permeates all modernities. Berman associated this dynamism primarily with capitalism, but showed that it was also a feature of communism. The dynamism of development is the constant search for newness, a tyranny of novelty, a dynamic of creative destruction.

There is a need to consider Berman's use of the term 'environment'. I understand him to mean not only the physical environment, but also the organisational and institutional environment. Thus I understand him to refer to the organisational and institutional forms within societal, cultural and locational development. That is to say the collective and communal attributes surrounding the individual and within which the individual is an actor.

However, Berman's modernity appears to be aspatial. He writes 'Modern environments and experiences cut across all geographical boundaries . . . in this sense modernity can be said to unite all mankind' (Berman 1983: 15). I assert that modernity is experienced differently in different places, and different places shape the experience and meaning of modernity differently. Place and modernity are reciprocal and dialectically interactive.

Capitalism as modernity and Taylor's Theory of Three Prime Modernities

Berman points out that modernity has been strongly associated with capitalist development and its dynamic of creative destruction. Berman refers to and dismisses 'modernisation theory' as developed by Rostow (1961) as an American imperial strategy for the Third World. Like Berman, Taylor (1989) rejects the 'developmentalism' of modernisation theory. He advocates Wallenstein's world system approach to capitalism as modernity.

Drawing on Braudel and the neo-Marxist Gunder Frank, Wallenstein developed the idea of 'historical systems' to conceptualise societal change (Taylor 1989). Wallenstein recognised three broad categories of system defined in terms of modes of production. **Mini-systems** depend on 'reciprocal lineage' between producers, are made up of kinship groups and are local in scale. **World empires** are based on agricultural surpluses produced by large groups of agricultural workers. These surpluses are in part reciprocally exchanged with artisans but mainly appropriated by a military–bureaucratic, aristocratic–church élite. The **world economy** system is based on the capitalist mode of production. It is driven by profitability and the accumulation of surpluses in the form of capital. The system has a single market, multiple states and a three-tier structure. The geographical dimension of this structure is made up of core, semi-peripheral and peripheral zones that are differentiated on the basis of production systems. It is important to realise that 'world' does not mean global; world means an undetermined spatial extent that is greater than local.

Later, Taylor (1991) developed this theory into a theory of historical regions and then (Taylor 1999) combined these theories with the idea of hegemony derived from Gramsci and the idea of 'modernisation of modernity' derived from Beck, to produce a theory of three prime modernities as geo-historical periods. These were associated with Dutch, British and American hegemonic power, innovation and lifestyle that other societies sought to emulate. These hegemonies were projects that attempted to impose order onto or tame the maelstrom of modern life. The extension of this theory in terms of ordinariness and comfort emphasised the attractiveness of material prosperity and mass cultural forms. It also broke away from the idea of industrialisation as modernity that, according to Taylor, had held social science in thrall. Furthermore it went some way to redressing the balance between the overplayed historical dimension and the underplayed geographical dimension in the debate about modernity.

Taylor's is a powerful theory of modernities, but it is incomplete. It is incomplete because of its reliance on equating capitalism with modernity and because of its weak spatial elaboration. I shall show that communism is a world system and a category of modernity, and misinterpreted by Taylor when he sees it as a 'movement'. I shall also elaborate the spatial dimension through spatiality as outlined above.

Communist modernity – deviant or different?

Inkeles (1976) identified seven traits of 'modern' man that he stressed were universal and associated with modernisation irrespective of cultural or historical setting. They applied to capitalist and socialist societies. Field (1976: 8) commented that these ideas brought developments in the communist world into line with those in 'other advanced industrial countries around the world'. These were: a shift in allegiance from traditional authority, such as family, local community and church, to governments and secular public organisations; increased involvement in public activities; a broadening of individual horizons from parochial to national affairs; an openness to new approaches to life, natural, technical or personal; a belief in knowledge, especially science to overcome nature and solve problems; an awareness and mastery of time – an orientation towards the future; a maximisation of personal development and improved opportunities for the education of his children. These traits anticipate the dynamic of development stressed by Berman and emphasise their applicability to communism as well as capitalism.

Berman pointed out that the dynamic of the developer, the key driving and unifying force of modernity, is central to twentieth century socialism. In what Anderson (1984) calls 'a very important subtext' Berman discussed the idea of communism as modernity. He showed that Marx was a modern thinker and communism a 'modern' ideology and one therefore that was subject to 'melting into air'. Berman recognised that 'Marx's own analysis of the dynamic of modernity undermines the very prospect of the communist future he thought it would lead to' (Anderson 1984: 99). Berman asks 'If all new relationships become obsolete before they ossify, how can solidarity, fraternity and mutual aid be kept alive? A communist government might try to dam the flood by imposing radical restrictions, not merely on economic activity and enterprise ... but on personal, cultural and political expression, but insofar as such a policy succeeded wouldn't it betray Marx's aim of free development for each and all?' (Berman 1983: 104). This question arose in theory. As we shall see, in practice, communist governments transformed from the humanity of Marx, the Dreamer, into the inhumanity of Stalin, the Developer, would impose such restrictions. Paradoxically it was those restrictions that led to ossification and collapse, albeit aided and abetted by the dynamism of capitalist modernity.

Z (1990) referred to Sovietism as a deviant form of modernity. This raises the question of whether there is a standard form of modernity or simply different forms. Holmes (1997: 41/42) points out that whilst 'In theory communists were quintessentially modern', in practice communism was 'atypical regarding actual structures and processes normally associated with "modern" states'. For me this so-called atypicality is a misreading of historical difference. Taylor (1999) did not interpret communism as a modernity at all, but as a movement against industrial capitalism of the British hegemonic modernity. This seems to me a misinterpretation resulting from his over-commitment to the capitalist world economy model in his conception of modernity. Though the Soviet Union as a state was heir to the ideas of Western European civilisation, in its practices it was heir to the trajectories of Eastern European civilisation. Communism is a different mode of modernity. Communism driven by the idea of world revolution had its face firmly set towards the future and was positive, optimistic and humanistic in outlook. In its spatial practices it reproduced some of the most dramatic characteristics of Berman's Developer as the communist project imposed a remote-controlled order on the maelstrom of life.

Communism can be thought of as a world system, because it has a distinctive mode of production that extended beyond the local scale. In contrast to the capitalist world economy communism was not driven by the accumulation of surpluses as capital. Surpluses were accumulated for social redistribution by a party élite. The organisation of production was in marked contrast to that of capitalism. Capitalist private ownership was replaced by public ownership: capitalist devolved decision making was replaced by central decision making; and the market was replaced by planning. The system had multiple states, no markets, and a tiered social and geographical structure.

The communist world system began in the Soviet Union in 1917. It was enlarged to Eastern Europe after 1945, creating Eastern Europe and the Soviet Union as a particular international modernity. It spread to China and was adopted in some Asian and African countries following independence from colonial rule, and in Cuba. At first it depended on the leading role of the Communist Party of the Soviet Union and was an attempted Russian hegemony. From the late 1960s in some European communist countries new forms of evolutionary continuity occurred. The collapse of communism in the Soviet Union and Europe in 1989/91 terminated these continuities and initiated a discontinuity. A second hegemonic power was China. In China emerging market mechanisms are a form of evolutionary communist modernity.

Communism can thus be interpreted as compatible with and an extension to Taylor's theory of geo-historical modernities. In the twentieth century communist modernity competed with capitalist modernity at the local, regional, state, continental and global scales. Communism as a modernity failed in Eastern Europe and the Soviet Union when it was not able to evolve effective societal–cultural forms as it struggled to maintain its essential element – political control by the Party. Its end was hastened by the dynamism of the competing American hegemonic modernity that evolved global forms. Forms that are even more remote controlled than communist practices.

The idea of Europe and modernity

History is constructed in the present. The history of modernity has been constructed in 'the West' purporting to be a global history commencing sometime between the fifteenth and the seventeenth centuries (Tomlinson 1999). The dimensions of this history are so European that Agnes Heller (1992: 12) equates the idea of 'modernity' with the idea of 'Europe': *'Modernity, the creation of Europe, itself created Europe'* (Heller's italics). Habermas regarded modernity as a project. If it is, then it is clearly a European Project for shaping locational, societal and cultural relationships. However, as Therborn says, it was in 'Western Europe' that the definite victory of modernity was first registered. Europe as a cultural category is not the same as Europe as a continental category when viewed from a historical perspective that regards the idea of Europe as the idea of modernity. Continental Europe has various historical trajectories, different routes through modernity in different areas.

Bugge (1993) offers a useful division between 'perceptions of Europe' and 'projects for Europe.' The former are ways of thinking about and understanding the term Europe. The later implies attempts to unite the continent into a single societal and cultural domain. Habermas's reference to modernity as a project conveys a sense of assertion, a promotion, even imposition of a particular structuration of locational, societal and cultural dimensions that shape the meaning of place and press it onto others. Thus the view of Europe as modernity becomes a project to promote a particular idea of Europe, selective in its characterisation of 'Europe'. A discourse on this theme has been pursued among historians and in cultural studies and has recognised that there have been many attempts to unite continental Europe within a single societal and cultural domain. The most successful Project of this kind has been the European Union of the present day. I will return to this theme in Chapter 2.

The idea of Project

The idea of a 'Project' in this book starts with the writing of the German philosopher Frederick Nietzsche. With him I descended into the abyss of nothingness that confronts the mind when all the thought-holds that are used to cling to conventions are stripped away. Then came the realisation that Nietzsche's message was 'construct your own world' and furthermore convince everybody else that yours is the only 'natural' world. Thus invention and assertion lie at the heart of spatiality. Nothingness, I conceptualise as a maelstrom of people, events and ideas in a physical world; a flux of phenomenom; an absence of form; without conventions. Invention is the imposition of order, form and convention onto this mass. Assertion is the declaration of the 'common sense' of that form, the 'hegemony' of that invention as a world view. Inventors may claim that they have discovered the relationship, or that a relationship is revealed to them, and assert it as inalienable truthfulness possessing universal validity. The imposition of order is myriad, it can be literally individual and remain personal. More usually assertions gather disciples who promote, develop and elaborate them. Determined leaders will seize their opportunities to destroy existing conventions and construct the future in thought and action. In this way inventions and assertions emerge as Projects. Grand Projects are attempts to transform inventions by assertion into 'taken for granted', 'self-evident', 'natural', forms of universal relationship that impose order on the maelstrom or the flux. Religion, science, aesthetics and emotion, common sense are examples of such order (Abler, Adams and Gould 1972). A narrower compass identifies Projects with more limited objectives in the field of moral, cultural, societal and spatial relationships – attempts to capture ways of thinking in senses of identity, moral values and attitudes or in economic, political and social relationships. However, since the maelstrom or flux has no permanent form the invention and imposition of order or convention becomes a continual recombination of different elements, a continual reinvention, a continual assertion of reinventions. This ensures that competition between inventions and assertions, and the imposition of order in the form of Projects, is a constant process.

In this book the term Project is used to convey the sense of a set of ideas and conventions that is promoted and elaborated in the arenas of cultural, societal and spatial relations. These Projects each stand for a distinctive course of action. A Project asserts a particular vision of the world, has a strategy for achieving that vision and a number of actors sharing a particular platform. Projects as ideologies are often very specific and inhabit a real world of cultural and societal practices that is conditioned by broader projects claiming universal insight or the revelation of universal truth. All Projects have, and seek to impose, directly or indirectly a distinct spatiality.

Space/time relations

Space/time conjunction, time/space distanciation and space/time compression

Theorists of modernity and of geography have been interested in the relationship between space and time. Places are transpositions of space. In the earlier discussion of place the element of time was hardly referred to. This concentration on space in time/space relations is *time/space conjunction*. It emphasises the conjunction of forces in particular space at a particular time (Allen, Massey and Cochraine 1998). It leads to the idea of space being structured by place and scale. It emphasises space as a particular time slice of locational, societal and cultural relationships that, as experienced, perceived and conceived, is transposed into places. As such it is a vital concept in this book. Places are also situated in time. How long do places last? Societies and cultures come and go and perceptions and conceptualisations of places come and go with them. The practices, attitudes, norms and values by which societies and cultures are structured change over historic time. Space/time is transformed too. An alternative view of time/space relations emphasises time. The idea of *time/space distanciation* is most strongly associated with modernity through the work of Anthony Giddens. It emphasises change in communications technology through time that stretches out social relationships over space. This implies a communications theory of location as the principal explanatory framework operating in geographical space. In historical time, connectivity has been multiplying and speed of connection has been increasing. Improvements in transport technology have been superseded by improvements in communications technology. Movement-free interlocational connectivity has become dominant in some experiences of time/space. This process leads to ideas such as *space/time compression* (Harvey 1989) and the notion that today's electronic networks have collapsed space/time completely producing a borderless, deterritorialised global space.

Geographers have thought of space/time relations in other ways too that are relevant to this book. Place and space as the present in the past; the past in the present; the idea of continuity and change – the interplay of antecedent conditions and current processes; spatial diffusion, the space/ time sequence of the spread of ideas, artefacts and practices. These discourses will be woven into the fabric of the book, sometimes playing a prominent part in the analysis and sometimes more muted. The shifting of relationships that this implies underlines the argument of the book in the broad sense and in the more detailed elaborations.

The production of place, spatial modernity and geo-historic periods

I have now assembled all the main theoretical ideas that underpin the task of this book and so come to the unifying conceptualisation – **spatial modernity.** Geographical space and modernity are separate categories constructing an order from a maelstrom of people, ideas, events and physical environments. Geographical space imposes order as constructed, contested conceptualisations of *place* as a multi-scalar, meaningful experience of time and space. Modernity is a frame of mind looking to create the future. It is an experience of time and space as an arena of individual and environmental development. In neither of these ordering categories is anything fixed; there are no permanent structures. Place is an experience of time and space; modernity is an experience of time and space. There is a dialectical relationship between modernity and geographical space, from which arises the idea of spatial modernity.

Spatial modernity theorises an experience of time and space produced as multi-scalar place ordering, by individual, organisational and institutional development in thought and action expressed as competing spatialities in geo-historical periods.

From the perspective of spatial modernity the key relationship is development, which produces places identified as modern. Place development is interpreted as a process linking spatiality and modernity. Spatiality is a process of development that produces space and place. Spatial modernity is embedded in socio-cultural processes of development. The production of place is a specific development pathway through spatiality. Modernity is mediated by spatiality. Modernity is experienced differently by all individuals because of their situation in place and place relations. Modernity may become universal, but it is experienced through place particularities and place relations.

The nub of my argument is that geographical spatiality brings a new insight into interpreting modernity, by emphasising a hitherto neglected set of place particularities of meaning and multi-scalar place relationships of individual and institutional development that I call spatial modernity. Spatial modernity directs attention to the relational and multi-scalar properties of place to offer a balanced, empirically informed whilst theoretically underpinned discourse. Spatial modernity asserts that modernity is an experience of time and space through place as development. Place is here and now, it is bounded, multi-scalar, interconnected and meaningful space. Spatial modernity draws attention to the thoughts and actions, the ideas and practices, that produce modernity as multi-scalar place relationships of development in the here and now. To the spatiality of individuals, organisations and institutions as locational, societal and cultural actors in imagined, perceived and experienced place/space transpositions. To the constructed, contested production of place that imposes order onto a maelstrom of people, ideas, events and artefacts. Different modernities produced different places and the experience of modernity is different in different places. Places are experienced, perceived and imagined as modern.

Spatial modernity will be used to identify, describe and explain the geographies of Central and Eastern Europe in the years from 1920 to 2000. The First World War is interpreted as a break with the past that inaugurates the modern period in Central and Eastern Europe. Three distinctive geohistorical periods (space/time conjunctions) are identified that are produced and articulated through and by three ideological projects of modernity: the Nationalist Project, the Communist Project and the Neo-liberalist Project. These three Projects produce multi-scalar place in practices and discourses as particular forms of development. Set in a framework of antecedent conditions and current processes spatial modernity offers a dynamic geohistorical theory of multi-scalar renegotiation of spatiality and modernity.

2 Marches and disputed borderlands

What and where are the lands of which we speak?

What and where are the lands of which we speak? To ask this question is to open up a Pandora's box of conflicting and contested spatialities. These lands are marches. Marches are disputed borderlands. Disputes over borderlands cause fierce rivalries, military conflict and bitter distress. The eastern marches of Europe have been disputed by religions and branches of religions; by internationalism and nationalism; by powerful neighbouring states; and by political ideologies. 'Everything about the area continues to be disputed' (Banac 1990: 142). This chapter will explore the production of Central and Eastern Europe as Marchlands of Europe, and Marchlands of Modernities by reference to six discourses:

(a) Scholarly perception and representation of the ideas of Central Europe and Eastern Europe.
(b) The meaning of Europe, as continent, culture area and modernity.
(c) Modernity and the cultural invention of the east as an imagined place contrasted with the west.
(d) The theory of prime modernities.
(e) Russia, Europe and the communist project of modernity – the view from the east.
(f) The global arena of conflicting political and military ideological modernities.

It will stress that there is no fixed bounding of the Marchlands in the realm of thought or the realm of practice and action. It will conclude with an outline of the key features that have produced and reproduced the meaning of Central and Eastern Europe as Marchlands, as a geo-historical place.

Scholarly perceptions, representations and bounding

Are there two Europes or three? This question underlay a discussion of the historic regions of Europe (Szücs 1988). It was asked when the predominant, apparently solid, division of Europe into East and West was becoming hazy, and beginning to melt into the air. The haze was descending in social and spatial practice through events in Poland and Hungary. These events had complex internal causes, but were also related to the new Soviet policies of *perestroika* and *glasnost*. The question itself was part of an intellectual discourse that was reviving the idea of Central Europe. In practice and in discourse, the permanence of Western Europe was taken for granted. The point at issue was whether or not there was a division in Eastern Europe. This question went beyond the immediate situation of the late 1980s, which was a reconfiguring of an old socio-political and cultural discourse on Europe. At the root of the question was the idea of civil society and the nature of the zone of contact between Western European civilisation and Eastern European civilisation, overlain by the spread of western modernity. It is, in fact, the question of the distinction between Central and Eastern Europe. As such it lies at the core of this book and its central theme of spatial modernity and the Eastern Marchlands of Europe.

The scholars who have turned their attention to the study of the history, geography and culture of the eastern area of continental Europe have found it very difficult to agree on an answer. They do, however, accept that first there are no clear physical boundaries; second, that the location of their studies is broadly bounded to the west by Germany, and to the east by Russia and Turkey; third, that no satisfactory single name has been found to describe the area. The British historian H. Seton-Watson (1943: xv) declared that his subject was 'the region lying between Germany and Russia . . . It is unfortunate that no single expression exists which satisfactorily describes the area in question.' Historians C.A. Macartney and A.W. Palmer (1962) also studying 'that part of the continent which lies between Germany and Italy in the west and the USSR in the east' regretted that 'the wit of man has not yet devised for this area any term which is not objectionable on grounds of accuracy or euphony'. Not surprisingly, therefore, scholars often think of their definitions as arbitrary, when it is a trait of their personal spatiality, and define the extent of their study area in terms pertinent to their particular interest and use different terms to describe it. Seton-Watson and Macartney and Palmer conceptualised their area of study as *Eastern Europe*. The Polish writer Czeslaw Milosz (1991: 18) defined *Central Europe* as 'all the countries that in August 1939 were the real or hypothetical object of a trade between the Soviet Union and Germany. This means not only the area usually associated with centrality, but also the Baltic states.'

The idea of Eastern Europe in scholarly discourse

Eastern Europe is an idea widely used in academic discourse, but it has conveyed different meanings to different scholars at different times. Rugg (1985: 2) thought of Eastern Europe as 'the eastern portion of the continent . . . primarily an historical concept, evolving from the particular course of its history'. Okey (1982: 9), a historian, delimited his 'Eastern Europe' as 'the lands between Central Europe and Russia . . . what once was the Habsburg monarchy, partitioned Poland and Turkey in Europe and is now the states of Poland, Czechoslovakia, Austria, Hungary, Romania, Yugoslavia, Bulgaria and Albania'. In the period following the Second World War for many scholars 'Eastern Europe' referred to the group of countries outside the Soviet Union that had communist governments. The geographers N.J.G. Pounds (1969),

R.E.H. Mellor (1975) and D. Turnock (1978), for example, used the term in this sense. It is in this sense too that the term was widely understood by political scientists, within the framework of the Cold War division of the continent into East and West.

Tiltman (1934) called the lands east of Vienna *Peasant Europe – the other half of Europe*. Rupnik (1988) applied the idea of *The Other Europe* to the eight communist states of Europe. In the media, Cable News Network (CNN) Business News also refers to the 'Other' Europe in its transmissions discussing business affairs in the region. The same area, comprising the same group of countries was referred to as *East-Central Europe* by the geographers R.H. Osborne (1967) and Gy. Enyedi (1990b). The eleven volume *History of East-Central Europe* edited by Suger and Treadgold covered the same countries but also included Greece.

Mitteleuropa versus Zentraleuropa – the Central Europe debate

'Central Europe is not a universally agreed geopolitical or geocultural expression, but very largely what we want to make it' (Willett 1991: 1). Sinnhuber (1954: 15) believed that the idea of Central Europe, Mitteleuropa, Europe centrale, as used in geography, was so imprecise that it 'gives rise to misunderstanding of geography and even its ill repute among scholars of other subjects'. From my point of view Sinnhuber provides an excellent illustration of the contested production of place through competing spatialities. Central Europe as conceptualised and bounded by six celebrated geographers, of different nationalities, at different times in the twentieth century, is shown in Figure 2.1. The most extensive area conceptualised as Central Europe was that of Albrecht Penck, who named it Zwischeneuropa in 1915, and the smallest was the Central Europe of Violette Rey in 1996. All constructed their Central Europe from a personal mixture of experience, perception and imagination of Europe in their own present. Four constructed Central Europe from the perspective of 'insiders', and two from the perspective of 'outsiders.'

The idea of '*Central Europe*' as used today has been shaped within two main contending traditions. There is the German tradition of 'Mitteleuropa', drawn from geo-politics. In contrast there is the Austrian, Polish, Czech, Hungarian tradition revived in the 1980s and called by Fehér (1989) 'Café Zentraleuropa – the debate on Central Europe', drawn from a cultural discourse. However, the term all but disappeared from academic and cultural usage between 1945 and 1986.

German geographers have traced several strands in the idea of 'Mitte' in German philosophical and political thought since the mid-nineteenth century (Sandner 1989). Heffernan (1998) points out that the idea of Mitteleuropa was originally an economic space. It was also discussed as a federal structure before its transformation into an imperialist project. He suggested that the idea of Mitteleuropa was elaborated in part as an alternative to the geo-political 'Heartland' theory of Sir Halford Mackinder. The idea of Mitteleuropa was a complex one and was often used by different groups for different reasons. Penck used the term 'Zwicheneuropa' implying a middle zone in Europe

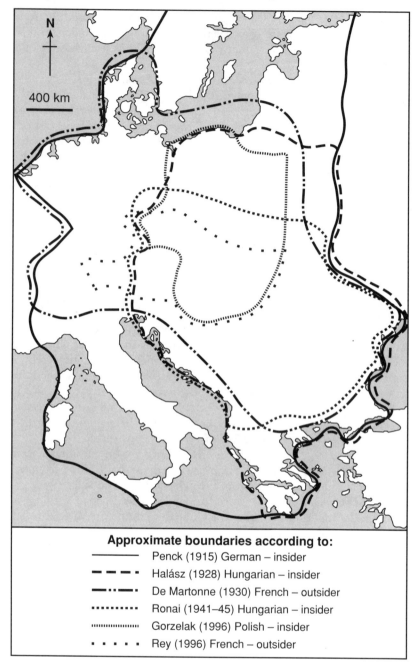

N

400 km

Approximate boundaries according to:

——— Penck (1915) German – insider

– – – Halász (1928) Hungarian – insider

–·–·– De Martonne (1930) French – outsider

············ Ronai (1941–45) Hungarian – insider

⁗⁗⁗⁗⁗ Gorzelak (1996) Polish – insider

· · · · · Rey (1996) French – outsider

Figure 2.1 Where is Central Europe?

Sources: Sinnhuber (1954) *Central Europe – Mitteleuropa – Europe Centrale*; Gorzelak (1996) *The Regional Dimension of Transformation in Central Europe*; Carter, Jordan and Rey (1996) *Central Europe After the Fall of the Iron Curtain*; Halász (1928) *Atlas of Central Europe*; Rónai (1993) *Atlas of Central Europe*.

under German patronage. F. Neuman developed the 'Mitteleuropa' idea after 1916 when it was seen by many as an attempt to justify German imperialism. The idea became totally discredited when, following Haushoffer, it became associated with the idea of Lebensraum and was adapted by Nazi Germany in an attempt to justify conquests in the east.

The Hungarian geographer A. Rónai (1993) conceptualised Central Europe as inner Europe. Rónai upheld a conceptualisation of Central Europe that perceived the Carpathian Basin as the settled territory and natural culture area of Hungarians. But after 1945, the term 'Central Europe' practically disappeared from the academic and political vocabulary, until, as T. Garton-Ash (1986) declared, 'Central Europe is back.' For three decades following 1945 nobody spoke of Central Europe, at least not in the present tense (Schöpflin and Wood 1989). 'Zentraleuropa' was a place that existed only in the minds of elderly survivors of the Austro-Hungarian monarchy. It was used only by 'consenting adults in private' (Garton-Ash 1986), until its revival in the mid-1980s. Enyedi (1990b), an insider, rejected the idea of Zentraleuropa as backward-looking, sentimental and nostalgic. But, for its leading protagonists, the Czech exile writer Milan Kundera, Vaclac Haval (Czech), Gyögy Konrád (Hungarian) and Adam Michnik (Polish), it was concerned as much with the present and future as with the past. Konrád wrote that 'Compared with the geographical realities of Eastern Europe and Western Europe, Central Europe exists today only as a cultural political anti-hypothesis' (cited by Garton-Ash 1986). It was certainly an anti-Russian project and it has also been interpreted as an attempt to distinguish Central Europe culturally from Southeast Europe which was regarded as inferior (Delanty 1996). In the 1990s it became incorporated into the world of practical politics and economics in the Visegrád Association and the Central European Free Trade Area, co-operative international institutions established by the first post-communist Polish, Czechoslovak and Hungarian governments (Dingsdale 1999b). Even so Rey (1996: 46) could state that 'In 1995, without doubt, Central Europe is still more of an aim to be accomplished than a reality.' The revival has been noted by geographers and became the regional framework of study for Carter, Jordan and Rey (1996) and Enyedi (1998a).

In so far as the idea of Central Europe was used as an imagined place to distinguish it from other parts of the Marchlands, it contributes to the recognition of subdivisions in the centre and east of Europe. Geographers and historians have recognised other subdivisions too. The term the 'Danube lands' was used between the wars (Seton-Watson 1943). The Baltic States, excluding Sweden and Denmark, have also been recognised as a distinctive group of countries (Wanklyn 1941). Other subdivisions have been identified and discussed separately. Among them Southeast Europe, often also referred to as 'the Balkans', has been prominent (Hall and Danta 1996; Royal Geographical Society 1997; Todorova, 1994). There have also been many studies of individual countries. All are examples of contrasting and competing spatialities.

In the 1990s the terms 'post-socialist', 'post-communist' and 'post-Soviet' have been used by geographers and economists to describe the east of Europe. Central and Eastern Europe are frequently used together in current academic writing, by Smith (1997), for example. These concepts are clearly inadequate and I suggested a new division of the eastern territories of the continent into Central Europe, Balkan Europe, Baltic Europe and Eastern Borderlands (Dingsdale 1999b).

The meaning of 'Europe': continent, culture, civilisation and modernity

The question of where and what are the Marchlands also demands consideration in the wider question of the meaning of 'Europe'.

We are accustomed to thinking of Europe as a continent. Defining a continent as a major territorial unit of the earth's surface presents no problem for Europe in the north, west or south. On each of these margins the presence of major seas provides a sharp surface edge for the land. Offshore islands can be acceptably accommodated in this way of thinking. To the east, however, there is a problem. The traditional definition of the eastern continental boundary of Europe is the Urals mountains. This range is hardly a significant physical divide. It does not stretch across the whole of the land mass and it has never been a major obstacle to the movement of peoples.

The idea of Europe as a continent, separated from other land masses, arose from the misinterpretation of the extent of the eastern rivers and seas. Classical scholars, like Heridotus and the geographer Hecataeus, thought that the Caspian Sea was the southern section of an ocean that separated Europe from Asia. Later the River Don (Tanais) was seen as the continental divide in medieval 'T in O' maps. Knowledge gained from exploration challenged these views and the Black Sea and the Caucasus and Urals mountains became regarded as the continental boundary. Maps have played a critical role in defining Europe as a continent (Vujakovic 1992). Early maps portrayed the world views of Greco-Roman scholars and their medieval Christian successors giving them a permanent form. Despite their heavy symbolism these maps continued to influence thinking even when their lack of topographical accuracy was well known. Thus the idea of Europe as a continent was constructed by cartographic representation of Christian dominated cultural conceptualisations.

Europe's eastern boundary arose from cultural origins. This raises the difficult question of defining European culture and how the ideas of Central and Eastern Europe fit into it.

European culture and civilisation: east and west

Bugge (1993) points out that Edgar Morin distinguishes between European culture and civilisation. Culture means the Judeo-Christian and Greco-Latin foundations of Europe. Civilisation means humanism, science and freedom.

European culture originates in Judeo-Christian theology and the Roman Empire out of which came Christendom – 'a community of peoples and a geographical area, as distinct from "Christianity" which is a religious faith' (Seton-Watson 1985). Christendom was more or less co-extensive with the continent. In the west the idea of Europe and European civilisation emerged out of Christendom through the Renaissance and Reformation of Catholicism, the secularisation of social and political power, the Counter-Reformation and the Enlightenment. Particularly important were a series of 'Revolutions' beginning in the late eighteenth and nineteenth centuries that combined these cultural antecedents into a practical transformation of societal and cultural relationships. European culture was given its modern trajectory by the French Revolution, the capitalist urban/industrial revolution, the demographic transition, the nationalist revolution and the technological revolution. From this sequence:

> Liberal democracy, industrial capitalism and nationalism comprise the ideological trinity of Western European civilisation's modern political culture. They have come to be institutionalised in the West's unique political creation – the nation state. Because all four are products of Western European culture – the west's perception of political reality – few of us in the west recognise just how distinctive they really are.
>
> (Hupchick 1994: 112)

A distinctive Eastern European civilisation also emerged from Christendom, but in this case from the eastern branch, the Byzantine Empire and the Orthodox Church. This historical sequence had no reformation of the Church and no conflict for authority between the Church and the state, instead the Church and state were united in the person of the ruler. The princes of Muscovy and czars of Russia saw themselves as the only Christian rulers and their territories as Christendom. Peter the Great sought to westernise Russia; the Russian dynasty intermarried into western royal families and the Russian court took on western forms. Russia became involved in western and Central European political and military conflicts (Seton-Watson 1985). However, the impress of the west was never deep. In other Orthodox countries too, such as Romania and Bulgaria, the association of Church and State remained a strong centralised institution. Traditional societal relationships were hardly touched by capitalist urban industrialisation, instead it was the communist form of urban industrialisation that occurred.

> Autocracy, absolutism, centralisation and divine sanction – such are the terms that historically and consistently describe the political culture of Eastern European civilisation for well over a thousand years.
>
> (Hupchick 1994: 118)

Thus Eastern European civilisation emerged from Christendom, differently from that of the west but with an equal or stronger claim to historical continuity (Figure 2.2).

Within these two civilisations there were more particular fault lines of cleavage that separated ethnic groups. The lands between Germans and Russians are settled to the north by several Slavic nationalities: Poles, Slovaks, Czechs, Ruthenes. The south is also settled by Slavic nationalities: Bulgarians, Serbs, Slovenes, Croats. Between these northern and southern Slav nationalities are non-Slavic nationalities, Romanians, Greeks and Hungarians (Figure 2.3).

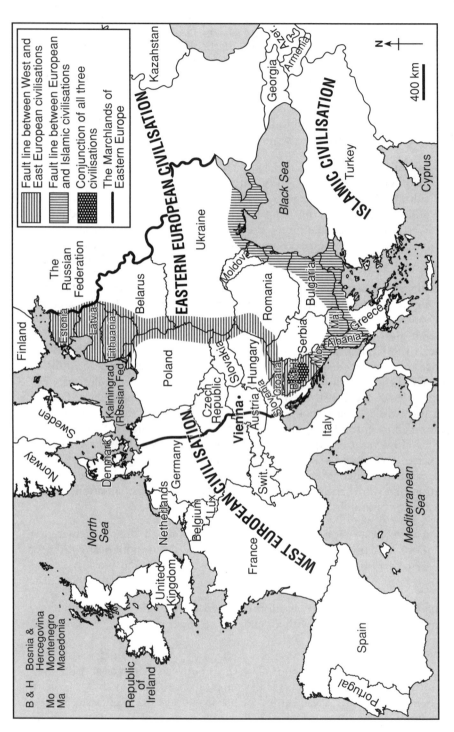

Figure 2.2 Marchlands of civilisations.

Source: Adapted from Hupchick and Cox (1996) *A Concise Historical Atlas of Eastern Europe*, Macmillan, Basingstoke.

Continental Europe was also touched by Islamic culture brought by Arab and Turkish invaders. The Iberian Peninsula was ocupied by Moorish Arabs. The Ottoman Turkish empire's incursion into the southeastern territory of continental Europe lasted for over 500 years. At its greatest extent the Ottoman power spread to the gates of Vienna. The city was besieged by Turkish armies more than once. The defeat of the last Turkish siege of Vienna in 1683 signalled the beginning of the reconquest of Islamic Europe for Christian Europe – a reconquest achieved by Catholic and Orthodox Christian efforts.

Modernity and the European Culture Area: measures of Europeanness

The contrast between this Eastern European and Western European civilisation is no less than the history of modernity and the idea of Europe. This 'modernity' was invented by Europe and is the principal defining feature of Europe (Heller 1992).

The American geographer T. Jordan has made an attempt to set out essential components of the European Culture Area in a plain and comprehensible format. He developed a list of 'European traits' from which he derived measures of Europeanness that could be used to define the European Culture Area. Whilst much of this approach is contestable it does provide a simple and useful empirically based 'checklist' of Europeanness. It is clear that religion, language and race are seen as important cultural dimensions. European civilisation is strongly represented with attributes of modernity being prominent features of the list. By assigning scores to each 'European' country Jordan analysed the European Culture Area in terms of a core and a periphery. Clearly the 'core' of the European Culture Area as he defined it is located in the north west. The east and the south of the continent of Europe are seen as peripheral to the culture area. They form transitional zones merging with other culture areas. This map of Europe reflects the historical zone of conflict between Western European, Eastern European and Islamic civilisations and the impulses of modernity from the west (Figure 2.4). Jordan's methodology was criticised by Taylor (1991) because the traits he recognised were also found in America. Taylor seems to have forgotten the location factor. Europe and America may share many societal and cultural features, but they have different global locations. These three elements combined to make them different places

Modernity and the cultural invention of Eastern Europe

In the first chapter I referred to the association of the idea of Europe and the idea of modernity being for some scholars one and the same thing. In Jordan's European Culture Area cultural traits are merged with modern ones, and the modern ones predominate. I will now take this question of Europe as modernity further in relation to the east of Europe. Throughout the discourse on the idea of Europe there has been the persistent but background question of the territorial dimension and the place and nature of the east in the meaning of Europe.

Figure 2.3 The broad patterns of ethnic nationalities in the Marchlands.

Source: Adapted from *Times Atlas of World History.*

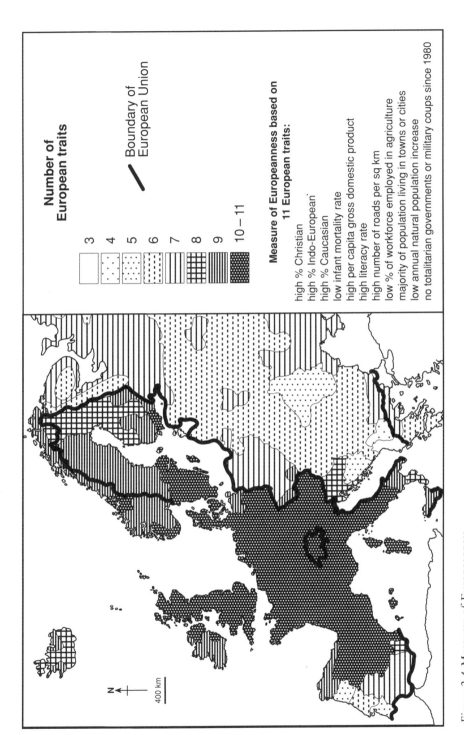

Figure 2.4 Measures of Europeanness.

Source: Adapted from Jordan (1973, 1988, 1996).

This literature identifies the east as a part of Europe's 'other'. The outsider against whom 'Europe' must protect itself and against which it must be contrasted. De Rougement (1966) points out that projects for European unity have all had a strong sense of crusade against an outside foe. The Turks and the Russians have been assigned this role. The Turks, against whom Christendom ('Christian' Europe) waged a 'crusade', and also the early Russian states were labelled 'infidels' and 'barbarians', different from the 'civilised' Europeans. However, contemporary princes of Muscovy and early czars of Russia saw themselves as defenders of 'Europe' against the Mongol hordes and the only true Christian princes (Seton-Watson 1985; Tchoubarian 1994). In the eighteenth century Russia was given this role of outside enemy by western powers, whilst at the same time being considered part of the 'European' state system. In the twentieth century the mantle was passed to the Soviet successor of the Russian empire.

According to Wolff (1994) the idea of Eastern Europe was invented at a time when the states in Western Europe were looking westward, out across the Atlantic. He argues that 'Eastern Europe' was 'invented' by the French Philosophes during the Enlightenment. Eastern Europe was a cultural creation designed to show the superiority of Western Europe. It depended on the linking of asserted similarities binding together Russia, Poland, Hungary and the Ottoman lands giving it a strong territorial dimension. The Philosophes deliberately promoted the idea that Eastern Europe was characterised by 'backwardness' as compared with Western Europe's 'civilisation'. Like the 'Orient' it was alien and inferior (Said 1978). Eastern Europe was, however, a sort of halfway house between the 'Occident' and the 'Orient', invested by the Philosophes with a 'demi-Orientalism'. Eastern Europe was not entirely alien to 'Europe' so, despite its backwardness, it was higher on the scale of development than the Orient. Wolff argues further that this Enlightenment idea of Eastern Europe has had a strong influence on how 'Europe' has been conceptualised since the eighteenth century. It reinforced the notion that the Iron Curtain of the Cold War was a cultural divide. It even played a subtle part in perceptions of Eastern Europe in the 1990s.

Eastern Europe was invented as an attribute of the Western European mind by the first moderns in their project to construct Europe as modernity. It was the first 'other' place used to bound this Europe, to assist the Enlightenment thinkers to destroy Christendom. It was the first imagined place to be defined in order to promote Western European modernity's appropriation of the appellation 'Europe'. This has culminated today in the idea of the European Union (EU) as 'Europe'.

However, the idea of Europe as a cultural community extending from the Atlantic to the Urals has remained alive. In the post-Second World War period General de Gaulle's concept of Europe was one of a 'l'Europe des patries', extending from the Atlantic to the Urals and reflecting French culture and leadership. President Gorbachev also imagined a 'common European home' from the Atlantic to the Urals when he addressed the United Nations in 1988. It was also invoked by the then British Prime Minister John Major, when the enlargement of the EU into Central and Eastern Europe came onto the political agenda in the early 1990s (see Chapter 16).

Central and Eastern Europe in the Theory of Prime Modernities

In his Theory of Prime Modernities, Taylor's (1999) spatiality of dynamic equilibrium defines a core, semi-periphery and periphery experiencing a long cyclical evolution of economies made up of logistic and Kondrotieff cycles. These occurred within a sequence of hegenomic modernities (Dutch, British and American). According to Taylor's account, the capitalist world economy was established in the second logistic wave from 1450 to 1750 and extended from Eastern Europe to the New World. Eastern Europe was part of the periphery. It experienced the 'second serfdom', that witnessed the ruling aristocracy reimposing the feudal obligations of the peasantry in order to take advantage of the market for grain in Western Europe at the time of Dutch hegemony. Eastern Europe was characteristic of a world empire. Central Europe was interpreted as part of the semi-periphery. Its cities underwent a relative decline. Central Europe remained part of the periphery throughout the Kondrotieff cycles of the British and American hegemonic periods. Russia entered the semi-periphery in the 'A' phase of the Second Kondrotieff (1850–95) during the British hegemony. Eastern Europe is noted as part of the semi-periphery in the A phase of the Fourth Kondrotieff (1948–2000), the period of American hegemony, the B phase of which occurred with the collapse of communism in Eastern Europe and the USSR. In this sequencing there is no clear territorial delimitation of Central or Eastern Europe and the Theory of Prime Modernities treats communism as a movement that is fossilised as a reaction to British industrial modernity. Communism collapsed because it remained in a time warp of economic organisation and could not compete with the form of the American post-industrial modernity.

Russia, Europe and the communist project of modernity – the view from the east

Just as Eastern Europe, including Russia, was invented as modern Europe's 'other' so Europe was Russia's 'other'. The meaning of Europe in the Russian and Soviet debate was Europe as western modernity. Europe was thought of as developed and Russia as backward. From the time of Peter the Great, however, opinion among the Russian élite was divided, between those who admired this Europe and those who rejected it. In the nineteenth century the debate hardened into the 'Westernisers' – promoters of European material and secular modernity – and the 'Slavophiles' – upholders of Russian culture, spirituality and national traditions. The Soviet debate contrasted a 'true', progressive, socialist Europe with a 'false', capitalist, reactionary Europe (Neumann 1996).

The debate, like that among the Europeans, was ambivalent towards Russia's place in or out of Europe. Like Western Europeans, Russians never decided, thinking of Russia as in Europe and out of Europe at the same time. This confusion arose because neither the Russians nor the Europeans distinguished clearly between the meaning of Europe as a continent, as a culture area or as modernity. Russians entered the debate about where to locate the eastern boundary of Europe. The Urals boundary was selected in the

eighteenth century and challenged in the nineteenth century as an artefact in
the ideological conflict over Russian self-identity, in the Westernisers versus
Slavophile debate (Bassin 1991). Some Russians did perceive the Urals as their
cultural boundary as well as one of physical geography (Mikkeli 1998).

For Russians, as for Europeans, the process of 'othering' was complex
and confused. In spite of this, and the division of opinion within the Russian
élites, the state maintained a sense of its own superiority over the western
idea of Europe. Neumann expresses this as a sense of a 'true Europe':

> During the nineteenth century, the Russian state represented itself as the
> 'true Europe' in a situation where the rest of Europe had failed the best
> in its own traditions by turning away from the best values of the *anciens
> régimes*. During the twentieth century, the Russian state represented itself
> a 'true Europe' in a situation where the rest of Europe had failed the
> best of its own traditions by not turning to the values of socialism.
>
> (Neumann 1996: 194)

That is to say Europe as western modernity was not truly Europe at all!

Russian and Soviet thinking polarised the debate between Western Europe
as modernity and Russia. How did the Russians and the Soviets view their
nearer western neighbours? The Russian imperial view had a mixture of
great power politics and cultural kinship. The Balkans were perceived as
part of Russia's vital strategic interests in controlling the Black Sea fleet's
access to the Mediterranean. Russia also felt a dual cultural mission: to lead
and protect Slavdom – the lands of fellow Slavs, protect it from German
domination, and also to lead and defend fellow Orthodox Christians from
Islamic, Turkish injustice. The Soviet vision was formed from three perspectives. First, the Soviet rulers aimed to promote a world revolution. The
next centre of this revolution they thought would be Germany, so they
paid little regard to the lands in between. Second, they wanted to consolidate Soviet power, and aimed to regain the lost territories of imperial
Russia. Third, they continued the great power strategic perspective with its
emphasis on the need for control, particularly of the Balkans. These perceptions persisted into the post-World War Two period (see Chapter 16).

Eastern Europe in the global arena of competing modernities

We can lift our gaze beyond the perimeter of Europe and explore the global
horizon in our search for Central and Eastern Europe and recognise them
as Marchlands of Modernities in the twentieth century.

The British geographer Sir Halford Mackinder's *Democratic Ideals and
Realities* (1919) theorised a global geo-political perspective around the idea
of the World Island. A 'Pivot' area recognised in his paper of 1904 was
enlarged to form a 'Heartland'. The territorial extension creating the
'Heartland' was made up mostly of what he called East Europe. East Europe
had emerged as a distinct region, as 'real' Europe, extending from the
Atlantic to the Russian steppes, had diverged into two antagonistic parts,
west and east. The broad boundary between west and east ran southwards

from the Baltic Sea to the Adriatic Sea. The western edge of East Europe was a transitional land of buffer countries preserving 'real' European civilisation in France and Britain. Mackinder feared that a German–Russian alliance dominating the 'Heartland' would threaten West Europe. Despite the crude cartography, his illustrations showed the location and territoriality of the 'buffers', East Europe and the Heartland.

In Mackinder's early theorising the 'Pivotal Area' was hedged in by a Rimland and beyond this was the rest of the world. When American global strategists sought a theoretical underpinning to their policy of containment against the Soviet Union, in the 'Cold War' Mackinder's theory fitted the bill well. American global strategy was to contain communism by forging a series of alliances. The North Atlantic Treaty Organisation (NATO) and South East Asia Treaty Organisation (SEATO) were created to form an encircling zone around the heartland of the communist world system. Eastern Europe was within the European section of the Soviet hegemonic territory that was thus contained.

The above follows the traditional geo-politics account, but other theories have also situated Eastern Europe within post-World War Two global geo-politics including: a post-Stalinist/Atlanticist duality; a Balance of Power approach; and a World Systems theory. These will be discussed in Chapter 9.

The production of Central and Eastern Europe as Marchlands – key defining features

The production of Central and Eastern Europe by differentiating them from each other and from other Europes has been a complex process of contested spatialities. This chapter has shown, first, that the ideas of Central and Eastern Europe convey distinctive meanings that differentiate them from other Europes. Second, that whilst the two ideas are often set together, they are differentiated from each other in societal and cultural conceptualisation. Third, that the meaning of Europe and the project of western modernity have been constant bedfellows and have been most intensely contested in the east.

Central and Eastern Europe have been produced as places in material and discursive modes as constructed and contested places, without sharp or constant boundaries. Rugg (1985) identified in his Eastern Europe four very broad, cultural, historical and geographical periods: feudalism, multiculturalism, nationalism and socialism. From the discussions in this chapter six key features can be recognised that have given Central and Eastern Europe geo-historical definition and identity through these periods. The reworking of these features between 1920 and 2000 has given the region its distinctiveness in relation to spatial modernities. These can be summarised as follows.

First, it is on the margin of the 'European Culture Area', invented in the western 'European' imagination as a backward demi-Orient (Wolff 1998). It is Europe's first 'other' and the 'Other' Europe. It forms a borderland of European civilisation (Helecki 1952), where the varieties of Christianity have sought to spread from their southern European bastions and have come into conflict with 'barbarians' and Islamic civilisation (Hupchick 1994). It is the disputed borderland of 'European' cultural attitudes, values and norms.

Second, it is a zone of political and military conflict fought over and controlled by powerful, outside, states. Ottoman Turkey and Russia/Soviet Union from the east have contested the area with Habsburg and German states from the west (Turnock 1989). In the twentieth century the two world wars were shocking testimony to this military frontier character. These outside powers have always dominated the area. Russia, Prussia and Austria removed Poland from the map in the late eighteenth century, partitioning the country between themselves, three times. The 1939 Nazi–Soviet Pact defined spheres of influence for these outside powers. The failure of the west to assist the Hungarian Revolutionaries of 1956 can also be read as part of a tacit recognition of spheres of influence among outside powers.

Third, it is an economically backward region, peripheral to outside cores of economic innovation and power (Enyedi 1990b). The influence of new ideas of economic organisation and political economy originating in Western Europe have competed with those from the east, to 'modernise' the region out of its economic backwardness. Historically, the internal social structure was dominated by a conservative aristocracy and a numerous gentry. Liberal economic innovation was the project of socially and nationally outcast groups. Western European capitalism arrived belatedly and was soon replaced by Russian communism, which itself proved short lived.

Fourth, it is an ethnically and culturally diverse region made up of small nations surrounded by more populous ones (Pearson 1983). As the 'outsiders' have spread across the region as settlers or conquerors, they have claimed specific places and revised the map of socio-cultural diversity and threatened the national identity and culture of the inhabitants (Seton-Watson 1985).

Fifth, the indigenous peoples have always adapted 'foreign' ideas to their own ends, modifying and rethinking them. The Western European Enlightenment and the first onset of capitalism were moulded to meet local conditions (Okey 1982). The process of adaptation continues today (Smith 1997).

Sixth, as a maelstrom of peoples and ideas, there have been constantly shifting political and cultural boundaries; nothing seemed to last for long. In alternate phases ideas from the West have displaced those of the East, only for the pendulum to swing again and ideas from the East displace those from the West. The whole historical geography of the area is well summed up in the phrase 'the lands between' (Palmer 1970). In the twentieth century Central and Eastern Europe have been the lands disputed between competing modernities.

These key features have shaped the processes that have constantly produced and reproduced places across the region. From time to time a semi-permanent set of relationships halts the movement to mark out a distinctive geo-historical period, redefining meaning and territoriality. A new phase has often marked its beginning in dramatic fashion – a war, a revolution – but the mixing of new societal–cultural innovations into the existing conditions has moved at a slower pace. It is this mixing that has created and recreated places to form the defining geo-historical identity of Central and Eastern Europe as Marchlands of Modernities.

Part 2

Spatial modernity and the Nationalist Project

3 The Nationalist Project

The assertion of ethnic nationality in modernity

This chapter introduces the main features of the Nationalist Project. It will present a broad description of social and cultural change in the Marchlands as a whole, following the ideas and values of the Nationalist Project, as it competed with other projects. Some comments will touch on contrasts at different scales across the region, but the purpose of this chapter is not to provide a geography of the Nationalist Project. The chapter is intended to discuss the socio-cultural changes in which the geography of the Marchlands is embedded. The other chapters in this part of the book will then narrate the variety of experiences of the nationalist period as spatial modernity through a detailed discussion of multi-scalar place particularities and place relations, ranging from the local to the global.

The totality of the First World War provided a historical marker for the onset of modernity in the Marchlands in the way theorised by Therborn (1995). The defeat of the Central Powers and the successful Russian Revolution provided a break with the past that would permit new societal and cultural ideas, especially those favoured by the victorious allies, to take shape. Whether or not they could take permanent root was an entirely different question.

The Nationalist Project began to take shape in the Marchlands in the 1840s, but achieved its final success in the aftermath of the First World War. The idea of nationalism, originating in Western European modernity, encountered very different societal conditions when it was asserted in Central and Eastern Europe. The intelligentsia as a class played a significant part in constructing this new nationalism. Western nationalism was reinterpreted to produce two competing forms. One was associated with liberal middle-class concepts of democracy and capitalism, the other with romantic irrationality and exclusivity. The former looked to the future from contemporary realities, the latter was an invention from pre-modern myths and adorned the future with nationalist ambitions (Sugar and Lederer 1969).

The boundaries of the Marchlands in the inter-war period

The bounding of the Marchlands can never be sharply defined. Nations and states were never precisely aligned. Figure 3.1 shows the boundaries that define the area in the following discussion. This area was produced as a place of conflict and compromise, of contention and co-operation, a struggle for power among societal and cultural élites. The war weakened traditional élites, but not quite terminally. They still contended for power with the new nationalist and capitalists élites who promoted the ideas and practices of modernity. The old aristocracy inhabited a world, now withering, in which landed property, family prestige and the Church had shaped the practices and meaning of place. Nationalists and capitalists challenged this with increasing confidence and success. In the end a fascist modernity took hold and plunged Europe into the Second World War. In the east, across the Marchlands, the war was fought with the greatest ferocity and brutality, between the competing ideologies of Nazism and Communism.

Nationalism, the state and democracy

Nationalists imagine a world made up of nations, and assert that their nation must enjoy cultural, political and territorial sovereignty, in their own homeland. The Nationalist Projects thus set about inventing new imagined communities to replace religious and dynastic communities, as the primary source of identity and so invest places with new meaning. Under the influence of the victorious western allies, they adopted Western European and American style liberalism as a means of destroying the former ruling conventions, and constructing new futures.

The Nationalist Projects were founded on a duality that stressed nationality as the basis of cultural identity, and political statehood with liberalism as the organising institutional framework for societal relations. Statist and capitalist projects were interwoven with the nationalist project, creating an amalgam that shaped the particular features of the ideology that promoted 'modernity' in Central and Eastern Europe. However, this ideology was refashioned by local conditions, and was challenged by revisionist authoritarianism and fascism. Both authoritarianism and fascism rejected 'modernity' as promulgated by liberal democracy. Fascism promoted a complex and contradictory ideology. Heffernan (1998), drawing on Herf's idea of 'reactionary modernism', and Griffin's 'form of populist ultra-nationalism', argues that fascist ideology was shaped by contradictory notions derived from corrupted eighteenth and nineteenth century ideas. In fascism, the concepts of nation, culture and race were fused with modern forms of militarism. When viewed through the lens of Berman's 'developer' concept of modernity elaborated in Chapter 1, fascism can be seen to be an extreme form of nationalist modernity shaped by a particular mélange of rational and irrational ideas, very different from that associated with democratic

Figure 3.1 The Marchlands, 1920 to 1939.

ideals. Communism also challenged nationalist liberal democracy, but with a different cocktail of ideas – a different project of modernity.

The spatiality of Nationalist Projects centred on independent nation-states, with democratically elected governments providing, in theory, freedom of speech, and of action for citizens under the law. Each country, as an independent state, with its own economic and political system, its own aspirations for international relations and its own sense of cultural identity, was responsible for shaping its own internal, local and regional geography and finding its position within continental and global space. The nation-state was a primary differentiating place of the experience of nationalist modernity.

The overthrow of the old ruling order was viewed optimistically by contemporaries, and democracy had a distinctive populist and nationalist tone. Liberal democracy was perceived by contemporaries as having a clear and comprehensive modernisation programme. Politically, it stood for full-blown parliamentary democracy; socially, it stood for land reform, the expropriation of the large estates, and their redistribution among the peasantry; economically it promoted measured industrialisation; diplomatically, it stood for the formation of international alliance networks (Okey 1982).

In all the new countries of the Marchlands, except Hungary and Albania, democracy was given a try. It was, however, an highly centralised democracy; local independence was weak. Local government was stronger in some growing cities than in smaller towns and rural areas. Most states adopted extremely democratic constitutions that were not well suited to the prevailing socio-political and territorial conditions. The complex ethnic and social fragmentation made the formation and working of democratic institutions difficult. The Projects were plagued by excessive nationalism, leading to expulsions and discrimination against minorities. The strength of nationalist feeling bedevilled the progress of democracy and civil society. At the same time, international tension, and the failure of the great powers to properly support the international state system they had created, undermined democracy. The Great Depression of 1929 also sapped the will of democrats.

In these circumstances, authoritarian regimes became established in the late 1920s. As Nazi strength grew in the 1930s, German influence returned. The tendency to adopt authoritarian, fascist or Nazi models of political organisation and ideology increased. To many, the Nazi authoritarian system came to be seen as a more attractive way forward to a prosperous and secure future. It seemed to offer an orderly and efficient philosophy for economic modernisation. It was a strong alternative to liberal or social democracy.

In the 1930s international trends polarised the socio-economic possibilities for many of the new countries. On the one hand, schemes for Danubian co-operation were proposed, by the countries themselves and by Britain and France, as attempts to overcome the problems of economic stagnation, and to spread prosperity. On the other hand, central and southeast, Balkan, Europe came into the German economic sphere, as Germany moved towards an aggressive military economy. The great German enterprises, such as Siemens and I.G. Faben, intensified direct and portfolio investments. The incorporation of the lands lying to the east in Germany's political and economic spheres of influence became an objective of German foreign policy. The role that this German plan assigned to its eastern neighbours inhibited the progress of modernity in the region as I shall show in later chapters.

Nationalism and capitalist societal modernity

Like nationalism, capitalist transformation aimed at replacing the old power nexus and introducing new societal relationships. Capitalism's primary focus was, however, economic rather than political or cultural. The coexistence of nationalist modernity and capitalist modernity was full of societal contradictions. Capitalist modernity tended to be inclusive, open to any successful entrepreneur or industrial businessman irrespective of background. The dynamism of capitalist success was difficult to control socially or politically, and it stressed novelty and innovation. In contrast nationalist modernity was exclusive, with group membership determined by ethnicity not organisational or financial ability. Nationalist identity owed more to romantic reinterpretations of the past than economic success in the present. Nationalism, furthermore, forged a new and particularly strong relationship between the state and the economy. State sponsored economic development was intended to strengthen the state economically and politically. Extreme forms of nationalist economic and political attitudes permeated policy making. In the 1920s for reasons of political nationalism and the economic consequences of the war, the new governments increased state involvement in the economy. During the 1930s, following the Depression, the need for state involvement was increased, but state activity in the economy was also given a theoretical underpinning by the new Keynsean economics.

> Previous views that state intervention in the economy meant socialism were no longer valid; but earlier ideas of laissez-faire went the same way. It became widely accepted that governments could, or should, play a decisive role in transforming static economies into growing ones ... the views on the relationship of the state to the economy had in practice changed permanently.
>
> (Berend and Ránki 1974: 35/36)

Capitalism had to adapt to these circumstances. A specific form of capitalist modernity would emerge much influenced by government activity.

In the empires that dominated Eastern Europe before the First World War the ownership of landed estates gave the nobility and the Church control of all the levers of societal power. Land was the currency of power. The new nationalist governments had to institute significant land reform in order to break the power of the nobility and establish their own credibility with the new electorates.

Industrialisation was a key to capitalist modernising processes. Eastern Europe is well endowed with industrial raw materials and energy resources.

There are deposits of hard coal, brown coal, oil, natural gas and a variety of non-ferrous minerals. Some, such as Poland's Silesian coal reserves, Hungary's Transdanubian bauxite fields and Romania's oil and natural gas reserves, were abundant enough to rank as important in the world scene. New resources were discovered by surveys undertaken in the 1920s and 1930s. These resources, however, were not as effectively exploited as they might have been because of the state territorial division after the war and the prevailing political and economic conditions.

The governments of the new post-war states were under considerable pressure to be proactive in managing their economies and assisting private capitalist, industrial development. The consequence of this was a particular form of capitalist industrialisation. The state took over the assets of many pre-war entrepreneurs and became highly active in regulating and investing in industrial enterprises. Industry made a significant contribution to economic growth despite various hindrances. High levels of foreign direct investment were encouraged and went, mainly, into banks and industrial enterprises. Economic development led to shifts in the branch structure of industry. There was a tendency to increase the size of industrial production units. Cartels were formed in many branches of industry, at first by voluntary agreements, but later by legal enforcement.

Governments hardly developed any urban policies and urbanisation resulted largely from spatial processes related to demographic and economic trends. Much of the industrialisation occurred in rural and semi-rural settings, growing out of the long tradition of small-scale workshop industries. It turned villages into small towns as the growing population created a demand for professional, administrative and commercial services. Urbanisation was also channelled through the existing network of small towns in the northern and eastern districts, elaborating the urban fabric. The growth of capital cities, however, as administrative, social, political and cultural centres attracted most industrial investment as they stimulated demand for manufactures. These cities housed some of the wealthiest groups as well as a mass of workers and were a variegated and substantial market. City regions concentrated investment in a circular and cumulative process that dominated national urbanisation. The international urban hierarchy grew out of the peace treaty redistribution of territory, was underpinned by national government policy and further shaped by independent private sector economic activity. The development of resource-based urban–industrial agglomerations was much weaker, but became, nevertheless, a feature of urbanisation that formed new types of urban regional settlement systems. Urbanisation through town and city growth created a new and dynamic setting for economic and social relationships to be reshaped into distinctive morphologies of nationalist, capitalist modernity.

The technology of communications is a vital aspect of modernity affecting economic development, the spread of prosperity and societal relations. The governments of the new states invested heavily in new transport infrastructures within their countries, promoting economic development and personal mobility. Telecommunications were encouraged by licensing telephone services and national radio broadcasting. Governments were much less willing to assist the development of foreign travel. They introduced passport and visa requirements, hindering personal contact. They promoted tariff régimes curtailing international trade. Despite these government restrictions, international rail and air services expanded. These trends increased the potential connectivity between people and transformed the time/space environment in which they lived, but the impact of these changes was strongly conditioned by locational and territorial circumstances.

These socio-economic trends led to occupational restructuring and lifestyle changes. However, so intensively did the feudal heritage lie on Eastern Europe, and so numerous was the agricultural population, that the shift of male working population from agriculture to industry and commerce proceeded extremely slowly.

Nationalism and socio-cultural modernity

Nationalist political policies affected a sense of identity and social relations. The development of capitalist social relations was interwoven with these. Social transformation and standards of living were also affected by demographic dynamics. For the most part, demographic trends were left to natural processes, though some right-wing governments pursued pro-natalist policies. The population of Eastern Europe rose by about 25 per cent between 1920 and 1939. The general patterns of demographic change fit well with the features of the demographic transition model from traditional to modern patterns. The high rates of natural increase were sustained by falling mortality rates and not through unusually high birth rates. Infant mortality remained high, never falling below 100/1,000 for the area as a whole. In the 1930s there was a general reduction in fertility rates. The high birth rates and declining death rates produced a bimodal age distribution with exceptionally high proportions of younger and older age groups that had implications through strained dependency ratios. Whilst general trends were similar, particular national experience was related to different combinations of several factors. Among these were the displacement of populations following the peace settlement, and the longer-term effects of the war that affected different countries in different ways. This factor influenced, and was influenced by, the ethnic and religious make-up of the country that was a significant element in population patterns. At the same time patterns of international and intercontinental economic migration affected trends within particular countries in different ways.

Education is a vital ingredient of modernity as an experience of time and space. Education was an important weapon on the side of the nationalist and capitalist assault on the old régimes and in promoting new ways of thinking. Education encouraged national community building and a national consciousness by teaching a nationalist history. It could provide a workforce able to follow instruction and undertake semi-skilled and skilled jobs for which training is essential. It provides a channel for self-development, broadening personal experience and a gateway to personal development and advancement, enhancing a personal perception and conception of place. It is a vehicle for generating social and spatial mobility. The nationalist programme of modernisation in most countries included compulsory elementary education, but low public expenditure on education resulted in too few and inadequately trained teachers, a scarcity of textbooks, poor, overcrowded school buildings and large classes. Educational radio broadcasts were an innovation in Finland. Over half the national schools heard educational broadcasts (Lingeman 1938). The Churches also retained a strong position. Curriculum development was weak. This inevitably produced school leavers who were ill-equipped to take up the semi-skilled and skilled jobs created by industrialisation. During the inter-war period the provision of elementary education expanded but progress in secondary and vocational education was generally slow. Governments did, however, seek to increase the numbers in education right from the elementary school to the university. The university expansion seemed to be motivated more by prestige than a commitment to modernity, reflecting a priority of national identity over economic modernisation. Governments never really decided whether education was to be provided for the benefit of individual students or was intended to be a long-term public investment to underpin the shift to a modern occupational structure. In this situation the former tended to predominate and the education system did not generally provide a workforce trained for modern industry and commerce (Hauner 1985).

The single most salient sign of the move to modernity was the defeat of illiteracy. The meaning of modernity as self-development and institutional development depends on the widespread achievement of literacy. The frontier of modernity in the Marchlands as measured by the spread of literacy among the population aged over ten years, during the mid-1930s, shows a marked west–east gradient (Figure 3.2), but it is not a simple pattern. There is a complex zone of transformation. The trend appears to be independent of national boundaries, but the western and northern boundaries of Bulgaria seem to mark a sharp localised divide. This suggests that national education policies did make some difference. There is some suggestion that higher levels of literacy are found in urban centres, but this is not especially marked across the whole of the area, when broad bands are considered. The remoter rural areas, often in mountainous terrain inhibiting communications, also have a shadowy presence in the pattern. There is some hint too that ethnic factors may be important. There is a salient extending eastwards between the northern and southern Slavic settlements. Historical factors may also play a part. The former Russian areas of Poland seem to have lower levels of literacy than the Austrian or German areas. The areas long occupied by the Turks also seem to have achieved lower levels of literacy than elsewhere. However, as noticed in Bulgaria, recent policies can make a difference. None of these factors seems in itself to be crucial, but as layers of space they construct distinctive patterns of places.

As the next chapters will narrate, the varied experience of modernity was etched into the mosaic of locality, region and state. However, on the basis of an analysis covering 1,500 localities, four broad shades are discernible on

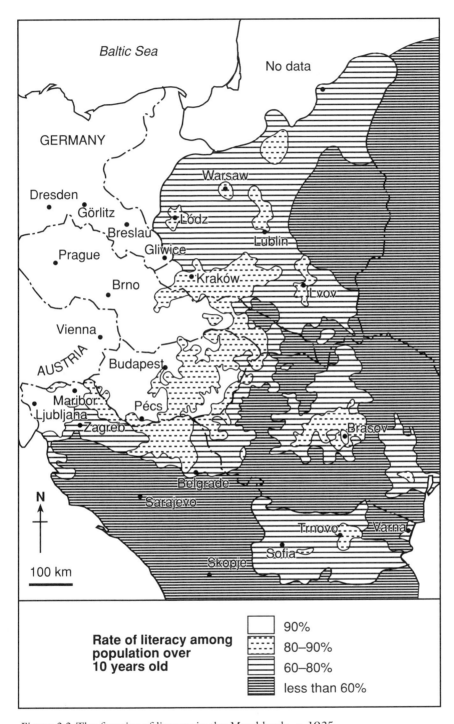

Figure 3.2 The frontier of literacy in the Marchlands, *c.* 1935.

Source: Adapted from Rónai (1993) *Atlas of Central Europe* and Pándi (1997) *Kösztes-Európa*.

the map of the Marchlands of Central and Southeast Europe, when the east-ward motion of western modernity is freeze framed in the mid-1930s (Figure 3.3). In the west, traced out by a line intertwined with the borders of the Czech lands and Austria, modernity is almost fully coloured in. Just to the east, the shade is that of lands on the threshold of a mature modernity. They stretch from Slovenia, through the western and northern edges of Hungary and western Slovakia, to the very western margin of Poland. Beyond this to the east two separate lobes of territory have a lighter shade. They are but finely brushed by the practices and ideas of modernity. The southern lobe swathes over the great Pannonian lowland, central to the Carpathian Basin. Its eastern edge steps out from Zagreb eastwards to Belgrade, from whence it traverses northwards through the Hungarian–Romanian border country, skirting the towns of Temesvar/Timisoara and Nagyvarad/Oreada to loop around Debrecen. The southern edge of the northern lobe follows the upper Vistula Valley eastwards, then turns north to the east of Lublin. Beyond these threshold lines, the shades that signal the early experiences of modernity are as yet absent, except for isolated, scattered clusters.

Cultural geography was shaped by the conflict between competing concepts of culture. Nationalism favoured the strengthening of a narrow 'pure' ethnic folk culture. The old Austro-Hungarian monarchy left a legacy of cosmopolitan multi-culturalism, championed by many intellectuals. Capitalist inspired commercialisation created a modern style of popular culture in which newer forms of entertainment, sport and recreation predominated. This was enhanced by metropolitanism concentrating the cultural avant garde in the capital cities. The mass media were growing in importance for entertainment and news information. Newspapers, a variety of magazines and comic books were published in large numbers and increasingly read by a better educated public. Cinema became a popular cultural experience. In the 1930s narrated newsreels were shown as part of cinema programmes alongside feature films. Foreign films often dealt with contemporary and romantic themes and gave the audience a glimpse of far-away, exotic places. Domestic feature-film production was encouraged, especially when its inspiration was derived from classical national subjects. Radio broadcasting was the latest communications technology and was reaching increasing audiences. Unlike the telephone it was not two-way personal communication, but one-way mass communication. It had great power in creating a sense of place and community, and in providing publicity for individuals and propagandising their ideas, as well as compressing and virtualising space. As noted above it was used for educational broadcasting. At the same time a rich, vigorous and distinctive national and also cosmopolitan high culture of music, literature, art, sculpture and architecture was strengthened by an infusion of modernist interpretations. Nationalist hostility led to the emigration of some artists and writers, whose ethnicity or individual style of expression was unacceptable to the extreme nationalist notions of culture. The cultural representation of place was framed within these competing forms and their creation of a new genre of expression.

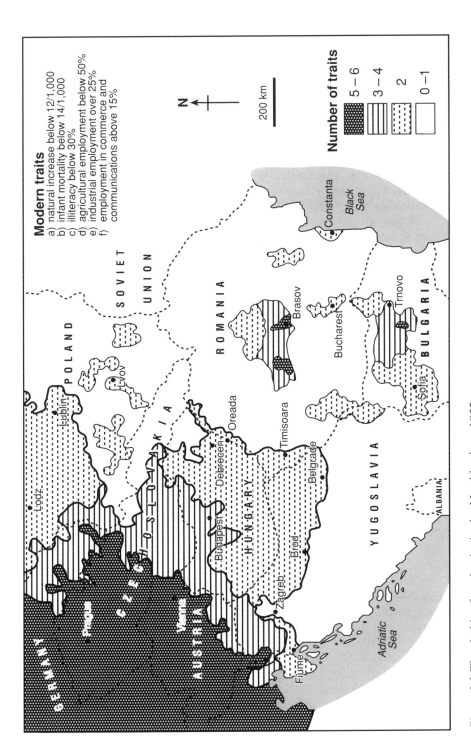

Figure 3.3 Thresholds of modernity in the Marchlands, *c.* 1935.

Source: Adapted from Rónai (1993) *Atlas of Central Europe.*

Conclusion

This chapter has reviewed some of the main socio-cultural trends in which the spatial modernity of the Marchlands between the two world wars was embedded. The Nationalist Project was forming opinion and, with capitalism, driving individual and environmental development. The chapter has suggested that the First World War provided a break with the past that allowed the societal nexus of the imperial period to be successfully challenged by liberal and social democratic, nationalist and capitalist ideas and practices. Western development culture was embraced by new political and economic élites, perceived as a pathway to modernity. Eyes became firmly set on the future. However, it was a fragile world of tense relationships between nation building, state building, societal modernisation and traditional attitudes and values. Liberal nationalist attitudes were defeated by romantic irrationality. There was excessive and disabling political, cultural and economic nationalism. Reinventing the past to serve the present, in order to make a nationalist future, hindered capitalist modernity. Nationalist Projects promoting liberal democratic ideas failed everywhere, except Czechoslovakia, to completely dislodge older societal values or to fend off authoritarianism. The awakening of national consciousness and identity was highly successful, but it paved the way to extreme nationalist policies that caused strife and conflict culminating in the Second World War. The Marchlands were reproducing their geo-historic characteristics as a zone of conflict, catching up economically with the west, modifying outside ideas in a maelstrom of tense ethnic relationships and adapting new ideas to meet local needs. These trends were conditioned by a nationalist spatiality that stressed ethnicity as the arbiter of societal and cultural inclusion and exclusion. The focus was on the nation-state with private enterprise and economic development controlled by the state, within an international state system of European great powers whose ambitions were global and imperialist. These elements were constructed into a distinctive spatial modernity as multi-scalar place particularities and place relations, to which, in the rest of this part of the book, I now turn.

4 The production of localities in nationalist modernity

This chapter first explores the production of localities in nationalist modernity, distinguishing between rural and urban localities as differentiated cultural, societal and locational attributes of nationalist spatiality. Localities are the places in which people spend most of their lives, enact their daily routines and with which they are most familiar. Localities in the Marchlands in the 1920 and 1930s were usually associated with communes, the smallest scale of local government, but counties also conveyed the sense of the local for some societies. In this chapter particular attention is given to Central Europe as defined by Rónai in the *Atlas of Central Europe 1945* (for which the 1993 digital facsimile edition has been used). In this atlas communes generally form the framework for analysis, and I have adopted them as localities. Through the example of Budapest, the most rapidly growing city of the region, I look at the ambivalent status of capitals, as symbols of national identity and as places of conflict between traditionalist, nationalist and modernist ideas. The urban–industrial conurbation was a relatively new settlement form. The Upper Silesian coalfield was the most developed urban–industrial, multi-centred, conurbation and is presented as a case example of resource-based urban–industrial development in the particular conditions of nationalist modernity.

Rural localities

The inhabitants of the Marchlands, in the 1920s and 1930s, were predominantly peasants. The places they experienced, through which they made sense of the world and that they invested with meaning, were their localities, their safe place. Their places were bounded by the estate, the farmstead or the village. They visited the local town, even the county town occasionally, but the state capital was an imagined place. Rarely did they step out of their daily experience and rarely did outsiders, except itinerant peddlers, step in. Life's experience was far from the styles of modernity. Yet, there was a restlessness across the peasants' lands. Some had enjoyed a little education and minds were beginning to open.

> Those who could, escaped. With the despair of drowning men, the farm servants now began to look even beyond the villages to the towns.

Budapest! In their imagination the capital gleamed afar like a wonderful fairy palace shining high above the curse-ridden morass of the Puzsta.

(Illés 1971: 265)

The peasants would forsake their safe place in ever increasing numbers. They would search for a better life in new places, with new jobs, make new friends and join new communities. The excitement of modernity was taking hold.

Gyula Illés, the Hungarian writer quoted above, was born on the puzsta. Yet it was not the romantic imagined place of the lyric poems of Petöfi. Gyula Illés was born in the countryside, but he knew nothing of the farmstead or the village. His place was the great landed estate of Transdanubia, western Hungary. He knew the elegant mansion house surrounded by parkland, with its tennis court, artificial lake and orchard, behind the fence that kept *him* out. The ox barn was the next most impressive building, followed by the manager's house then the bailiff's. The farm labourers lived under one roof in a single storeyed building like a barrack. When, at the age of eight or nine, he visited a village he was filled with stark astonishment, amazement and terror. He was petrified by the din of the carts, people, cattle and children in the narrow street, bordered by the lines of houses in a terrifying orderliness and congestion. Because the village was inhabited by Germans he believed it an alien, German, invention not spontaneously Hungarian (Illés 1971).

Outside Poland, Transdanubia, the Great Plain of Hungary, Bohemia and Moravia, the land reform policies of nationalist governments had mostly swept away the great estates and their, often absentee, landlords. The first practice of land reform redistributed around 70 million acres, created 2 million new farms and transformed 1.5 million tenants into owners. In Bulgaria the average size of some 800,000 individual farms was only 2 acres (Beaver 1940). With the large estates went some of the more recently introduced commercial and mechanised farming practices. The large-scale, consolidated holdings had been well suited to the steam technology of agricultural equipment, but large landowners had not always been in the forefront of technical innovation. Now there were millions of new, peasant small-holdings that were not commercial or mechanised. The cost of equipment, the conservatism of the peasants, the alien idea of commercial management, the small extent and the scatter of non-contiguous plots constrained farm modernisation. Local co-operative movements were formed among peasant farmers in most countries that allowed the small surpluses from each farm to be marketed more efficiently and also assisted in raising credit. Modernity, subtly differentiating rural localities, often rested on the enterprise of individual peasant proprietors and their acceptance of commercially based co-operative ventures. Many peasants remained suspicious of wealthier villagers believing that sudden wealth could only have come from devious or even unlawful practices (Illés 1971).

The peasants of the predominant rural–agrarian localities looked back to time-honoured custom and the rhythm of the seasons to count the measure of their daily lives. They walked the paths their forefathers had trodden with pride in a landscape of vernacular architecture that reinforced the meaning of shared ethnic traditions. Rural settlement reflected the mosaic of historic colonisation and the imprint of cultural experience and sequent occupance. The 'Dutch' villages of Poland, for example, were founded by Dutch and, later, German colonists moving east along the line of the glacial moraine and reclaiming swampland as they had done in the Netherlands (Maas 1951). The Hungarian recolonisation of the Great Hungarian Plain began as the Turks withdrew. The colonists built small-holdings known as Tanyas. They created a landscape of Tanya township localities with widespread low density occupancy in a process that continued into the 1920s (Duró 1992).

Everywhere distinctive village morphology and architecture reflected ethnic identity. It also reflected social stratification. Whilst land reform removed the great landed estates and thereby dented the power of the landed aristocratic and capitalist owners, there remained considerable differences in wealth among the village community. A large class of landless and day labourers remained extremely poor. A class of wealthy peasants commanded respect and deference from their poorer neighbours whom they often treated harshly (Illés 1971). The local doctor, schoolmaster and clerk were still respected by many peasants and retained a leading position in village life.

Rural industrial localities picked out the scatter of mineral deposits worked on a small scale by colonising mining communities of the past. Metal working often developed in these villages creating a tradition of skilled craftsmanship. Local initiative and foreign capital then transformed them into prosperous industrial enterprises and sometimes formed the basis of small town development such as in Bohemia and Moravia. In these, Czech, lands in the 1930s some rural localities had over 50 per cent of their inhabitants engaged in industrial activity (Rónai 1993).

The daily toil of the peasant and the craftsman was set within the traditional rites and customs of religious observance and ethnicity. Ethnicity and religious affiliation often went together. Localities had subtle distinctiveness in these cultural traits. For the Hungarian Illés the German village seemed alien. The new state boundaries drawn by the peace treaties had changed the ethnic composition of some localities. Voluntary and forced migration of some ethnic groups was followed by the settlement in these localities of new ethnic groups, often as new governments followed colonisation policies. Most localities were inhabited almost totally by just one ethnic group, but Rónai (1993: 107) mapped some localities with two groups accounting for over 10 per cent each, and in a few extreme cases where the inhabitants were from as many as five different ethnic groups, each counting more than 10 per cent of the population.

> The mingling of two nationalities is rather general along the language borders. Three–four nationalities are mingling in vast areas on the eastern part of Poland, in Galitia, South-Bessarabia, Dobrudja, Transylvania and in the Balkans, in Bosnia-Hercegovina, but in the most remarkable manner in the Banat and Bácska.
>
> (Rónai 1993: 104)

There was often, however, separation of the ethnic groups within a single locality. In the dispersed settlements of the eastern Carpathian mountains, for example, Fleure and Pelham (1936) found Hungarians and Germans living in nucleated villages in the valleys, whilst Romanians lived in hamlets on the upper valley slopes. Rónai (1993) noted:

> Sometimes even within one village the streets and houses of the different nationalities are segregated: e.g. in Transylvania it is a characteristic phenomenon that the upper end of the village is Hungarian, the lower end Roumanian or German. They have two churches, two schools, two pubs.
>
> (Rónai 1993: 104)

When nationalism stressed ethnic identity there arose the danger that ethnically mixed localities could become places of conflict and violence.

Rónai (1993) suggested that there was some connection between the adoption of modern techniques, the development of modernity and ethnic composition. His maps show locational associations from which this assertion arose but it may be only tenuous and conditioned by the official nationalism of the time.

Some support for Rónai's comments came from Gaffney (1979) who relates the story of Kisker, a German village in the Bácska region. He suggested that the varying prosperity of villages of the Bácska between the wars was related to ethnic and religious attitudes and values nurtured by history. The most prosperous villages were German. German settlers came dreaming of prosperity in a spirit of free enterprise in the eighteenth century. They had responded to Austro-Hungarian government offers of free land and material goods in return for settling lands recently conquered from the Turks. In contrast, the non-German farmers of the region, mainly Serbs, were descended from serfs emancipated in 1848. They had very small, fragmented holdings though just as high soil quality. Unlike the Germans, they had neither the attitudes nor the farm layouts to successfully organise commercial activities. Whilst the Serb farmers looked to their farms to provide subsistence, the German farmers were motivated to raise their standards of living and produce surpluses for sale. The Germans were equipped to participate in the capitalist international market for agricultural produce from which the Bácska was made into a bulk supplier of grain to Germany, Hungary and Austria. The exceptional success of Kisker, which stood out even from other German villages, was assigned to its particular historical trajectory. In the 1930s nearly all the villagers of Kisker enjoyed imported 'wristwatches, radios, bicycles, sewing machines and fine English suits' (Gaffney 1979: 140), commodities that inhabitants of other villages could not afford. The particularities of Kisker and its place relations made the spatiality of these German villagers more complex than that of others.

Agricultural districts were overpopulated. Rural birth rates remained high. After an initial wave of land reform between 1918 and 1920, further government redistribution of land moved at a slower pace. Some governments envisaged a slow transfer of land taking a decade or more. Because of the land hunger, governments sometimes promoted colonisation clearing state owned forests to create extra cultivatable agricultural land. The subdivision of peasant holdings continued, following the traditional partable inheritance laws. This provided some land for new generations to work, but reduced the size of holdings. In many districts peasants relied on communal land and forest for grazing and foraging. Transhumance was widely practised in mountain areas. Rural localities remained tied to subsistence farming but those located near to small towns or bigger cities had the opportunity to sell specialised produce at their markets. Small towns traditionally held fairs, generally twice a year, providing an opportunity for socialising and commercial exchange, attracting country folk from up to 25 km (Fleure and Evans 1939). In a few areas the commencement of local bus services enabled the journey time to the nearby market town to be reduced. These changes were influential in promoting the adoption of ideas and practices of modern spatiality.

Urban localities

Life in the towns of the Marchlands was still parochial and local issues and identities were predominant. The locality of the town filled the minds of most residents. The regional cities and particularly the capitals were beginning to experience the opportunities of modernity. The advent of mass communication, new technologies and competing cultures that produced the uncertainties of modernity were most evident here.

Naming transposes place from space and differentiates the hundreds of urban localities in the Marchlands imparting identity and meaning for their inhabitants. The multiple naming of places reflects the ethnic and linguistic mixed occupancy of urban localities that was historically widespread, but it does not imply a cosmopolitan intermixing of ideas and practices. The renaming of places reflects the sequent occupancy and competition for control that characterised settlement. What's in a name? Identity, meaning, and in the Marchlands of the 1930s, socio-ethnic conflict are in a name. In the 1930s Poles gave the name Wilno; Lithuanians Vilnius; Germans and Belarusians, Wilna; to a city most of whose inhabitants spoke Polish or Yiddish, whilst some spoke Lithuanian, Belarusian or Russian; where Catholics predominated, but Jews, Calvinists and Russian Orthodox also lived (Rupnik 1990).

The ethnic make-up of urban localities was mixed and fluid. For example, in the towns of the Carpathian Basin the ethnic balance of the inhabitants was complex, and constantly changing in the inter-war period (Kocsis and Kocsis-Hodosi 1998). Table 4.1 shows the ethnic composition of eight towns in the Carpathian Basin from 1910 to 1941. All of these towns were in Hungary before the First World War, but were incorporated into new states about 1920. Some were regained, temporarily, by Hungary in the Second World War. These particular features affected the ethnic balance, but the diverse ethnicity and mobility is still clearly apparent. This selection also suggests that the more populous towns tended to have a more even balance than smaller ones.

Table 4.1 Ethnic composition of selected towns, percentages, *c.* 1930

Date	Total pop.	Slovaks	Hungarians	Germans	Other
Kassa–Kosice (Slovakia)					
1910	54,331	25.1	66.5	6	2.4
1921	60,063	63.7	19.6	3.4	13.3
1930	81,802	64.7	14.3	4.1	16.8
1941	79,855	19.2	75.6	2.1	2.9
Beregszasz–Berehove (Transcarpathia)					
		Ruthenes/Ukrainians			
1910	14,470	1.6	96.4	1.0	1.0
1921	15,376	11	60.9	0.7	27.4
1930	20,897	10	51.3	1.9	36.8
1941	21,554	4.3	91.8	0.3	3.6
Huszt–Hust (Transcarpathia)					
1910	10,292	50.8	34.1	14.9	0.2
1921	11,835	56.9	7.7	3.5	31.9
1930	17,833	52.2	7.8	4.1	35.9
1941	21,118	49.7	24.6	2.0	23.7
Ujvidék–Novi Sad (Jugoslav–Vojvodina)					
		Serbs			
1910	33,590	34.5	39.7	17.6	8.1
1931	56,585	36.5	30.0	15.0	18.5
1941	61,731	28.4	50.4	12.4	8.8
Obecse–Becej (Jugoslav–Vojvodina)					
1910	19,372	34	64.5	1	0.5
1931	20,519	34.4	60.7	1.5	3.4
1941	21,200	28.8	68.8	1	1.4
Brasso–Brasov (Romania Transylvania)					
		Romanians			
1910	41,056	28.7	43.4	26.4	1.5
1930	59,232	32.7	42.2	22.4	2.7
1941	84,557	58.5	17.9	19.2	4.4
Nagybanya–Baia Mare (Romania Transylvania)					
1910	16,465	33.7	64.8	1.2	0.3
1930	16,630	50.8	39.2	1.8	8.2
1941	25,841	24.8	72.1	0.5	2.6
Poszony–Bratislava (Slovakia)					
		Slovaks			
1910	104,896	21.7	35.9	38.0	4.4
1921	122,201	42.6	21.4	26.7	9.3
1930	170,305	51.2	15.8	24.3	8.7

Source: Kocsis, K. and Kocsis-Hodosi, E. (1998).

The uniqueness of place as the conjunction of space/time experience of modernity is given the brilliance of crystal in Prague as expressed in the contrasting lives and literary work of the Langer brothers. 'Yet it is precisely Prague were the common denominator must be sought, in the unique, unrepeatable circumstances of that city, where three cultures, – Czech, German and Jewish – lived side by side, three peoples living under the constant pressure of symbiosis' (Dragen 1991: 189).

Jewish communities tended to be prominent in towns, because having been for centuries excluded from traditional occupations, they worked in trade, the professions and as industrial entrepreneurs, activities that were generally associated with modernising urban centres. Jewish communities were present in towns from the Baltic to the Black Sea and the Mediterranean. Rónai (1993) estimated that about half of the 3.5 million Jews living in Central Europe as he defined it, in the 1930s, lived in eight cities – Budapest, Vienna, Łódź, Lwow, Kraków, Bucharest, Kiev and Odessa. In Christianu, Iasi, Lwow, Lublin, Łódź, Odessa, Ziloniv and Vinnici more than 40 per cent of the inhabitants were Jewish. In Budapest, Nagyvarad, Galati and Kiev (Ukraine SSR), over 30 per cent of the inhabitants were Jewish. Jewish settlement in the towns and cities of the Balkans was usually more in the order of 10 to 20 per cent, but there were substantial communities living as far south as Thesiloniki in Greece (Rónai 1993). Many Jewish communities were still confined within a distinctive ghetto district giving a physical symbolism of their social exclusion in nationalist spatiality. That these communities were so easily removed, when the scourge of Nazism covered the land, is a testament to their 'otherness' for all the societies of the Marchlands.

There was a dense network of small towns in Bohemia and Silesia, an extension of the Saxon and Thuringian patterns. In other regions, cities dominated skeletal hierarchies of small towns and rural settlements. Capitals were the main destination for refugees forced out of other countries. Domestic migrants fleeing agricultural overpopulation, flawed land reform schemes and economic depression also made for towns, especially capital cities. Several capitals were the subject of large-scale planning schemes influenced by French, German and British visions of city planning, given a local character by native, but foreign-trained city engineers (Lampe 1984). Modernism in architecture made its appearance in several notable buildings erected in the 1930s in these cities. New symbols of nationalism were constructed such as the University in Sofia, the National Assembly building in Prague and the National Museum in Warsaw. New government buildings were constructed in Riga, Latvia, as part of a plan, prepared by the municipality and national government, to completely rebuild the central districts of the city (Hobson 1938). However, modern infrastructure was often restricted to central areas and commercial premises. Housing, especially single family dwellings of one storey, constructed on the edge of the cities, lacked piped water or sewerage systems and streets were unpaved. Housing provision could not keep pace with the level of demand from immigrants. Overcrowding was a serious problem in outer districts, so that some cities were little more than shanty towns. Even so, by the late 1930s the inhabitants of the major cities were beginning to experience the qualities of modernity in their lives. In Sofia, the capital of Bulgaria, perhaps the least modern of the bigger cities, 'By the late 1930s, the adult population typically read one of a dozen newspapers once a day and saw a film, typically foreign, once a week. The state radio station broadcast three hours daily' (Lampe 1984: 43/44). Capital cities also had stronger, more democratic local government than elsewhere.

The local collapsing of space had begun in the capital cities in the late nineteenth century with the development of public transport services. All the capital cities had a public transport tram system and suburban railways operating in the 1920s. In Warsaw the old Russian broad gauge system east of the city was converted to standard gauge and the electrification of suburban lines undertaken in the 1930s (Beaver 1937). Budapest's public transport system included an underground line that had been in operation since 1898. All the capital cities had local radio stations. Telephone services were available for the better-off. In the Hungarian capital Budapest, for example, the number of telephones in use rose from 26,903 in 1920 to 63,537 in 1936 (Tinar 1992). An inter-urban telephone connection linked Berlin, Prague, Vienna, Bratislava, Budapest, Szeged and Temesvar before the war. The most intensive use was between Vienna and Budapest. In 1927 a new cable link was installed offering 210 lines for simultaneous calls between Vienna and Budapest (Tinar 1992). Daily air services linking the capitals and provincial cities significantly reduced journey times between them for the wealthy élite.

Economic conditions did not encourage town growth, but urbanisation through medium sized towns occurred, especially in the north and west. Rapid urban industrial growth in some places in the nineteenth century had created urban landscapes of stark contrasts. In the Polish city of Łódź 'The dark insanitary housing of the vast majority of people was punctuated by flamboyant "palaces" belonging to the mill owners . . . next to the squalor stood the Scheibler "Kingdom" – a model village showing how industrial towns should have been planned' (Dawson 1979: 379). The foundation of new towns was rare, but the port town of Gdynia in Poland and the industrial town of Zlin in Czechoslovakia were new. Whilst urban facilities were often primitive, some enjoyed newspapers, radio broadcasts and telephone services. Provincial cities, such as Lwow, Kraków (Poland), Brno, Kosice (Czechoslovakia), Tartu (Estonia), Thessaloniki (Greece) and many others, had radio stations, broadcasting for a part of most days. The principal Hungarian regional cities all had local telephone services. Between 1920 and 1936, telephones increased from 905 to 1,802 in Debrecen, from 1,143 to 1,881 in Pécs and from 1,587 to 2,764 in Szeged (Tinar 1992).

In the south and east, urban networks were poorly developed. Many towns had very scanty urban infrastructure. This included primary and vocational secondary schools and a general hospital in even some of the smallest towns, as in Cutea De Arges (Romania) whose population was 6,831 in 1930 (Fleure and Evans 1939). Regional centres such as Brasov (Romania), Stara Zagora and Varna (Bulgaria) had radio stations.

The variety of urban life was thickening for those men and women who perceived it as insiders. It was a complex mixture of opportunity and restriction at the root of which was ethnic nationality and social class identity. Seen through the eyes of outsiders, such as Western European travellers, from the seventeenth to the twentieth centuries, modernity, conceptualised as Europeanisation, was a much simpler affair. Modernity came to the towns and cities of Central Europe and the Balkans with the expulsion of the Turks (Jezernik 1998). As the Turkish tide ebbed, so the wave of European modernity flooded over the urban scene and changed the form, experience and meaning of urban life. Turkish towns were perceived as places of transience, like military camps. Islamic, religious architecture was the predominant symbol of their landscape. Turkish towns had neither the function, nor the form, nor the secular architectural quality of European towns for these observers. Where Turkish rule lingered longest, towns were places of wilderness and savagery, the opposite of western civilisation. It seemed to these travellers that when towns were freed from Turkish control, they began a transformation that removed and remodelled the congested, filthy streets, mean buildings and irregular form. New, wide thoroughfares, fine new buildings and an orderly plan were introduced influenced by the western perspectives of style, technology and urban design. Thus the symbols of Turkish occupancy, capture and control of place, individual buildings and whole areas of the town, were 'cleared out' and replaced by new buildings and new town design, symbols of the presence of western, European, modernity.

Whilst social and economic change was spreading to many rural and urban localities, local political rights remained undeveloped. The introduction of parliamentary democracy in most countries did not have a marked local democratic component. Capital cities often had strong municipal councils elected by limited franchise but elsewhere local democracy was muted. The old societal nexus continued to influence local conditions in many places. Despite the introduction of new administrative practices in many places, the local power of the old élites lingered on. Formal and informal networks of deference and authority were maintained. Local officials still paid attention to the views held by the land owners, priests, teachers and lawyers. In this way members of the old local régimes retained their informal networks of governance.

Modernity through the lens of locality in Central Europe

The onset of modernity in different localities of Central Europe can be shown by adapting information from A. Rónai's *Atlas of Central Europe 1945*, which compiled data for 1930–35. I have developed six indicators to provide a profile of the impact of modernity in localities (Figure 4.1). These indicators were mapped for 1,500 localities in Germany, Austria, Czechoslovakia, southern Poland, Hungary, Romania, Bulgaria and Yugoslavia.

1   A rate of natural increase below 12/1,000, indicating that the demographic transition was well advanced.

2 Infant mortality below 14/1,000, indicating the improvement in medical and health care associated with modern patterns of life.
3 A rate of illiteracy below 30 per cent for the population over ten years old, indicating progress in education.
4 An agricultural population below 50 per cent.
5 Industrial employment more than 25 per cent.
6 Employment in trade and communications above 15 per cent.

These indicators define modernity as more than just industrialisation, but the last three indicators show the extent of the occupational shift associated with modernity.

Each of these indicators or traits was mapped separately for each locality. They were then synthesised to construct a mosaic of modernity in the localities. Finally, the pattern of each synthesised category (0–7) was mapped separately. From this procedure it became clear that the thinner the veneer of modernity the more extensive is the area affected. By beginning with the localities that have all the attributes and then peeling them away one by one, the geography of the impact of institutional modernity in the localities can be revealed.

It becomes clear that three broad zones can be identified. First, there is a 'core' zone characterised by a continuous area of localities that are 'modern'. Around this is a second, inner fringe zone, with a greater number of isolated 'modern' localities. Around this is an outer fringe zone that has fewer isolated 'modern' localities.

Figure 4.1 shows a modernity within localities that is most intense in the north and west and decreases southwards and eastwards. The contrasts between modern and non-modern localities are often sharp, but clusters of modern localities are also identified. Almost always the earliest 'modern' districts in an area are the urban localities. The biggest cities always show most features of modernity. Modernity diffused generally from the west but further east the detail of transformation was punctiliform and diffusion began in numerous urban centres.

Case example: Budapest and modernity: iconography and representations of constrained metropolitanism

From 1873, when it was founded by the amalgamation of Buda, Óbuda and Pest, to 1914 Budapest was a place of modern dynamism, on the eastern edge of the European core. Its civic leaders strove for its entry into the élite company of Europe's metropolitan world cities. Budapest's growth was matched only by that of Berlin. The city grew twice as fast as Vienna and three times faster than Paris and London. In 1869 it ranked seventeenth in the European urban hierarchy; by 1910 it was seventh (Bender and Schorske 1994).

Budapest was developed as a symbol of nation building, but was constantly at odds with nationalist conservatism that contested its Hungarianness. The metropolitan values of modernism, liberalism and cosmopolitanism were repressed by the persistence of aristocratic nationalism. By the mid-1930s,

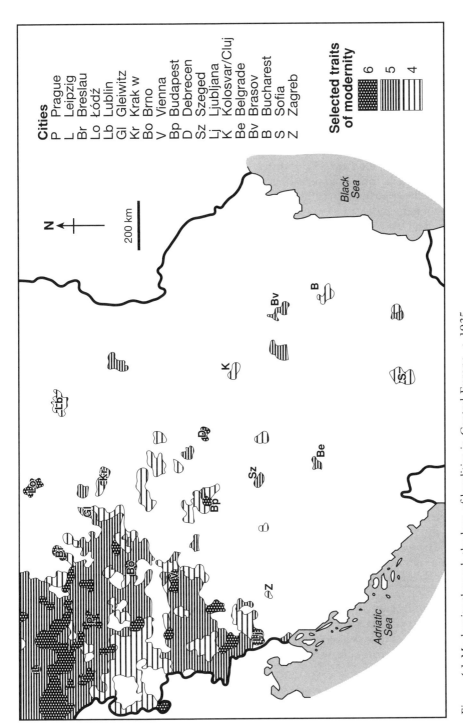

Figure 4.1 Modernity through the lens of localities in Central Europe, *c.* 1935.

Source: Adapted from Rónai (1993).

however, the dynamic agents of metropolitan culture had left the city to participate in the scholarly, artistic and entrepreneurial communities of London, Paris, New York and Los Angeles (Bender and Schorske 1994).

Modernism and tradition: the symbolism of the urban fabric

Budapest's rapid growth was closely regulated. A Master Plan was drawn up for Pest in 1872, for Buda in 1876 and for Óbuda in 1883. In 1886 four zones were delimited and within them plot size, building height and building materials were strictly controlled. Throughout the period of rapid growth the so-called Eclectic architectural style was predominant. Modernism made little impact against opposition from the authorities (Enyedi and Szirmai 1992). In the 1930s some individual modernist buildings were constructed, the best examples being Gyula Wälder's Madarch St apartment house and Alfred Hajós's Margaret Island Sports Swimming Pool. The Budaörs Aerodrome terminal is another example.

Modernist thinking in the mode of town planning sought to 'clear out' two old districts in Buda: Tabán or Raczváros (Serbtown) and Viziváros (Watertown) (Figure 4.2). The Tabán (Turkish for foothill) district was probably founded by Turkish settlers and in the Middle Ages Serbs settled there. It is situated at the foot of the Castle Hill. The small one storey houses, narrow winding streets and primitive sewage facilities were completely outdated. A plan of 1931 designed modern style apartment houses around courtyards and villas on the hillside. Some important Serb buildings were kept; 750 dwellings were demolished, but nothing was actually built. Sadly Joszef Vágó's 1936 modernist conception was never realised, but in 1938 a park was laid out. Viziváros is situated near today's Margaret Bridge. A plan to redesign Viziváros was prepared in 1937. Once the quarter of the well-to-do bourgeoisie, some elegant baroque houses were to be retained. Single storey dwellings were demolished and three to four storeyed blocks in gardens were envisaged for letting (Horváth 1980).

In contrast to the modernist clearing out of the Tabán and Viziváros was the construction of the St Imre Garden City. This 'City' was created in 1929 to rehouse Hungarians displaced from Transylvania and other territories lost after the First World War, many of whom were state officials and their families. The vision of the founders was to recreate the 'secure, tradition bound communities they were forced to abandon' on 170 acres of farmland in the Budapest suburb of Pestlörinc. It was named after Prince Imre, son of King Istán, founder of the Hungarian state. The street names recalled the places left behind. The residential architecture recreated the old family mansions and country houses they knew. Their houses were to be family fortresses of 'genteel officials suddenly snatched from the security of a feudal state, battered by the storms of history and the crisis of capitalism' (Teplán 1994: 176).

Artistic representation of Budapest

Hungarian artists did not celebrate the modernity, innovation, dynamism and technology of the Budapest landscape. Few had any interest in it. As the pace

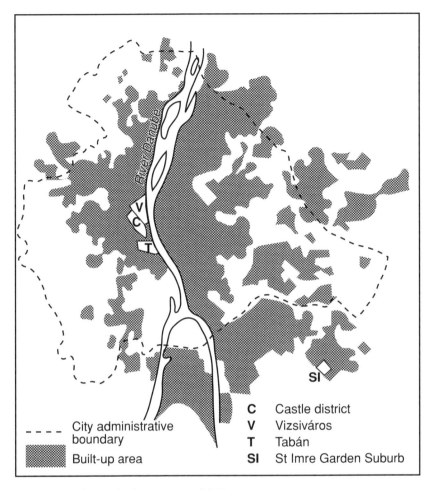

Figure 4.2 Budapest: built-up area, *c.* 1930.

of life increased, it was not the optimism of the vibrant future city that attracted them, but the melancholy and nostalgia of the past countryside. Artists were concerned with the tension between Hungarianness and Budapestness. When they did pay attention to Budapest, they painted narrowly focused cosy personal scenes. Only Hugo Schreiber's Factory (1925) is touched by a sense of urban dynamism. Armand Schönberger's Tabán (1920) is interpreted by Forgács (1994: 237) as 'betraying nostalgic feelings for old Budapest', but seems to me to represent something of the restlessness of urban life.

Case example: the Upper Silesian industrial district

The complex interconnection of nationalist ideology and capitalist development is shown by the experience of the Upper Silesian urban–industrial conurbation. Conflict between the post-war German and Polish governments split the district and divided the ethnic population leaving Germans in Poland and Poles in Germany. The two sections of the district that had earlier developed as a single unit followed separate pathways of economic and social development . Upper Silesia provides a good illustration of the problems that stemmed from the impact of the particular combination of nationalist and capitalist modernity on a culturally and politically divided remote border region.

Upper Silesia is a historic province that was divided between Germany, Austria and Russia in the Third Partition of Poland in 1798. The industrial district occupies only a small area of this province, measuring 25 miles long by about 13 wide. The district has no dominant central city.

The Upper Silesian Industrial District was made by capitalist investment in mining, heavy industry and concomitant housing, infrastructure and transport facilities. Feudal agriculture and trade routes had little interest in the area. There were no pre-industrial cities. In 1756 about 14,000 people lived in the area and there were four small towns (Hartshorne 1934). By 1932 there was a population of one million, six cities each with about 100,000 inhabitants and numerous mining and industrial towns and villages. Two cities, Gleiwitz and Katowice, had radio stations. Each had airports, Gleiwitz had two daily direct flights to Breslau and one to Kraków by Deutsche Lufthansa. Katowice had one daily service by LOT, Polish Airlines to each of Warsaw, Kraków and Brno (Rónai 1993).

The district's industrial structure and urban form emerged without control. The process of town and village extension by agglomeration around mines and factories was a formless scramble. Individual, state and corporate capitalist enterprise produced a landscape independent of any co-ordinated economic, physical or administrative framework. Mining and heavy industry were dominant. There was no investment in any other manufacturing. It was a landscape of close physical contrasts and great social distance. It juxtaposed workers' barracks, factories and mines in close proximity, whilst feudal landowners and capitalist entrepreneurs, often the very same families, lived in splendid palaces in semi-rural locations (Hartshorne 1934).

Following the First World War the Upper Silesian question arose because both the German and Polish governments wanted the economic wealth of the district. Should the new Polish state get it, because the majority of the population were Polish speaking, though not necessarily Polish 'feeling'? Should Germany keep it, because it had been separate from the historic Polish state for centuries?

The boundary dispute between Germany and the new Poland was not resolved until 1922. The final border was settled only after a diplomatic argument in Paris, an armed conflict locally and a plebiscite held in 1921. The plebiscite was conducted in the whole of the province of Upper Silesia whose population was 2.08 million, of which 60 per cent were Polish. The turn-out was 97.3 per cent and 59.6 per cent voted to be in Germany, 40.4 per cent in Poland. Subsequently, the line of the border gave about a sixth of the province to Poland, 29.9 per cent of its inhabitants being German. Five-sixths were awarded to Germany, with 49 per cent of the area's population being Polish (Pándi 1997).

The new state boundary ran right through the middle of the district cutting across established interplant linkages, tramlines, mineral narrow gauge and passenger standard gauge railway lines, roads, overhead electric cables and underground water and sewage pipelines. Landscape change on either side was small. The two sides remained connected by transport services, but these were much less used than before. They were disrupted by the eighteen new customs posts built where roads crossed the new international border.

Development immediately after 1922 was largely separated into two halves, yet the coalfield was of major world significance. In the mid-1930s the district was producing 80 million tons of coal per year, 7 per cent of the world's output. The value of output from individual mines was vastly in excess of that of others in the Marchlands. Developments continued to be shaped by the district's marginality and by distant events. It was an 'industrial island' region in an agricultural area on the margins of Europe's manufacturing belt and on the borderland between commercial Western Europe and subsistence Eastern Europe. The French occupation of the Ruhr led to extra demand for the output of coal from the German side. Support for the German side also came from German governments' aiming to free eastern Germany from dependence on Polish coal. The Polish side suffered from the German embargo and the Polish government's policy aimed at developing industrial districts away from the country's borders. On the other hand, the English coal strike of 1926 opened up Scandanavian markets to Polish coal exports. Polish government lobbying and cheap freight rates on Polish state railways, offsetting the long haul costs to Danzig and Gdynia, helped to retain this market.

Conclusion

Modernity in the guise of its nationalist, capitalist, liberalist structuration differentiated localities one from another and produced them as new places. There was a rich local mosaic that distinguished urban from rural localities, and picked out capital cities as particularly dynamic centres of modernity. At the same time, a locational dimension reflected the diffusion of ideas, methodologies and technologies from the west. The western districts were more affected by recent change than eastern districts. Most rural localities remained isolated, but the local transport services and localised commercial activity intensified rural urban links in some regions. Clearly most rural dwellers lived in a pre-modern mind set. Old attitudes and values still prevailed in rural agrarian villages, peopled ubiquitously by peasant communities, but cultural and societal problems were conspiring to denude rural areas of population, and migration to towns and cities proceeded apace. The extension of primary and secondary education into even small towns began to open new possibilities for the younger members of society. The development of industrial and commercial employment, the construction of urban transport networks, the provision of radio broadcasts, the increase in telephone services and the availability of cinema entertainment was beginning to shape the lives of urban residents. Urban populations were beginning to experience the opportunities of modernity. In cities there was a localised collapsing of space for the better-off. In capital cities a single urban culture with mass media was emerging.

5 The production of states and regions in nationalist modernity

Place, modernity and the Nationalist Project have their most complete expression in the idea of the nation-state. This chapter looks at the form and process of modernity through nationalist production of place in the practices of nation building and state building in the formation of the nation-state. It first considers the way that the boundaries of the new states were constructed, destroying the territorial order of the imperial system. It then examines the cultural, societal and locational practices that were used to modernise the new state division of space in relation to capitalist societal modernity. Regional modernity is discussed in terms of the approaches adopted by contemporary geographers and current theorists. Yugoslavia and its ethnic make-up and regional policy in Poland are discussed as case examples.

The delimitation of the new state boundaries

For the nationalists at the Paris Peace Conference, and immediately following it, the first act of nation-state building was the delimitation and consolidation of the new states' territorial boundaries. The struggle to delimit these boundaries was a competition to appropriate as much territory as possible for each nationalism. The stance taken by each nationalist protagonist invoked a mixture of cultural and societal attitudes and values, aided and abetted by manipulative practices to claim and capture as much territory as possible. Underscoring the territorial demands of each protagonist nationalism was a sense of place-constructed identity. This was encapsulated in the idea of 'homeland' conceptualised as 'fatherland' or 'motherland', that was portrayed as conveying a deep sense of cultural meaning for a particular tract of land. Kristof (1994) argues that the 'fatherland' idea, at least in the case of Poland, was culturally constructed by amalgamating an 'image' of the past and a 'vision' of the future. This conceptualisation of 'fatherland' was projected onto the nation-state and became highly politicised. Thus, the territorial claims of a cultural 'homeland' became embellished by the projecteers with political, strategic, economic and territorial aspirations. The political competition for territory in the aftermath of the First World War was dominated by the victorious powers. Whilst it was the hope of the American President Wilson that new, independent nation-states should be founded on the basis of national self-determination, the reality proved different. The process of boundary delimitation involved diplomatic lobbying and negotiation during the war, and after it at the Peace Conference in Paris, local armed skirmishes, annexations and wars between states, and, in some restricted localities, plebiscites.

The Paris Peace Conference was permeated by hostility and its overall tenor was to favour the claims of nationalists over those of the former, now defeated, imperial powers. Over large tracts of the Marchlands there was territorially consolidated ethnic settlement. There was relatively little difficulty in identifying these 'core' areas of settlement and including them in the new state boundaries. Difficulties arose on the edges of these tracts in which ethnically mixed settlement was often very complex (see Chapter 4). Resolutions were influenced by several factors. Among these the claims to locations declared by romantic writers of 'national' histories to be places of meaning, which created national identity, were prominent. The more recent imperial administrative boundaries were also significant. At the same time, economic and strategic considerations and the straightforward territorial aggrandisement of the new states played an important part.

The administrative provinces of the Habsburg monarchy often had the core settled 'homelands' of a particular ethnic group. Claims for statehood by this ethnic group tended to demand the whole of the provincial territory, not just the national core. Thus, for example, Czechoslovakia claimed all the provincial territories of Bohemia and Moravia (the Czech lands of the Habsburg monarchy) although the Czechs were only the absolute majority in central districts of these territories. Romania claimed the whole of Transylvania, although in many districts in the province Romanians were not in the majority as the later discussion of minorities in the majority will show.

Wars, annexations and local armed conflict also played a part in the delimitation of boundaries. There was full-scale war between Poland and the Soviet Union over the border line between them shifting the Peace Conference proposed 'Curzon Line' boundary eastwards and giving the Poles more territory. Polish and German armed groups skirmished in Upper Silesia. Romania invaded Hungary and annexed Transylvania and also seized Bessarabia by force from the Soviet Union.

Plebiscites were held in six border districts. Along the East Prussian–Polish border the inhabitants of two districts confirmed by a 96.7 per cent vote in favour, their inclusion in Germany (Pandi 1997). The districts of Teschen, Avra and Szepes, along the Polish–Czechoslovak border, chose between these two countries. In 1921 the German–Polish border in Upper Silesia was finally decided by plebiscite. The Burganland, the western border district of Hungary, was awarded to Hungary in the peace treaties, but was subsequently claimed by Austria. After the ratification of the Treaty of Trianon, the question was reopened. Finally, the Burganland became Austrian, except for the town of Sopron, whose inhabitants in a plebiscite voted to remain in Hungary (Pándi 1997).

National identity and nation building within the new territorial boundaries

When the boundaries were finally agreed, in no case was the bounded territory of the new states coextensive with the settled territory of a single ethnic group. This fact of geographical space exposed the fundamental weakness of the nation-state idea in the Marchlands. It introduced an ambiguity between the nation and the state as the source of civil rights and the sense of belonging in the new political spatial order. It thus brought two specific attributes of modernity into conflict. Nationalism, in capturing space for the 'nation' and for the 'state', had stored up some fundamental contradictions of place identity. Ethnically based nations were exclusive, whilst citizenship-based states were inclusive of individual membership. The new ruling élites were compelled to deal with this issue. Would they choose the 'closed' ethnic, national route or the 'open' citizenship route? How would they seek to build the future 'national' identity within the new 'state' bounded territory, the place they were constructing?

These were not simple questions and they go to the core of the nationalist spatiality of modernity in the Marchlands. Neither the idea of 'ethnic nation' nor the idea of 'state citizenship' meant much to the peasantry, who made up the vast majority of the population. The historic idea of 'nation' in Poland and Hungary was as a community of nobles. The peasantry was excluded. The peasants themselves had only a vague idea of nationality – their national consciousness was ill-formed. Political and cultural modern nationalism had to grapple with the aristocracy in extending the idea of nation to the peasantry and at the same time make nationality a vital dimension of the peasantry's self-identity. The idea of citizenship as conferring political and civil rights on the individual was equally alien. Once more the nobility had this sense of belonging to a polity, but the peasantry had no such sense or experience.

The slowness of these transformations during the nineteenth century was related to the entrenched power of the nobility, who steadfastly defended their feudal privileges, coupled with the illiteracy of the peasantry. The importance of printed, vernacular language in the breaking of traditional dynastic (family) and religious community values has been attested by Anderson (1991). Not until they transposed space to place through the territorial nation-state did the modern nationalists have a strong position to press upon the peasantry, or the nobility, their construction of national identity and national community.

Nationalist modernity emphasises the difference between ethnic groups and plays on suspicion, mistrust and arrogance between them. The key spatiality of nationalism is the ethnic test for inclusion/exclusion. Corollaries of the dominance of nationalism in the nation-state are the idea of national minorities and geographies of exclusion. During the boundary disputes and when the final delimitation had been made, displacements of enormous numbers of people got underway. Between 1919 and 1924 around five million people moved across the new international borders. There were 750,000 refugees from the Bolshevik revolutions in Russia. Some 480,000

Germans, 400,000 Hungarians, 350,000 Turks, 220,000 Bulgarians and 200,000 Latvians, Lithuanians and Estonians migrated during this period. The biggest groups to move though were the 1.36 million Greeks and 1 million Poles (Barraclough 1986). Some movements were in theory voluntary. Reciprocal resettlement agreements were signed, for example, between Greece and Turkey and Greece and Bulgaria (Pundeff 1969). In reality these agreements forced Bulgarians out of Aegean Macedonia and Thrace to make room for Greeks resettled from Asia-Minor (Figure 5.1). Thus, the forced international migration of people who found themselves on the wrong side of the new borders considerably redrew the ethnic settlement geography of the Marchlands.

But the mismatch of the mosaic of ethnic settlement and the new state boundaries in some regions was not resolved by these migrations, no matter that many millions of people were forced out of their homes. The boundary delimitations left some districts of the new states in which minority ethnic groups were the majority population (Figure 5.2). A second phase of nationalist nation building was enacted through the minorities and settlement/colonisation policies of the new states' governments. The policies of the Greek government are noted above. The Belarusians and Ukrainians settled in Poland's eastern borderlands were seen as in need of protection by Poland, but at the same time a threat to the state as their national consciousness was taking shape. They had to be assimilated, and so their schools were closed and their language supressed in favour of Polish (Frankel 1946). The attitudes of many governments is reflected by the Polish national democrat politician S. Grabski, expressing the sentiment in 1921 that 'today, the transformation of the state territory of the Republic into a Polish national territory is a necessary condition of maintaining our frontiers' (Tomaszewski 1993: 299). In other words, all those living in the territory of the state of Poland would be required to conform in self-identity and action to the idea of ethnic national Polishness. The state territory must become the place of Polishness as constructed by the Nationalist Project.

In their desire to shape patriotic and cultural identities attached to the new state territories, governments found themselves in one of two general situations. In one situation, as in Poland, Hungary, Romania, Bulgaria and the Baltic States, the new state boundaries encompassed one predominant ethnic group. This group accounted for 70 per cent or more of the inhabitants. However, in Poland, Ukrainians and Belarusians, and in Romania, Hungarians, were settled in particular districts where they formed the majority population. The other situation, as in Czechoslovakia and Yugoslavia, governments in fact ruled multi-ethnic states and several substantial ethnic groups were settled in the country (Figure 5.3).

It is also useful to think of a distinction between territorial 'gainers' and 'losers'. Gainers often incorporated minority populations as a result of their acquisitions and their 'national question' was internal. Losers became more homogeneous because of their loss and for them the national question was externally focused, though it had internal consequences. The territorial gainers, Poland and Romania, adopted discriminatory and assimilation

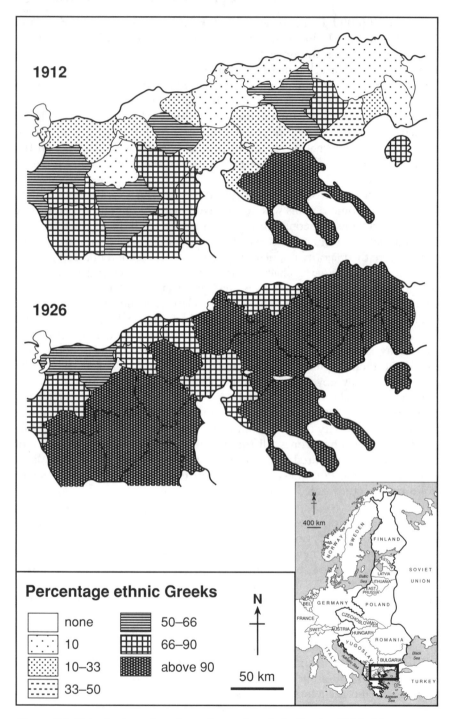

Figure 5.1 Colonisation of northern Greece by ethnic Greeks, 1912–26.

Source: Adapted from Pándi (1997) *Kösztes-Európa*, Osiris, Budapest.

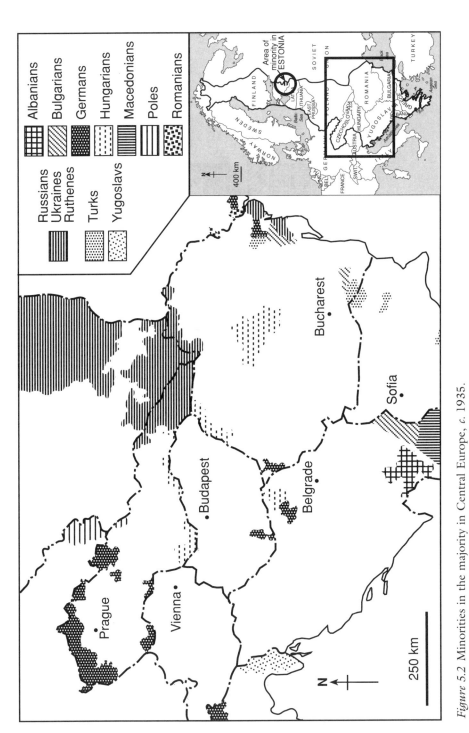

Figure 5.2 Minorities in the majority in Central Europe, *c.* 1935.

Source: Rónai (1993) *Atlas of Central Europe.*

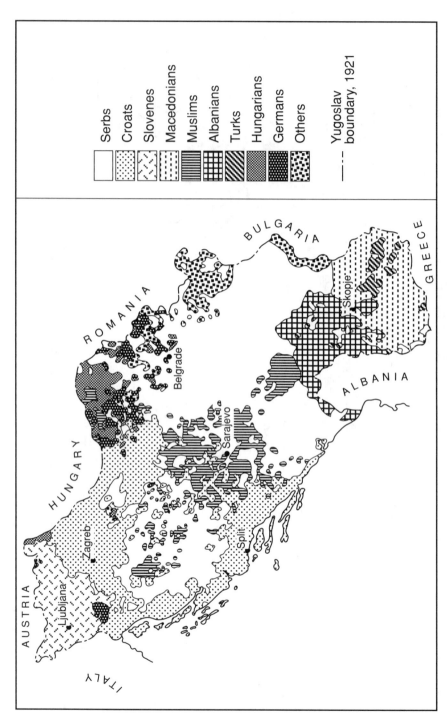

Figure 5.3 The ethnic settlement pattern of the kingdom of the Serbs, Croats and Slovenes, 1921.

Source: Adapted from Kocsis (1993) Jugoszlávia: Egy felrobbant etnikai mozaik esete.

measures against minorities in their strategy of national state building. Local populations were displaced. Refugees, veterans and repatriates were settled in new territories in order to change the ethnic population balance. Local jurisdiction and self-government, where they had existed, were removed by centralising state powers. Schools teaching in local languages were closed. Nationality was coupled with ethnicity. Citizenship rights were fully advanced only to those of the predominant ethnic group.

Jews were discriminated against everywhere. Jewish communities were excluded by the predominant nationalities. At the same time Jewish communities often excluded themselves from society by following their religious practices very strictly. This emphasised the difference between themselves and the particular national societies within whose state they lived. Many Jews having amassed substantial capital were economically powerful. Jews were also prominent in the medical, legal, journalistic and banking professions, despite restricted access to higher education. Thus Jews often espoused commercial and capitalist values of modernity alien to the predominantly peasant population. For this reason anti-semitism often expressed a reaction of traditionalist attitudes to modernity.

Case example: Place, national identity and the state in the idea of Yugoslavia

The foundation of the Kingdom of the Serbs, Croats and Slovenes in 1921 and the idea of Yugoslavia illustrate the complex interplay of locational, societal and cultural motives that were at work in the nationalist transposition of place from space.

The Yugoslav Project sought to give form and practicality to South Slav nationhood but this was a highly contested and tenuous concept. Yugoslavia was not a nation-state idea, it was a multi-national state idea.

The creation of Yugoslavia, it might be claimed, can be traced to the Napoleonic establishment of the French imperial province of Illyria, because this province incorporated Serbs, Croats and Slovenes in a single administration. At about the same time, the awakening of national consciousness began in these three groups.

Serbs settled in Vojvodina, located outside Serbia proper, and began to write the history of the Serbian nation. They constructed an identity in a Romantic style permeated with a sense of adversity and defined by means of places to which Serbs had a strong emotional–psychological attachment. Such places were widely scattered and Romantic nationalists used this to promote a sense of a Greater Serbian identity. Some Serb nationalists came to believe that all South Slavs were essentially Serbs and Serbia's destiny was to unite them in a Greater Serbian state (White 1996).

Slovenes gained from Illyria by the recognition of their language. In 1853 a map of Slovene place-names had promoted the idea of a single Slovene administrative area in the Austrian crown lands. This was never achieved. In 1921 a provisional government united the Slovene lands and expected to be accepted as such in the new kingdom. It was disappointed (Velikonja 1994).

In the later nineteenth century Croat and Serb intellectuals conceived the idea of Yugoslavia formed by a common Yugoslav identity constructed without a dominant ethnic group. This idea was not popular among the peasantry. Some Serbs interpreted the idea as a willingness of other South Slavs to become Serbs in a Serbian state. This was a misinterpretation.

In 1915 international diplomacy became involved. The London Yugoslav Council produced a territorial claim and a secret agreement delimiting a South Slave state was made with Britain. Another secret agreement made between Italy, Romania, Croatia, Serbia and Montenegro, but without the participation of the Slovenes, promised acceptance of a different border. The Serbs were the most committed of the South Slav ethnic groups pressing diplomatically for the formation of a South Slav state. In 1917 the Serb Prime Minister declared in Corfu the foundation of a South Slav kingdom headed by the Serb Karadjordjevic dynasty.

However, the final boundaries of the Kingdom of the Serbs, Croats and Slovenes were drawn differently from the wartime agreements (Pándi 1997). Added to the former territory of Serbia were the Vojvodina, Banát and Bácska from Hungary; Craine, Slovenia and Dalmatia from Austria; and Bosnia-Hercegovina formerly jointly owned in the monarchy. Montenegro was brought into the kingdom as were several small territories from Bulgaria.

This solution brought ethnic tension as eleven distinct ethnic groups were incorporated into the territory, but not all Slovenes. In 1920 a plebiscite in Carinthia favoured Austria, the peace treaties confirmed Slovenian Venetia in Italy and Slovene lands within the kingdom were split into two for administration. It also brought political tension. The Serbs strove for a unified state with strong central powers. The Slovenes and the Croats wanted devolved power structures. In 1929 the king declared a royal dictatorship in a deteriorating situation. The Constitution of 1931 signalled a drive for total Yugoslav integration; to create a homogeneous Yugoslavia. A Yugoslav language replaced Slovene, Croat and Serbian, only national political parties were allowed, provincial institutions were restricted and ethnic identities suppressed. This was interpreted by many as the imposition of a Greater Serbia.

Case example: Czechoslovakia

In Czechoslovakia the 7 million Czechs (about 50 per cent of the population) saw themselves as predominant in a politically and culturally unified Czechoslovak nation-state, the 'domain of the "Czechoslovak nation"' (Zacek 1969: 193). However, the 3 million Germans, 750,000 Hungarians and 80,000 Poles who had not wanted to become part of the state, refused to accept minority status and agitated to be united with their respective 'national homelands' in contiguous territories. The 500,000 Ruthenes began to recognise in themselves a Ukrainian identity. The 2 million Slovaks resented their lack of autonomy and nurtured their separate identity. However, compared with the other states, Czechoslovak treatment of non-Czechs and Slovaks was reasonably fair. Citizenship rights were extended to all its inhabitants and democratic practices were generally well observed.

The state and capitalist societal modernity

Capitalist modernity reached the Marchlands in its British industrial hege-monic phase, using Taylor's (1999) geo-historic periods. Gerschenkron (1966) emphasises the particularities of a country's mineral and human resource endowment, level of technology, infrastructure and the attitudes of ruling élites as strong influences on trajectories of industrialisation. Backward countries can create substitutes for the preconditions of industrial growth found in countries that had been industrialised earlier. In contrast, Rostow (1961) argued that there was just one pathway to industrialisation, each country must pass through a particular sequence of stages to achieve indus-trialisation. Dunford (1998: 63) argues that 'the timing and nature of mod-ernisation lies in the trajectories of social and institutional relations', a factor to which, in his view, neither Gerschenkron nor Rostow gave sufficient weight. The period of the Nationalist Project also conforms with the later part of the Third Kondrotieff cycle. This cyclical model of industrial devel-opment has been interpreted as concordant with sequences of technical inno-vation and theories of institutional regulation (Dunford 1998). All these economic or social institution-based theories understate the importance of spatiality, especially territory and location.

Berend and Ránki (1974) point out that the Marchlands experienced several different pathways to modernisation in the nineteenth century, but they had some common features. These included a reform from above, a relatively prominent role for the state and a failure to solve the land ques-tion or the problem of the peasantry, as well as slow progress to democracy and a strong nationalist colouring. By the end of the nineteenth century in some countries there was a free labour force, freedom of enterprise, secu-rity of property and a credit system, but a great shortage of up-to-date technology and infrastructure.

Nationalist spatiality and its new delimitation of state territorial bound-aries had a critical effect on capitalist societal modernity as mediated through the state as a place. Political reterritorialisation delimited arenas of interac-tion between capitalist and nationalist modernity, not only by apportioning the nature and amount of economic resources – human and material – avail-able for development, but also by apportioning the existing territories of modernised economic development. The new boundaries disrupted existing territorial patterns of economic space and brought together regions with contrasting levels of economic development. These features of economic space added to the tensions arising from mismatches of ethnically settled and politically bounded territories. Extreme forms of nationalist economic and political attitudes permeated policy making. However, the forces of capitalist/societal modernity themselves would help to cement the new state boundaries and the idea of national identity. Modernity brought new expe-riences to fashion and unified a sense of national identity within the new boundaries. It was within the boundaries of the new nation-states that modernity created new lifestyles and new opportunities.

The territorially delimited mix of nationalist and capitalist modernity was permeated by social and economic contradictions that affected the formation of state economic policy. The inclusiveness of capitalism contrasted with the exclusiveness of nationalism. Whilst remnants of the aristocracy and the ubiquitous peasantry opposed nationalisation as a state economic policy, the intellectuals drawn from the middle class and petit bourgeoisie supported it. For them it represented an opportunity for security and prosperity as the administrators in state enterprise (Haynes and Husan 1998). As foreign investment increased other actors were added. Foreign investors played an ambiguous role in influencing policy. Whilst foreign capitalists may have been natural opponents of state involvement in enterprise, many foreign capitalists approved of state aid, as long as they were the beneficiary of that aid (Haynes and Husan 1998).

State regulation of capitalist dynamics was difficult to achieve, but nationalism was able to forge a new and particularly strong relationship between the state and economic development. State sponsored economic development was intended to strengthen the state. As Haynes and Husan (1998) have pointed out a specific form of nationalist state-capitalist modernity emerged that reduced the importance of the market in economic development.

Whilst the relationship between the nationalist state and capitalist modernity had a number of common features, differences at the state level were such that each state may be regarded as following its own particular trajectory. State regulated capitalist societal modernity can be assessed by reference to trends in economic development, infrastructural and communications provision, and demographic characteristics (Table 5.1).

The peace treaties permitted the allied governments to expropriate property belonging to aliens by what was termed 'nostrification' (Berend and Tomaszewski 1986). Early state intervention in capitalist economic development occurred through 'nostrification' and land reform. In Czechoslovakia and Poland, but particularly Yugoslavia and Romania, nostrification was used to strengthen state and private domestic ownership of landed property and capital assets. Land reform attacked the aristocratic establishment. It was most radical in Romania and Yugoslavia for several reasons. First, the 'foreign' or expelled aristocratic landlords were powerless to prevent it. Second, land hunger was strongest. Third, alternative industrial employment was unavailable. In Czechoslovakia the need for land reform was eased by the growing demand for industrial labour, at least in the Czech lands. In Poland the landed aristocracy was strong enough to limit reform whilst the Hungarian aristocracy was so powerful that it kept reform to a minimum. Bulgaria, in contrast, was a country of peasant ownership and few landed estates, but some expropriation and redistribution did occur.

These processes created a weak undercapitalised agricultural sector in which capitalist transformation of economic and social relations was rendered difficult (Berend and Ránki 1974). Land reform consolidated the power of the peasantry. This was invariably used to defend traditional values and practices, slowing the introduction of new and innovative ideas and techniques of management.

Table 5.1 National economic, demographic and communications indicators, c. 1930–35

Country	Economic measures				Demographic measures				Communications measures 1938			
	GDP per capita 1937	% agriculture	% industry	% commerce	Infant mortality deaths per 1,000 live births	Natural increase per 1,000	% urban dwellers	Illiteracy % of population over 10 years	Cars per 1,000 population	Radios per 1,000 population	Rail identity km/km² population	Telephones per 1,000 population
Albania	—	—	—	—	100	17	12	70	—	—	—	—
Bulgaria	75	80	8	12	144	14	21	31	0.7	2.7	32	3
Czechoslovakia	170	28	42	30	110	8	48	4	7	61	100	14
Estonia	—	66	15	19	89	—	32	8	3	14	0.003	9.5
Greece	92	61	18	21	118	11	—	41	—	—	—	—
Hungary	120	54	24	22	132	9	43	9	25	40	93	15
Latvia	—	67	15	18	—	—	35	—	2	17	—	33
Lithuania	—	80	6	4	—	23	37	1	1	—	—	—
Poland	100	65	17	18	140	16	27	27	1	20	52	7
Romania	81	78	7	15	179	14	20	43	1.3	8.4	38	3
Yugoslavia	80	79	8	12	140	15	22	45	1	6.4	38	3
Finland	214	60	18	22	74	6	13	—	6.5	61	56	3

Sources: Economic measures, GDP: Ehrlich (1985); % employment: *Industrialisation and Foreign Trade*, League of Nations 1945 (facsimile edition Garland Publishing 1983); Demographic measures: Hauner (1985): 76, 83, 93, 98; Communications and transport measures: Rónai (1993); Mitchell (1975).

The sectoral structure of the economies shifted, albeit slowly, towards employment in industrial and commercial sectors and away from agriculture (Teichova 1985). GDP per capita rose and the contribution of industrial production to GDP was increased. The contribution of agriculture to GDP declined though it remained the largest component everywhere except in Czechoslovakia. Branch shifts occurred within industry. Teichova (1985) notes that the textile industry was particularly dynamic in Hungary and Romania, overtaking the food industries in employment, whilst the chemicals industry experienced especially rapid growth in Czechoslovakia. Rich mineral reserves in Hungary (bauxite), Romania (oil and gas) and Yugoslavia (non-ferrous minerals) were the basis of growth in heavy industry. Metal and engineering, food processing and textiles were the dominant branches in output and employment in most countries, but their relative positions differed between countries.

Industrialisation was promoted by government investment, protectionist policies and by foreign direct investment. This resulted in a tendency to increase the size of industrial production units. Industrial concentration of this kind was recorded in the numbers of workers employed, the energy consumed and the value of output. In the early 1930s, for example, enterprises with more than 500 workers employed 29.4 per cent of Czechoslovak industrial workers, 36.7 per cent of Poles, 32 per cent of Hungarians, 27 per cent of Bulgarians and 18.8 per cent of Romanian industrial workers (Teichova 1985). Concentration also promoted the domination of some industrial branches by a relatively small number of big companies. Soon cartels became predominant in the energy, metallurgy and building materials sectors. At first cartels were formed by voluntary agreements but later were enforced by state policy. The world economic crisis further increased these tendencies to monopoly conditions as companies and governments wanted to protect their market share. In the 1930s foreign direct investors owned 83 per cent of joint stock company equity in Romania, 61 per cent in Yugoslavia, 48 per cent in Bulgaria, 44 per cent in Poland, 29 per cent in Czechoslovakia and about 25 per cent in Hungary (Teichova 1985: 292). Despite these trends, small-scale enterprises remained important in industrial development.

Csernok, Ehrlich and Szilágyi constructed a complex index of infrastructural development for the Marchlands, encompassing transport, communications, health and education, housing and culture (Ehrlich 1985). This index can be utilised to assess the relative progress of countries to material modernity. The position of countries in the ranking and value on the index showed stability, mobility and different rates of change, all attesting to the individual experience of the states over the period between 1920 and 1937. The countries also showed strengths and weaknesses in different infrastructural sectors. Countries with the most developed infrastructure were Austria, Finland and Czechoslovakia. The least developed was Romania. Greece, Hungary, Yugoslavia, Poland and Bulgaria had intermediate levels, but their relative status changed over the period. Many countries followed a pattern in which levels of infrastructure declined in the 1920s, as measured by the index, but recovered during the 1930s. Exceptions were Austria and Poland whose index values rose steadily and Hungary which had a strong rise in the 1920s, falling back in the 1930s. The gap between the top and bottom countries increased steadily. Six countries had a higher level of infrastructure in 1937 than in 1920, but three had seen a decline. There were also differences between the least developed and the most developed components of infrastructure in the various countries. Education had been the least developed element in most countries in 1920, but by 1937 communications infrastructure was the least developed branch in the majority of them. In this vital dimension the gap between the best provided and the worst provided had grown, but all the countries had seen improvements over the period. In this fashion a west–east divergence was formed.

Capitalist industrialisation reshaped societal relationships within the framework of a transition to modern population patterns. Amongst the indicators of demographic modernity are particular characteristics of national vital statistics, rates of illiteracy and the proportion of the population living in urban centres. All countries had generally the same demographic experience as they passed through the demographic transition, but several factors affected the particular patterns of national demographic trends. The religious make-up of the population had a general effect. Protestants and Jews tended to have fewer children than Catholics, whilst Orthodox, Uniates and Moslems generally had more (Hauner 1985). The First World War also had an effect. Apart from direct military deaths, epidemics also took a heavy toll. Even in 1940 the military losses of the First World War and the loss of births were still reflected in population patterns.

National population dynamics were also affected by international and intercontinental migration. Industrial Western Europe had always attracted immigrants from the east. Immediately following the Peace Settlement, there were substantial population movements across the new international boundaries, as ethnic refugees moved into their new homelands, often following expulsion from territories that had changed hands as a result of the Settlement. Large numbers moved into Eastern Europe from Russia, fleeing the civil war and Bolshevik régime. Overseas migration had been a common trend of the pre-war years. In the decades 1921–30 and 1931–40 Central and Eastern Europe accounted for 21.6 per cent and 25.2 per cent respectively of overseas migration from Europe (Kirk 1946). However, in 1921 the United States government introduced restrictive quotas slowing emigration. Emigration was also severely curtailed by the Depression. The reasons for emigration remained the same though, namely the surplus agricultural population and, for Jews, anti-Semitic discrimination.

The Peace Treaty added greatly to the length of international borders in the Marchlands. The creation of new international borders, in the prevailing conditions of international tension, brought hindrances to movement such as the introduction of passports and the restriction of trade. This occurred at the time when the most recent advances in communications technology were opening up the potential of removing borders. Expanding air travel, increasing telephone services and the introduction of radio broadcasting all provided a means of contact that by-passed territorial borders and the hindrance they imposed on contacts and communication. It was important, therefore, from the point of view of governments that the state should successfully regulate these new means of communication and, where necessary, they should be organised through interstate agreements.

Technological development of telephone services and of radio broadcasting kept them primarily within the ambit of state boundaries. The telephone systems relied on a network of telephone exchanges for intra-urban calls. Several countries developed inter-urban networks; the Finnish network, for example, amounted to 141,800 kilometres of lines by 1938 (Lingeman 1938). International calls were only a small proportion of telephone usage, although inter-capital city trunk lines, between Berlin, Prague, Vienna and Budapest, had existed since before the turn of the century (Tinar 1992). An underwater cable provided telephone connection between Finland and Sweden from 1928 (Lingeman 1938). The range of radio stations was also restricted by technical factors that made it essentially a national means of communications. However, the establishment of radio stations near international borders, such as that between Germany and Czechoslovakia, permitted the reception of foreign stations in some regions (Figure 5.5). Thus telephone services and radio broadcasting effectively collapsed time and distance mainly within regional and national space.

The production of regions

The practising geographers of the 1930s perceived a world of 'natural' regions. These were bounded by natural features and within them particular relationships between the inhabitants and the natural environment were seen to form distinctive patterns of life. Hence the idea of the 'region' carried meaning beyond just 'area'. For geographers the idea of the region conveyed a sense of a particular relationship between natural and societal features that endowed territorial divisions with cultural meaning. To some geographers of today this seems a strangely static view of the world. Perhaps they forget that even sixty years ago the pace of life was much slower than today, and this was especially so in the Marchlands. Geographers of the 1920s and 1930s were interested in regional economic development. Regional economic specialisation within individual countries was a strong theme. Agricultural specialisation, for example in Yugoslavia (Shackleton 1925) and Bulgaria (Roucek 1935), was interpreted in relation to the specific physical, societal and historical conditions of the country. The prospects for economic development of border regions, places of special sensitivity at the time, attracted attention, as exemplified by Geddes' (1940) review of the Polish–Soviet borderlands.

As lifestyles were changed by advancing modernity, new approaches were needed. W. Christaller (1933) approached regional delimitation indirectly from the point of view of towns as providers of special functions for agrarian regional hinterlands. He stressed location, connectivity and movement in defining regions focused on a hierarchy of central places identified as market, administrative and transport nodes. His location theory was derived from empirical work in southern Germany, in terrain replicated in large areas of Poland, the Hungarian Great Plain and the plains of Walacia. This kind of regional analysis draws attention to an important aspect of regional development during the inter-war period, namely the increasing regional influence of cities, extended over hinterlands of widely different localities (Figure 5.4).

Geographers now theorise that capitalist modernity (a) increased contrasts within regions and redrew regional boundaries; (b) increased the contrasts between regions in terms of economic development; and (c) increased the contacts between regions. Out of this emerged the idea of uneven regional development and of regional systems. The principal underlying processes causing these transformations were local and regional concentration of circular–cumulative effects and core–periphery relationships in socio-economic interactions. The outcome was an emphasis on the classification of regions as economically dynamic or backward, and the relationships between them conditioned by 'spread' effects or 'backwash' effects. Some regions had an 'initial advantage' derived from earlier development that tended to ensure long-term continuity, more stable economic conditions and greater growth potential.

These points are exemplified by a synthesis of cities and their contrasting hinterlands during the 1930s (Figure 5.4a), constructed from the maps compiled by Rónai (1993). Prague was the centre of a densely populated dynamic industrial region. Prague's sphere of influence, as measured by the six hour railway isochrone, covered the historic province of Bohemia almost exactly. There were 7.1 million inhabitants of whom 42 per cent were employed in industry and 20 per cent in commercial and professional services with only 24 per cent employed in agriculture. The region was made up almost entirely of modern localities, to the reciprocal benefit of the region and the city. The rail network centred on the city was intensively used. The city had three radio stations, and within Bohemia there were a further two radio stations. In different parts of the region, radio broadcasts from Vienna and Dresden stations could be received. Prague also benefited from inter-national air services with four direct flights daily to Vienna and Brno, two to Munich and Leipzig in 1935.

In contrast, the spheres of influence around other capitals such as Belgrade and Bucharest were far less developed. Around Bucharest, for example, only the locality of Ploesti was 'modern'. Just 10 per cent of the population worked in industry and 13 per cent in commercial and professional services, with 70 per cent engaged in agriculture. The rail network was thin, the range of radio stations localised and daily direct air services few and mainly to national centres.

The spheres of influences surrounding regional centres such as Zagreb, Lvov and Kolozsvar (Cluj-Napolca) were much more limited (Figure 5.4b). Their spheres of influence covered agricultural regions, with few modern localities and thin rail networks that were only lightly utilised. Radio stations had a very small range, inhibiting the impact of the dynamism of city region

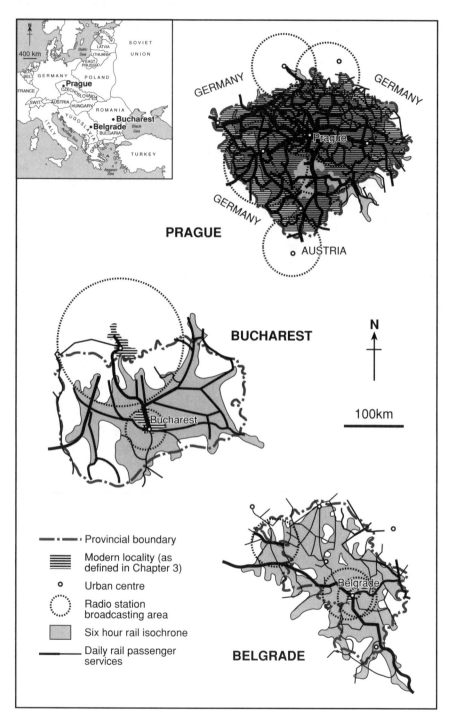

Figure 5.4a Modernity and regional development: Prague, Bucharest and Belgrade.

Source: Adapted from Rónai (1993) *Atlas of Central Europe*.

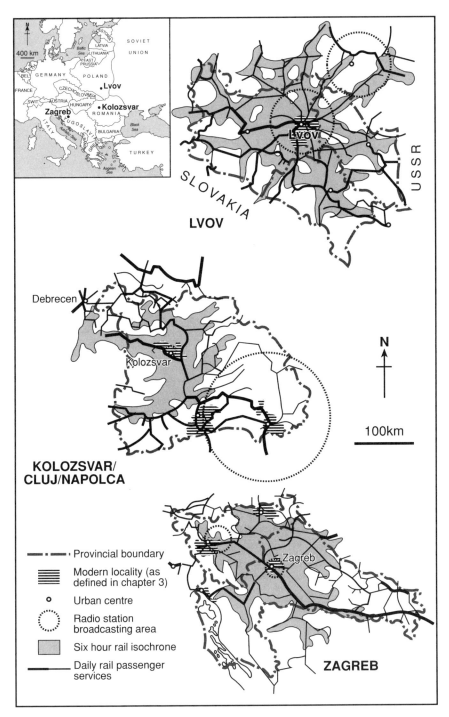

Figure 5.4b Modernity and regional development: Lvov, Kolozsvar/Cluj-Napolca and Zagreb.

Source: Adapted from Rónai (1993) *Atlas of Central Europe.*

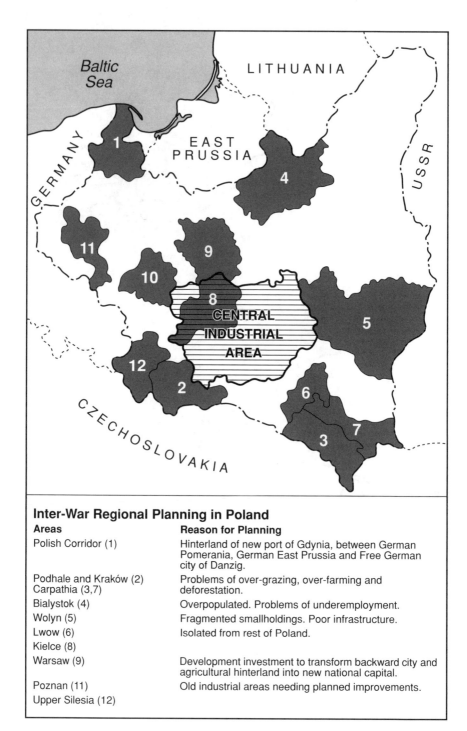

Inter-War Regional Planning in Poland

Areas	Reason for Planning
Polish Corridor (1)	Hinterland of new port of Gdynia, between German Pomerania, German East Prussia and Free German city of Danzig.
Podhale and Kraków (2) Carpathia (3,7)	Problems of over-grazing, over-farming and deforestation.
Bialystok (4)	Overpopulated. Problems of underemployment.
Wolyn (5)	Fragmented smallholdings. Poor infrastructure.
Lwow (6)	Isolated from rest of Poland.
Kielce (8)	
Warsaw (9)	Development investment to transform backward city and agricultural hinterland into new national capital.
Poznan (11) Upper Silesia (12)	Old industrial areas needing planned improvements.

Figure 5.5 Planning regions in Poland in the 1930s.

Source: Adapted from Fisher (1966) *City and Regional Planning in Poland* and Hamilton (1982) 'Regional Policy for Poland: a search for equity' *Geoforum*, 13, 121–32.

interaction. All had daily air services but networks were very limited. Zagreb had two daily flights to Susak on the Adriatic coast, but only one to each of Lubljana and Belgrade, whilst Lvov linked to Kraków, and was a stop on the route from Warsaw to Bucharest.

Some important regional centres that had developed in the Habsburg territories were adversely affected by the international boundary changes following the end of the First World War. Several regional cities, such as Szeged, Bratislava and Trieste, were cut off from their hinterlands. The new boundaries located the cities in one country, whilst their hinterlands were incorporated into other states, disrupting local and regional economic and commercial life.

These contrasting patterns of regional development emerged in the Marchlands as capitalist industrial modernity shaped regional evolution into a competition between different spatialities. Industrial capitalists evaluated places from the point of view of the richness of exploitable mineral resources or as peopled by inhabitants who formed a viable market for industrial goods and commercial services. This was a different perspective, not only from the former aristocratic élite, but also from their mercantile capitalist predecessors. In the inter-war period the advance of industrial capitalist modernity favoured capital cities and regions rich in mineral resources, such as coalfield and oil field regions, as locations for capital investment.

The redrawing of state territorial boundaries brought together regions of contrasting economic development. The states were significant investors in promoting industrialisation, but efforts were focused on sectoral structures. Governments did not generally embark on spatial policies for regional development directly. However, the government of Latvia, concerned at the concentration of factories in Riga, the capital, did undertake a weakly formed spatial policy. In the mid-1930s, further factory building in Riga was stopped and a dispersal programme proposed for new factory building. The construction of the Kegums hydro-electric power station was expected to assist the dispersal policy (Hobson 1938). It seems that only in Poland was there any large-scale attempt to formulate a regional policy to influence industrial location (Hamilton 1982). Elsewhere state investment was strongly sectoral and so the promotion of regional development was only indirect and through the framework of a sectoral profile.

Case example: Regional policy and planning in Poland in the 1930s: a unique policy initiative

Whilst all governments invested heavily to promote industrialisation and economic development, they did so by sectoral investment strategies. Only the Polish government introduced in the 1930s a directly regional policy – a specific spatial practice, as a means of promoting national development (Hamilton 1982).

By 1930 a number of important ideas that would underpin regional policy and planning had taken root in Poland. There was strong awareness that particular regions had specific development problems. It was realised that regional planning should be formulated within the framework of a national plan. Strategic needs were deemed to require some government direction of industrial location. Warsaw, if it was to become a worthy national capital and metropolitan centre, needed huge infrastructure investment.

These ideas were incorporated in the Four Year Investment Plan of 1936–40. Poland was divided into 'Poland A', west of the River Vistula and more developed, and 'Poland B', east of the Vistula and relatively un-developed. A 'Central Industrial Area' was delimited and control of its development vested in the Ministry of Defence. Ten other regions, each with specific development problems were designated as planning regions (Figure 5.5). Planning offices were set up in these regions and planning studies begun.

Hamilton (1982) saw these activities as creating the roots of a search for equity that continued into the socialist period. Zaremba's judgement (1966) was less positive. He contended that pre-war regional planning had no influence on investment trends or spatial change and had not been an instrument of spatial co-ordination of investments. None of the regional projects had been finished before the war. He conceded, however, that pre-war regional planning brought to light valuable regional resource inventories and many social, economic and technical problems that needed answers. It also highlighted the strong disparities in the socio-economic levels within regions and between regions that capitalist modernity was creating and that needed government intervention to ameliorate. The policy also pioneered new planning methods.

Conclusion

The practice of constructing the new state boundaries was very different from the utopian conception from which it began. The redrawing of state boundaries over the antecedent conditions of pre-war cultural, economic and social development created a serious problem of ethnic mixing in the context of virulent nationalism. The ideal of national self-determination was flouted amidst a conflation of historically fanned imagined places and spatial practices laden with vested interests. The vision of the Marchlands emerging from the control of tyrannical empires as harmonious nations at peace with each other was an unrealistic Panglosian myth. Despite huge population movements ethnic exclusion and discrimination were a marked feature of conflict in places formed by nationalist spatiality. This brought together the imported ideas of socio-cultural ethnic nation, and political statehood, locationally, in an area of complex ethnic settlement and feudal thought and practice. Old social élites clung tenaciously to power and the peasantry strengthened by reformed land ownership resisted change.

The new boundaries also created a distinction between countries economically. Their economic profile became predominantly industrial, semi-industrial or agricultural. The northern and western states, most closely attached to the Western European core of modernity, in thought and action, became more distinctly separate from the southerly and easterly countries. The placing together of regions with contrasting levels of pre-war industrialisation was a key factor. Regional transformation within modernity proceeded unevenly as favourable societal interactions propelled some regions forward, whilst others were held back. The resituation of regions in place relations was of distinct importance in shaping regional trajectories of modernity. The duality of the Nationalist Project, national cultural identity and state tempered capitalism produced a contradictory spatial territoriality. Nationalist spatiality produced states that followed their own particular pathway of modernity, but all experienced modernity through a vortex of intense cultural and societal tensions.

6 The Marchlands in European and global space

I have looked so far at the production of localities, regions and states in the western modernity of the Marchlands. This chapter looks at the production of the Marchlands and their resituation in European and global place relations of western modernity, during the period of the Nationalist Project.

Central and Eastern Europe as Marchlands in the European arena

In Chapter 2 I pointed to a confusion in the perception and meaning of Europe that had implications for the idea of European space. Ideas of Central and Eastern Europe were no less ambiguous and contested. Those peoples living west of the Urals, and those who wrote about them, came to think of themselves as European because they inhabited a continent they called Europe. But the eastern boundary of that continent was defined culturally. Most, but not all, of these people had a Christian heritage, spoke an Indo-European language and were of Caucasian race. This European culture area owed much to the idea of Christendom that was territorially almost coextensive with continental Europe. In the minds of many, however, especially intellectuals, scholars and ruling élites, to be European was to be 'modern'; to share a particular experience of time and space. The source regions of this Europe of modernity lay in the western half of continental Europe. Its history had been to emerge from and displace the truths and experiences of western Christendom.

In this modern Europe, Eastern Europe had been invented as an 'other' place, a demi-Orient, lying on its eastern borders. Its cultural purpose had been to hold up a mirror to the superiority of western civilisation. Its social practices were labelled barbaric and cruel. But its economies were brought into the domain of modernity as source areas for agricultural commodities. Its societal élites re-enforced the traditional feudal obligations of the peasants in order to benefit from the opportunities created by the material needs of capitalist modernity of the west. Elements of modernity spread eastwards in the nineteenth century making significant inroads in some places. However, the eastern half of continental Europe remained under the control of the traditional aristocratic and Church élites. Though Greece, Bulgaria and Romania had gained independence from Ottoman imperial rule, they had remained steeped in the culture of the past. Mustachioed, bearded emperors, dressed in operetta uniforms, headed societal hierarchies whose etiquette of behaviour conserved the past and viewed the future with suspicion. This was a western liberalist hegemonic spatial modernity that, aided and abetted by local élites, held its Eastern Europe firmly anchored to the past, to meet its own present needs. In the late nineteenth century the Marchlands stood in modes of production and patterns of trade in a peripheral relationship to the Western European core. Only the Czech lands were developing as semi-peripheral. Between 1870 and 1914 British hegemonic modernity was the core of the capitalist world economy and the dominant exporter of capital for economic development and of ideas and innovations. Little of Britain's capital export, however, was directed to the Marchlands. German and French capital on the other hand was invested in the area and made it one of the principal capital importing peripheral zones. German and Russian cultural practices competed for dominance. To the western viewer it appeared a golden age, marked out by optimism and progress. The Marchlands were an allotted place in the normal cultural and material practices of European experience.

'To European minds, World War One – the Great War as we call it – stretches across time like a curtain, beyond it is history, this side is the world we live in' (Pitt in Keegen 1971). The First World War changed the experience of being European. The optimism, sense of progress and dynamism of development was profoundly shaken by the carnage of mechanised total warfare. The pride in technical achievement, so prominent before the war, was replaced by a pessimism of shock, a depression of exhaustion that was not relieved by a much proclaimed commitment to 'never let it happen again'.

The Peace Treaty of Paris claimed to be the herald of a new beginning, but in reality it was permeated by old style thinking. The negotiations were dominated by the victors who sought to disable the vanquished by dividing up and redistributing their territory, by national self-determination and imposing financial and reparations burdens upon them. The treaties were written entirely in the interests of the victors in order to impose their world view. Self-determination was also the instrument for enforcing the destruction of the defeated ruling élites. It was the strongest factor in shaping place, yet, as I have shown (Chapter 4), it was accompanied by motives of aggrandisement and was manipulated or ignored if it did not suit the interests of the victors and their nationalist clients. Germany was excluded from the Marchlands because her pre-war cultural and societal influence had been strong there. Austria-Hungary was dismembered as the price of defeat. Russia, having failed the test of war, and become the Soviet Union, was excluded because of its weakness and from fear of its new government. The area was to provide a buffer against the potential of Soviet power – the spectre of communism. The newly independent states were modelled on their western patrons. However, those patrons showed little commitment to them.

Competing visions of Europe

The French and British governments retained the perception of Europe as an international system of states among whom alliances were to keep the peace, despite the failure of this system in the past. France in particular practised old style alliance diplomacy recruiting the new successor states of the Marchlands to oppose the revisionist policies of the defeated.

Yet there were alternative visions of Europe. Heffernan (1998) points to what he calls 'radical regionalism'. This movement asserted that a territorial system of nation-states would not ensure a united, democratic and peaceful Europe. Inter-war regionalist debate was vigorous in Britain, France, Spain and Italy. The idea of a European federation constructed on a framework of subnational regions or provinces was founded on the pre-war visions of men like Sir Patrick Geddes, H.G. Wells, C.B. Fawcett and Paul Vidal de la Blache. However, it never took practical form and in the nationalist sentiments of the Marchlands found little resonance.

A supranational challenge to the idea of Europe as a continent of nation-states also gained adherents. The Pan-European Union, the best known of several Europeanist groups established after 1918, was headed by Count Richard Coudenhove-Kalergi. Coudenhove-Kalergi's Europeanism refused to be cowed by current realities or intimidated by what existed as opposed to what should exist (White 1989). His cosmopolitan family experience, as the son of an Austrian diplomat father and a Japanese mother, led him to imagine a pan-European Union bounded on the east by the Soviet Union that on the west excluded Britain. The Union was to include areas outside continental Europe, among them Palestine, large tracts of northwest and west central Africa and parts of east Africa and Asia. The Pan-European Union founded in 1923 achieved support across the continent, from academics and influential politicians such as A. Briand and E. Herriot, both French government ministers, G. Stresemann, the German Chancellor, and E. Benes, the Czechoslovak Foreign Minister. Numerous local branches were formed, conferences held and offices were opened in most major cities. The idea of a single Europe had some appeal for the more liberal thinkers of Central and Eastern Europe.

Contemporary western perceptions of the East

Travellers, writers and academics who produced the eastern lands of Europe for the British between the wars dressed earlier images in the clothes of contemporary perception. Hall (1999) showed how the images of British writers constructed the Balkans from the themes of classical, oriental and elemental folk culture. Hall quotes comments on Albania made by Scriven in 1919, 'As beautiful as fairyland ... and as unknown as Africa', and Ackerman in 1938, 'A living museum of everything medieval' (Hall 1999: 162). The studies of Carpathian localities undertaken by Fleure and Pelham (1936) and Fleure and Evans (1939) for the Le Play Society of London convey a sense of the quaint. The approach, though certainly unconscious, is that of the superior westerner and is reminiscent of studies of African tribes. The emphasis is on the traditional fabric of peasant culture, with illustrations of dress, social practices and farm implements. References to change are present, but muted.

The sense of strong contrast with Western Europe, a sense of the unknown, of tradition, a very slow pace of change, and the uniformity of the east is given by H.H. Tiltman (1934) following his travels and conversations in *Peasant Europe – the other Europe*:

> Western Europe, preoccupied with problems of international relations, industry and the future of armaments, is sometimes in danger of overlooking the fact that over half the entire population of the continent is composed of peasants.
>
> (Tiltman: 1934)

He goes on:

> Meet the peasant lands. Meet them on the peasant highway that winds northwards from Bucharest ... That highway is as good an introduction as any to the lands of the peasant millions, for there is no intruding civilisation, and few villages to disturb the picture of ageless calm – the landscape is empty save for the wayside shrines erected by the pious when the Turkish tide ebbed southwards and the eternal peasant figures to right and left, men and women tilling the soil ... It matters not which highway you choose as long as it carries you east of Vienna, for in everything that matters the peasant lands are a single unit ... The fact that there are few large scale industries in Eastern Europe strengthens further the common bond. In western countries such as France and Germany, the state can safeguard its peasants ... But in the real peasant lands that stretch eastwards from Vienna, Prague and the frontiers of Germany until they merge into Asia, the peasant *is* the state.
>
> (Tiltman 1934: 13/14)

Europe as perceived by Poles, Czechs and Hungarians

Such are the images portrayed to the readers of accounts in English. What of the European perspective of writers and thinkers in Poland, Czechoslovakia and Hungary? They exhibited two principal strands. First, there was a wish to identify themselves as Central European and to distance themselves from Eastern Europe; none regarded themselves as Eastern European. Second, there was a sense of being in between, a suspicion of being seen by westerners as not quite European. Both stemmed from the ambiguity inherent in their sense of Europeanness. They felt part of Western European civilisation as distinct from Eastern European civilisation, but were aware of being viewed by westerners as part of the 'other' Europe. Poles and Hungarians perceived themselves as defenders of western Christian Europe against eastern barbarism. Poles had a strong sense of difference from Lithuanians, Ukrainians and Russians who they regarded as Eastern Europeans. Hungarians distinguished themselves from Romanians and Serbs in the same way.

On the other hand, there was anti-western sentiment too. Among Poles this was nurtured by a feeling that the west had not supported them in their heroic defence of western culture. Hungarians felt that betrayal by the west was written into the pages of the Trianon Treaty. The Czech writer Krejci (1931, cited by Bugge 1993) summarised Czech frustration, and spoke for Hungarians and Poles as well. All had a strong sense of being European, of knowing all about Europe, yet being completely unknown by Europe. Such sentiments also permeated the European idea of Slovenes and Croats.

Some recognised a distinctive Central Europe in the cultural experience of supranational cosmopolitanism inherited from the Habsburg monarchy and bounded territorially by it. Odon Von Horváth, writer and composer, summarised this sense of Central Europeanness in 1930 in reflecting on his own life experience.

> I was born in Fiume, I grew up in Belgrade, Budapest, Pressburg, Vienna and Munich and I have an Hungarian passport; but I have no fatherland. I am a very typical mix of old Austria-Hungary: at once I am Magyar, Croatian, German and Czech; my country is Hungary, my mother tongue is German.
>
> (Quoted by Rupnik 1988: 41)

At the same time a strand of German–Jewish culture permeates the uniqueness of the Central European experience. This strand became prominent at the turn of the century. From my point of view it is a key idea, because it spans the cultural and the material dimensions of modernity to reinforce a perception of Central Europe as a place distinguished from Western Europe and Eastern Europe.

Societal modernity

Capitalist, societal, modernity struggled to take root in the Marchlands. Its tentative steps needed to be encouraged by state sponsorship. Foreign capital was vital in promoting economic development. The Marchlands were part of the European space of the capitalist world economy as a peripheral place. Capital imports were, however, more or less confined to the 1920s. After the Great Depression foreign investments were reduced to a trickle.

The core, semi-periphery and periphery of the capitalist world economy contributed to the production of European spatial modernity. Figure 6.1, using data from the United Nations Economic Commission for Europe 1948, shows Europe west of the Soviet Union as five subcontinental zones constructed from the occupational structure of the gainfully employed population around 1930. It enlarges the field of vision and brings into sight other battlegrounds of conflict between the forces of modernity and conservatism in the societal shaping of contemporary perception and experience of place. My discussion is further elaborated by reference to other signs of perceptions and societal practices of modernity. Zones one and two of the map form the core of modernity, zones 3, 4 and 5 are peripheral.

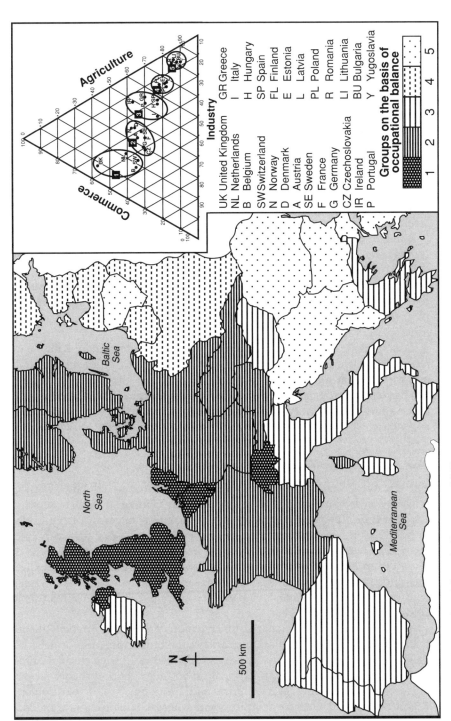

Figure 6.1 Core and periphery in Europe, *c.* 1930.

Zone 1 comprises the UK, the Netherlands, Belgium and Switzerland. Around 1930 these countries had the highest proportion of their labour force engaged in manufacturing and services. Agricultural employment accounted for below 20 per cent and farming was generally commercially orientated. They are the most prosperous as measured by GDP per capita and the most efficient as recorded by values of manufacturing output per head. The UK and the Netherlands derive as much of their national income from services as from manufacturing. Urbanisation and infrastructure are well developed.

In zone 2 the occupation structure is evenly balanced, showing the continued prominence of agricultural employment. Peasant communities producing for family consumption are still widespread. Prosperity and overall efficiency in production are marginally lower than in zone 1, as is the proportion of national income derived from services as compared with commodity production. Infrastructural development and urbanisation are at diverse levels in this group.

Zone 3 is a Mediterranean grouping, but also includes Hungary and Ireland. Agriculture remains a prominent employer. Radical projects have failed to dislodge the traditional societal power structure formed by the landed aristocracy, military and Church élites. Prosperity and efficiency are relatively low, infrastructure and urbanisation are weakly developed. Despite the relatively low level of industrial production national income, derived from services, is proportionally low.

Zones 4 and 5 are within the Marchlands, but distinguish the Baltic States from the Balkan States. As we have seen the societal relationships of independent peasant agriculture predominate. Though the measures adopted for this synthesis are incomplete those available signal the slowness of the permeation of modernity as a way of thinking and in societal practices and experience when viewed from the perspective of European space.

The return of German influence and the Nazi New Order

The Treaty of Versailles excluded Germany from the Marchlands, but they remained a prominent and distinctive place in German imagination, scholarly perception and spatial practices. The manifest injustices of the Treaty, from the German viewpoint, fuelled resentment and spurred geo-political revisionism culminating in the Nazi abomination. The Germans were the outsiders most interested in the Marchlands. This section illustrates how the lattice of imagination, perception and experience of development constructed the Marchlands as a particular place in inter-war German socio-cultural ideas and practice.

Mitteleuropa was a nineteenth century vision of Germany in a new Europe. It had no precise geographical territorial delimitation, function or meaning. In Weimar Germany scholarly critiques of Versailles Europe clothed Mitteleuropa with Völkisch German nationalism and geo-political theory producing a place with more precise form and meaning. Völkisch nationalism delimited a broad German Kulturboden, where German cultural practices predominated and a German Volksboden of historical settlement scattered

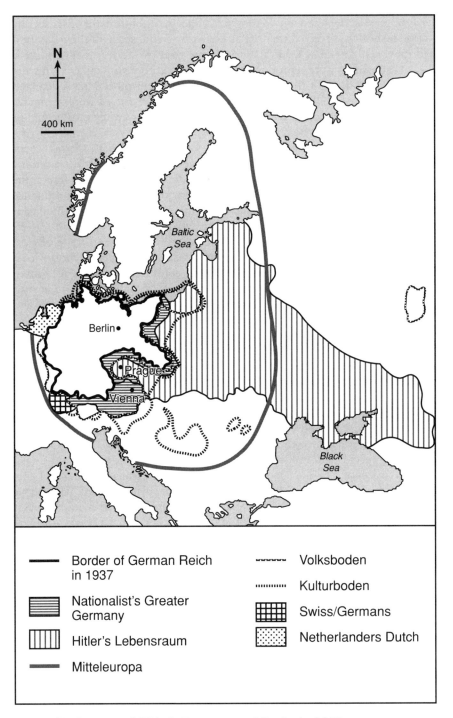

N

400 km

Baltic
Sea

Berlin•

Prague

Vienna

Black
Sea

	Border of German Reich in 1937
	Nationalist's Greater Germany
	Hitler's Lebensraum
	Mitteleuropa

Volksboden

Kulturboden

Swiss/Germans

Netherlanders Dutch

Figure 6.2 German and Hitler's European spatiality in the 1930s.

Source: Adapted from Heffernan (1998) *The Meaning of Europe*.

across the east depicted on Figure 6.2. The geo-political theorists led by Karl Haushoffer perceived the east as Germany's vital space in the formation of a new, self-sufficient Central Europe free from western capitalist materialism. These ideas were reproduced in Weimar textbooks and atlases. In what Heffernan (1998) calls the cartography of resentment official maps depicted Germany Mutilated, Germany's Enslavement and Germany's Imprisonment. These images conditioned young German minds that were later subjected to the pernicious nastiness of Nazi ideas.

German economic recovery from post-war inflation led to German economic influence steadily returning. The success of the German economy was a model for her neighbours. German trade, investment and preferential credit and tariff policies were all important factors in recapturing influence in the countries of the Marchlands. Germany soon played a significant role in the international trade of all the Marchlands countries. By 1925 Germany accounted for more than 10 per cent of the foreign trade of each country and more than 20 per cent of the trade of Austria, Czechoslovakia, Hungary and Romania. In contrast, these countries together accounted for only 11.5 per cent of German imports and 12.9 per cent of German exports. Furthermore, whilst they provided Germany with raw materials and foodstuffs, Germany supplied them with high value machinery, pharmaceuticals, electrical and consumer goods (Hálasz 1928). After 1934 trade recovery was closely tied to trade with Germany, especially for the smaller countries. German companies were important foreign investors. Germany in 1937 accounted for around 10 per cent of foreign equity capital invested in Czechoslovakia, Bulgaria, Poland and Yugoslavia. German preferential credits and tariffs further bound these countries, together with Hungary and Greece, into German dominated economic space (Drabek 1985).

The cartography of resentment, German economic strength and the Nazi project of 'reactionary modernity' culminated in a horrific vision of the east. Hitler's fusion of race and space transpositioned the east as a place of Lebensraum for Aryans. The east was to be won in a purifying conflict with Slavic and communist ideological enemies. The victories achieved would reinvigorate Aryan spirituality. Cleansed of inferior races, Hitler's Lebensraum was to be an integral place in the New Order of his 1,000 year Reich. A distinct place in a German dominated Europe. From 1939 to 1945 as a space of war the Marchlands were made up of places of horror and obscenity as Hitler tried to put his New Order into practice.

British and French perspectives: the idea of a Danubian Customs Union

The idea of the Danubian Customs Union illustrates the belated and weak attention that the British, and to a lesser extent the French, as outsiders, gave to the economic and political problems of the Marchlands. This in turn reflected their concentration on global horizons in constructing their European spatiality. The global issues, not European ones, were the salient features of British and French governmental spatiality. This was in marked contrast to the perceptions of insiders.

For many insiders the idea of reconstructing the economic unity of the Habsburg monarchy by means of international economic co-operation was a project of great interest. The economic disruption caused by the dismemberment of the Habsburg monarchy, especially the Hungarian lands, was ably demonstrated by the Hungarian delegation at the Trianon negotiations. An atlas prepared for Count Pál Teleki, head of the delegation and a geographer, detailed the extent of the economic integration within Great Hungary in 1910–13 and the disruption caused by the new territorial boundaries. This disruption had a significant influence on the nature of spatial modernity, as it developed within a multi-state nationalist project in the Marchlands. G. Hálasz, one of the compilers of the 'Teleki' atlas, showed in his *Atlas of Central Europe* (1928) that the new states were each others' principal trading partners in the mid-1920s. In fact the idea of reconstructing a single economic space from the nation-state place fragmentation of the old imperial territory had been alive almost from the moment of the Empire's collapse, among the insiders. The restoration of normal commercial relations and steps towards closer economic co-operation were discussed at numerous conferences between 1920 and 1925 (Pasvolsky 1928). The Portorose Conference, 1922, was attended by delegates from Austria, Hungary, Czechoslovakia, Romania, Yugoslavia, Poland and Italy. The Central European Economic Conference in Vienna, 1925, called for the establishment of a permanent commission to prepare the ground for a Central European Economic Union. The idea seemed, however, a utopian vision in the hostile nationalism of the 1920s. The Depression revived the idea and attracted outside support. The French Tardieu Plan advocated preferential agreements. The British in contrast proposed a full customs union creating a single trading space surrounded by a bounding tariff barrier. The motivation for the British plan, which had political, economic and financial aspects, was obscure and ill-thought-out. As a consequence it made little progress (Stambrook 1963). Thus, the Danubian Customs Union existed as an imagined place, invented to contribute to a solution to the perceived economic development problems of the Danube countries. As such it formed a part of the geo-economic and geo-political discourse for more than ten years. In political and economic practice it lasted for less than one year.

The Nazi–Soviet Pact

For the Polish writer Czeslaw Milosz (1991: 18), as an insider, the meaning of Central Europe as a place was most keenly delimited and defined as 'all the countries that in August 1939 were the real or hypothetical object of a trade between the Soviet Union and Germany'. The vulnerability of the Marchlands to the strength of outsiders, and their spatial imagination, perceptions and practices, is most emphatically illustrated in the way that the two powerful totalitarian states agreed to divide them in the Nazi–Soviet Pact.

The Soviet Union and communist modernity were absent from the Marchlands arena, except for the secret negotiations with Nazi Germany that culminated in the Nazi–Soviet Pact. The Pact was a momentous event for the Marchlands. It provided an opportunity for Hitler to overrun the small states of the Marchlands without fear of Soviet intervention, and to begin to put into practice his vision of a German dominated east. When Hitler reneged on the Pact and invaded the Soviet Union in 1941, he started a story that was to lay open the Marchlands to Soviet domination following the Second World War.

Spatial practices and international trade in the production of the Marchlands

The return of German economic influence and the British and French plans for Danubian states to co-operate were visions of the Marchlands that resituated them in competing conceptions of European economic space through the spatial practices of international trade. The success of German policy and the failure of British and French policy, much less vigorously pursued, is reflected in the shift in international trade relations between the 1920s and the 1930s. In the 1930s the Danubian states traded mainly with each other. In the 1930s, whilst they still traded with each other, Germany was the leading trade partner of each of them.

The spatial practices of international trade also resituated the Marchlands in European and global space of capitalist modernity as conceptualised by world systems theory. The contrasting modes of production in core, semi-peripheral and peripheral zones produced and were reproduced by the spatial practices of international trade relations.

Austria and Czechoslovakia were states within, but located on the edge of, the European core of modernity. An analysis of the information in the Rónai atlas shows that although Czechoslovakia's contacts were global and Austria's more confined to Europe they had similar trading profiles in the 1930s. They imported textile raw materials and exported textile manufactures. Trade with the peripheral states was characterised by imports of raw materials and exports of manufactured products and machinery. With core states they exchanged manufactured goods, but also provided foodstuffs and raw materials. The peripheral states of the south east, Yugoslavia, Romania and Bulgaria, imported textiles and manufactures whilst exports to the core were made up of raw materials, animal products and cereals. Hungary was in an intermediate position, reflecting its location and economic status. It exported foodstuffs, animal products and cereals to the core states, receiving raw materials for textiles and machinery from them. Its trade with peripheral states was to provide machinery and receive raw materials.

Case example: Czechoslovak trade

Czechoslovakian trade illustrates well the importance of intra-core trade for development, but at the same time shows the need for links with the global periphery as a source of raw materials.

Table 6.1 Czechoslovak global trade, *c.* 1935

	Imports (%)	Exports (%)	Total (%)
Global core (Western Europe and USA)	64.7	70	67.2
Eastern Europe periphery	18.6	18.9	18.8
Southern European periphery	2.5	3.0	2.7
East Asian global periphery	11	5.6	8.3
South American global periphery	3.7	1.8	2.7

Source: Figures calculated from Rónai (1993).

In the 1920s Czechoslovakia's main trading partners were its immediate neighbours. As its industrial production increased, its trading patterns changed and trade with the core rose, resituating it within the core–periphery system. Germany had a special place in the 1930s but Czechoslovakia unlike its neighbours was never dependent on Germany.

In the 1930s Europe was often perceived as divided into an industrialised west and an agricultural east (Hartshorne 1934). Czechoslovakia, as delimited in 1920, straddled this west–east divide. Bohemia, in the west, had a strongly industrialised economy, Moravia in the centre had localised industrialisation, but Slovakia and Ruthenia in the east had agricultural economies.

Czechoslovakia traded on a global scale (see Table 6.1). Its principal imports were raw materials for the textile industry. These raw materials came from the states of the global core and global periphery: Western Europe, the USA, Egypt, India, Australia and New Zealand. Its principal exports were textile goods that were sent to Western Europe and the USA. There was a small trade with South America, whereby animal products and food-stuffs were imported and machinery exported.

The production of the Marchlands in the global arena

The spatiality of the European victors at Versailles was constructed as the territorialisation of the, Eurocentric, imperialist imagination of western modernity. Imagined places shaped attitudes and actions. Eastern Europe, as revealed by Wolf (1995), was imagined as a demi-Orient. As such it was different from the Orient as revealed by Said (1979) and Africa. The peoples of the demi-Orient were European, almost like themselves and so superior to the peoples of the Orient and Africa who were definitely not. Being Europeans, though subordinate, they had the right to shape their own identity. The subordinate peoples of the other imperial territories of the defeated powers were regarded as 'objects', not allowed to shape their own identity. Nationalists had lobbied and negotiated, with the victors and amongst themselves, during the war in order to win the peace. Hence the Wilsonian ideal of self-determination could be applied in the Marchlands. It was also in most cases extended to the Germans. The German possessions in Africa were treated differently.

Britain took over German West Africa (as a League of Nations protectorate) and German East Africa; France took over Togoland. The Cameroons were partitioned between them. The Ottoman possessions in the Middle East were shared out between Britain and France. In all these territories there were local independence movements. In a relatively short time some of them made gains, but they were not offered self-determination in 1920. The collapsing Russian empire was reconstructed in different circumstances.

The Marchlands in the global economy

Around 1900 the capitalist world economy achieved a global reach (Taylor 1999). Imperialist cultural and spatial practices are evident in the shaping of the capitalist world economy at the global scale in the inter-war period, as it adjusted first to post-war reconstruction and then to the Great Depression. Britain and the United States of America were at the core of this global system, and were the main exporters of capital (Figure 6.3). France was the only other significant global capital exporter in the inter-war period. As multilateral trading broke down, a system of spheres of influence dominated by the major industrial powers evolved (Figure 6.4). These spheres were associated with the global empires of the major European powers. This global spatial system of separated spheres was marked by a tendency to bilateral trading, preferential credits and tariffs, and self-sufficiency. The newly independent states of the Marchlands were badly situated to attract trade and investment when compared with other peripheral and semi-peripheral zones.

As in the previous half century, the Marchlands was a relatively unimportant area for British capital or trade. British investors preferred to put their funds into colonial economies like Australia, New Zealand, India and South Africa, as well as the Far East and South America developing trade contacts in these global regions. Even so British investments in southeast Europe were estimated at more than £100 million in 1932 (Stambrook 1963), and British investment was a significant proportion of foreign investment in Czechoslovakia and Yugoslavia. French capital and trade contacts were also mostly directed at imperial domains, but France was an important source of investment in Czechoslovakia, Poland and Yugoslavia. The USA directed capital and trade to the Pacific zone and South America but was present in the Marchlands. Belgian investments and trade went to the Congo, but Belgian capital was a significant proportion of foreign investment in Bulgaria. Only Germany, seeking to build up influence in the area, looked upon the Marchlands as a major zone for investment and trade. Even so, German investment funds and trade links were also directed into South America. Thus, whilst from the point of view of the Marchlands, foreign investment and trade were significant in development, the principal investing and trading states had little interest in the area, and looked elsewhere for capital investment and trading opportunities.

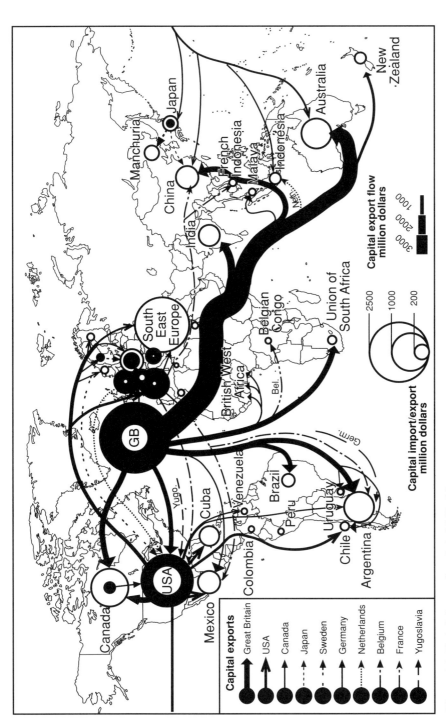

Figure 6.3 Global flows of capital, 1920–30.

Source: Képes Politikai és Gazdasági Világatlasz © Kartográfiai Vállalat, Budapest 1966.

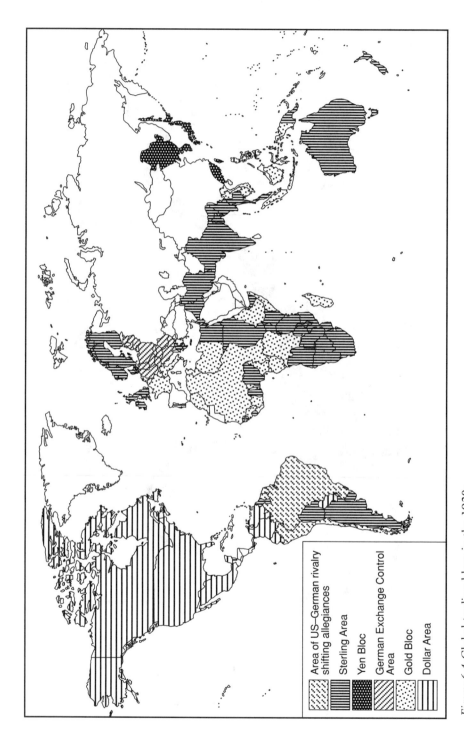

Figure 6.4 Global trading blocs in the 1930s.

Source: Adapted from *Times Atlas of World History*.

Air services and spatial modernity in inter-war Europe

Modernity is about novelty, the search for perpetual newness. Innovations in transport and communications technologies have a particular importance in shaping modernity. In the 1930s the technical progress in aviation captured the imagination and exemplified spatial modernity. Aviation offered freedom from terrestrial travel and opened up the seemingly boundless skies as a new medium of connectivity. The actual development of air services shows that spatial modernity produced places in the sky as well as places on the ground to structure a new time/space.

In the aerodrome, often an icon of modern architecture, the traveller entered the exciting uncertainty of a journey on the newest technical achievement – the aeroplane. Flying in comfort and style, at an unknown speed, at an unknown altitude, in an unknown direction, for an unknown duration along an imagined route, would bring the traveller to the intended destination. An air journey was a metaphor for modernity. It required a culture of faith in the future, borne on technology and the professional skills of mechanics and pilots. Once airborne, there could be no change of mind, there were few intermediate landing places and no unscheduled stops.

Air services produce space as isolated terrestrial places connected by invisible sky places, represented on air navigation charts that transformed time/space. However, governments transpositioned air space as places coextensive with state territory and controlled by themselves. The control of air space thus delimited as places gave governments a role in structuring domestic and international air traffic. The attitude of governments also conditioned the form and growth of the aircraft production industry. Thus, by the mid-1930s air services in different states were usually provided either by several commercial enterprises or by one state national flag carrier. For example, Britain had twelve private companies offering regular air services, whilst Germany, France, Belgium and the Netherlands, leading countries with air services, had just one state airline each. In Britain twenty-three aeroplanes were in regular service, in France there were 134 and in Germany 172 (O'Dell 1935).

The development of air services in the Marchlands, east of Vienna and Berlin, reflects all those locational and societal factors that affected the onset of modernity. The peace treaties forbade military air services in some countries (Rónai 1993). This hindered experimentation, aircraft construction and pilot training, indirectly affecting civil commercial air service development. By 1935 each country had developed a network of domestic air services with the capital as the focal point of routes radiating to regional centres. There were few direct interconnections between regional centres. International air traffic had been seriously curtailed by the hostility between nationalist governments. Access to air space transposed as places was often denied to airlines from neighbouring countries. Later under strict control designated airways and airports were opened to foreign carriers. The economic expansion of Germany in the 1930s stimulated international air services. The earliest international air services were operated by Deutsche Lufthansa, but soon air service providers from other countries joined in. By 1935, Vienna was established as the principal air traffic centre. It handled domestic traffic, international traffic and for KLM intercontinental services to the Far East. All the other capitals handled international flights as destinations and as transit points, boosting their traffic.

There was a European core area of intensive air traffic connections. Berlin, Amsterdam, London, Paris, Geneva and Vienna delimit this place. The most intensively served routes were London–Paris and Amsterdam–Rotterdam. Routes radiated from Brussels, Hannover, Munich, Stuttgart and most of all Berlin. Air traffic on the periphery was much sparser. Aerodromes here often participated in air services as staging posts for intercontinental flights occasioned by the limited range of the aircraft in service. The Soviet air traffic network was almost completely separated from the European services.

Conclusion

The Marchlands were produced in inter-war European and global spatial modernity by an array of contested spatialities. From the far west, Britain and France, the Marchlands were overlooked in a structure of thoughts and actions that privileged an imperial presence in global space. In a Europe perceived as a state system the new states were viewed as a buffer between Germany and Russia, but of little interest. Western travellers reproduced earlier images of a place culturally inferior and economically backward. From Mitteleuropa emerged a German spatiality focused on continental European space. For the German geo-politicians and for Hitler the east was envisaged as a place of German living space, settled by peoples of cultural inferiority and economically exploitable backwardness. A conception of Zentraleuropa among Poles, Hungarians, Czechoslovaks, Slovenes and Croats drew cultural and societal boundaries embedded in western civilisation that separated them from Eastern Europe, but they were uncertain about acceptance in the west. Only the Austrian–Japanese Coudenhove-Kalergi was bold enough to promote a cultural pan-European vision. The practices of international trade and investment flows linked and integrated more tightly the core of the European capitalist world economy. Air transport technology added a new binding free from the direct hindrances of terrestrial territorial borders. The western boundary of the Marchlands was thus not sharply drawn but was produced in thought and action as an eastern periphery, a demi-Orient, in a Western European modernity that overlooked them in its imperial global hegemony.

Part 3

Spatial modernity and the Communist Project

7 The Communist Project

The assertion of collective development and competing global modernities

The Communist Project was the modernity of politicised social and cultural revolution. In the terms of Berman's allegorical trilogy interpreting the Faust legend as modernity, I can see Marx as the Philosopher and Dreamer, Lenin as the Lover and Stalin as the Developer. Marx declared the bourgeoisie as the most revolutionary class in history and the trend of history to lead on to the hegemony of the working class. Lenin created the Party as the champion and leader of political, social and cultural revolution. Stalin was the remote, insecure Developer who lost sight of the purpose dreamed of by Marx, and loved by Lenin. The tragedy of communism was its development of western ideas as an heir to eastern statehood. The overwhelming humanity of Marx, the Philosopher, was transformed into the inhumanity of Stalin by communist modernity's internal propensities, the tradition of autocracy and absolutism inherited from czarist Russia, and the hostility of western capitalism.

Communism achieved little success in the Marchlands after the First World War. Communism evolved in the Soviet Union alone. Z (1990) argued that the Communist Project had matured in the Soviet Union on the foundation of: a utopian vision; the myth that socialism could be built; and the lie that it had been successfully built by Stalin. But in 1945 it appeared that Stalinism had been the foundation for victory in the 'Great Patriotic War' and was indeed the font of a new future. The war formed a break with the past that would permit new societal and cultural ideas and practices favoured by the Soviet Union to take shape. Would they take permanent root?

Just as after the First World War, when the Marchlands were the first place, outside its original hearth, that western modernity had been tried, so after the Second World War the Marchlands were the first place, outside its hearth, that Soviet communism was tried. The territorial enlargement of communist modernity ended the extension of the nationalist/capitalist modernity into the central and eastern areas of continental Europe. Communism was imposed on the Marchlands to create Eastern Europe in the wake of the victory of the Red Army over Nazi Germany, the agreements made by the victorious allies at Yalta, and subsequent international congresses. The idea of Eastern Europe took on its clearest conceptual and territorial meaning, became a place, as eight states with communist governments outside the Soviet Union (Figure 7.1). These states were Albania,

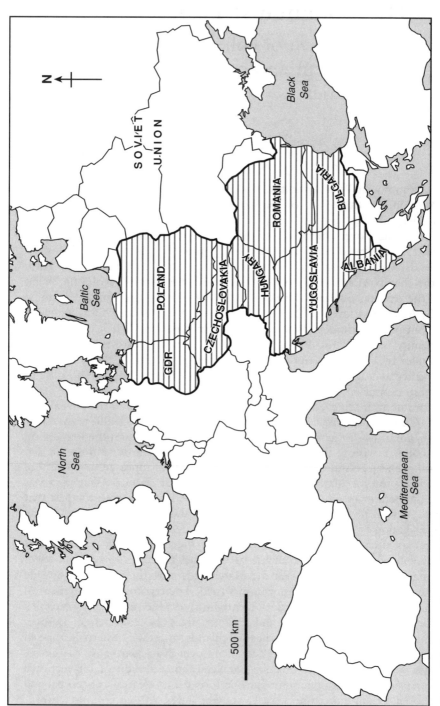

Figure 7.1 Eastern Europe, 1945–90.

Source: Dingsdale (1999b) 'Redefining Eastern Europe', *Geography*, Vol. 84(3), 205.

Bulgaria, Czechoslovakia, the German Democratic Republic, Hungary, Poland, Romania and Yugoslavia. With the exception of Albania and Yugoslavia, the governments of these states accepted the international leadership of the Communist Party of the Soviet Union. The external boundaries of this new subcontinental area were changed to fit the Stalinist, Soviet Project in the face of the collective spatiality of the western powers and inherited nationalist spatialities. The westward enlargement of the Soviet Union incorporated the whole of the Baltic States and large tracks of inter-war Poland, Czechoslovakia and Romania. This ensured direct territorial contact between the Soviet Union and its immediate neighbours. On the western edge, Yugoslavia gained territory from Italy, and the German Democratic Republic extended the boundary of Eastern Europe to its furthest western extent, when it was created in response to the acceptance of the Federal Republic of Germany into NATO.

Like all other modernities, Communism 'swept away ideas, values, institutions, solidarities and people, preserving only a few of those from the previous state of affairs, so that not much of it remained' (Schöpflin 1993: 256). The plurality of nationalist modernity was replaced under communism by a single organisational practice. The production of place, as the arena of conflict and compromise between competing spatialities' was fused in the communist model into a struggle between interest groups within the Communist Party, and the parallel state institutions that it dominated. All other contenders were dismissed as dissident. All aspects of society and culture were encompassed by the Party. Collectivism suppressed individual expression and drove it underground and into allegorical representations in the arts. The separation of society and state that characterised 'civil society' as understood in western modernity was abolished. The framework of thought and action thus created clearly laid a stress on the spatiality of the political élites. The spatial modernity of communism was constructed by the central decision makers, planning committees and ministries, and the party and state representatives of regional and local interests.

Communist modernity introduced new attitudes, values and practices that were intended to create 'socialist man'. The new faith from the east promised a radiant future. It matured in the Soviet Union under the dictatorship of Stalin into what became known as 'real existing socialism'. As such it had come to resemble, in its essential unification of thought and action, all the most horrific aspects of the Developer depicted by Berman in his interpretation of the Faust epic. Remote, insecure, absolute power, bureaucratic abstraction, gigantuan schemes and inhuman practices were all found in abundance. The Stalinist model of communist modernity was formed out of a deep tension between philosophical propositions, ideological convictions and the practical realities of communism in one country. It has been perceived as a variant of the 'internal colonisation' model of development, whereby urban areas looted rural areas of resources (Smith 1989). The guiding absolute of this Stalinist model was the political interest of the Communist Party; all decisions were motivated by the imperative of maintaining the political power of the Party (Z 1990). The evolution of communist modernity in Eastern Europe was the struggle, within the communist parties and their cadres, to find modes of societal and cultural change that could be implemented without loss of political control exercised by the national parties and approved by the Soviet leadership.

The Stalinist system was an imposition of Soviet hegemonic power that incorporated great power politics as well as ideological tenets of faith. It implied societal and cultural values that were Russian as well as communist. Responses among intellectuals and other social groups, to Soviet imposed governments, intermixed reactions to Russian, and to communist, hegemonies. In other words, for many it contained a historically conditioned ethnic, cultural dimension, as well as a current political and ideological one. The idea of building socialism in each country reinforced the ethnic dimension and it showed its presence in the modernisation of Stalin's modernity, when the reforming élites of the different countries followed different paths away from Stalinism. This was the root of the ambivalent attitude to the Soviet Union. The imitation of the Stalinist system implied a legitimisation of its ideology and practices, but the Soviet Union did not enjoy the prestige that is normally accredited by emulation, neither in official circles nor among the private citizens. Sharp (1979: 218) writing of collectivisation expressed a general truth that 'there was a kind of psychological block and a critical attitude even towards potentially beneficial change if its decrees are "made in Moscow"'.

The central unity of Berman's understanding of modernity is development. This takes the form of institutional development and self-development. These two forms are in conflict under Stalinism and produce its essential contradiction as a modernity. Institutional development all but obliterates self-development. The personal dynamic of self-development, the personal freedom of expression, which does not conform to the model of institutional development is cast out as dissidence. Institutional development pursues neo-Faustian gigantuanism. It imposes a strict stability on the instability of self-development. The 'existential fatalism' of limitless self-development is confined within the collective institutional fatalism of real existing socialism.

Communist material modernity

In Soviet policy Eastern Europe was conceptualised, perceived and treated as a group of separate countries, but they were all dealt with in a similar vein. Communist modernity did not spread westwards, as western modernity had spread eastwards (Figure 3.3), it was imposed in country packages, following the 'communism in one country model'. The Stalinist model was applied to each country irrespective of territorial extent, population, physical resources or existing level of development because it was the only model of socialism that governments had to draw on. This in itself ensured new place/space transpositions as new current processes interacted with different antecedent conditions.

When applied in Eastern Europe, Stalinism established a relatively coherent development scheme that I will call 'the National Development

Strategy'. This strategy was founded on an institutional framework of public ownership, central decision-making and planning within the unified societal framework of a totalitarian one Party-state. It was capable of mobilising and marshalling huge resources and directing them at specific objectives. Its practices favoured the gigantuan over the small scale. The objectives were set boldly as industrialisation, urbanisation and the collectivisation of agriculture in the economic sphere, and equality of access to employment, health and education in the social sphere. Regional and local development was to be equalised indirectly by the achievement of these objectives and directly by spatial measures that would ensure that living standards would become similar in rural and urban places all over the country. This strategy meant that all thought and action was framed in the institutions and practices of the Party-state and its public policy.

The institutional framework created conditions that were generally insensitive to local needs and circumstances. Central decision-making was the principal culprit. Public ownership, especially state ownership, of enterprises and comprehensive planning that formed the practices of action contributed because they encouraged remoteness and aloofness among decision-makers. The institutional framework also encouraged a culture that shunned accountability. The strictly hierarchical technical and political establishment, the nomenclatura, discouraged challenges to superiors. Deference was paid to higher authority, enabling each decision-maker to evade responsibility.

Remoteness of decision-making, and a lack of accountability, had two further significant consequences. There was a lack of awareness of the real effects of policies. Subordinates only told superiors what they wanted to hear. When information was provided as to the ineffectiveness of policy there was an unwillingness to recognise the meaning of the evidence. One further feature was derived from the general societal framework. It was the lack of concern for public well-being that was most tellingly revealed in the repression of public participation in the processes of modernity. A huge labour force toiled in the dangerous, modern occupations of mining and heavy industry, yet there was no concept of health and safety at work to protect them. The political experience of modernity, expressed in meaningful individual participation in democratic representation, was also absent. The political representation of the people was usurped by the Party.

The above factors stemmed from the general culture of communist modernity. Other factors were related to specific objectives. The focus on investment in industrialisation and urbanisation was intended to transform rural–agrarian economies into urban–industrial ones and peasant societies into proletarian societies. However, the Stalinist model imposed three other early features. First, there was a concentration of investment in heavy industry, especially energy, iron and steel and heavy engineering sectors, to establish an economic base of producer goods. Second, there was a condition of autarchy or self-sufficiency. This often had the effect of forcing the use of low quality raw materials. At the same time, for ideological reasons, as well as practical ones, resources were utilised to exhaustion, rather than output discontinued at the point of uneconomic production. Third, there was a lack of new technologies that resulted in inefficient utilisation of resources and low productivity.

The dynamism of communist development was the drive for economic growth, whilst preserving political control. Like capitalism, communism was a material modernity, but its mode of production was different and, it claimed, more effective. After reconstruction of wartime damage, the early years were characterised by extensive development. This relied on the application of extra inputs of labour and materials, rather than the investment of capital or technical improvements. The achievement of high annual growth rates depended on tapping new sources of labour, a stress on quantitative targets and reliance on domestic resources. Forced investment, directed through industrial ministries, enabled new power stations and factories to be built and all on the gigantic scale. There was rapid sectoral restructuring of the economy, but enormous sectoral imbalances were created, and there was little improvement in productivity. Occupational and social restructuring occurred rapidly. High levels of local and inter-regional migration were recorded from less developed rural–agrarian areas, generally starved of new investment, to rapidly growing urban–industrial centres.

The rapid pace of annual economic growth initially achieved by extensive development could not be sustained. Labour supplies and physical resources were finite and limited in each country. Isolated attempts at new practices occurred in the 1950s, but a general effort was made from the mid-1960s, when signalled by the policies of Kosigin in the Soviet Union. There began a shift to intensive development, laying stress on inputs of capital and technical improvements. These involved attempts to reorganise economic, financial and managerial practices, shift from producer goods to consumer goods production, introduce greater flexibility from plan rigidities and proposals to loosen central control. Circumstances were strained. There were shortages of capital for investment, but available funds were dispersed in a more balanced way than before.

Technical innovation proved difficult. Several new development models were tried. The first was a search for 'market socialism'. This idea had been proposed in the mid-1950s in Poland, and was revived in the 1960s and heralded the most optimistic route forward for communist régimes, but few were able to grasp it. Market socialism was politically risky, because economic decentralisation was potentially coupled with political decentralisation that was not acceptable to most régimes, and certainly not to the Soviet leadership. A second, new development model was to introduce more democracy through workers' councils and self-management. A third development model advocated turning to western corporations to acquire higher technology production systems and western banks for capital loans. Generally, these foreign sources were willing, under certain political and strategic constraints, to co-operate. However, imported production systems were often already obsolete from a western point of view, and bank loans were rarely used effectively. A fourth development model was to create vertically integrated cartels. This provided a relatively small number of large-scale enterprises whose role in economic development could be relatively easily controlled by planning, and did not threaten the Party's political power. With the adoption of these new strategies, social and occupational restructuring continued, but only very slowly. Inter-regional and local migration slowed as most countries adopted spatial development practices that, in some measure, ameliorated the worst effects of earlier investment decisions. Even so, investments in state-owned companies through sectoral ministries dominated the processes of change, central decision-making and planning remained the framework for action.

None of the reform models proved effective. The continuation of the old system, if watered down in some countries, was assured by the strength of its political lobby among the vested interests of the Party bureaucracy and nomenclatura. Growth rates became slower. Foreign debts mounted, but productivity remained low. Shortages permeated the entire economic system. There was no mechanism for change. Failing, poorly managed, firms continued to be bailed out by the government. Successful firms were not rewarded, but had surpluses removed to prop up the ailing enterprises. The economic system began to fossilise. New industrial investment was at a minimum. 'The model of the future failed because it was, in fact, a model of the past' (Rupnik 1988: 179). It was a model conceptualising economic growth as production when the dynamic of growth had elsewhere passed to consumption. However, where exposure to western influences was greatest, trends in the capitalist world economy began to have an impact. In the more market-orientated economies, employment in heavy industry and manufacturing fell, whilst tertiary employment grew. The transformation of the occupation structure continued, albeit very slowly. However, the incoherence of these changes threw the process of economic development into further crisis as job security disappeared and social welfare was undermined. Migration rates were very slow and urbanisation began to separate from industrialisation. The entire system was grinding to a halt.

The evolution of official development policies in each country was an ongoing outcome of a struggle within the local communist parties whereby Stalinist 'hard' communism was challenged by a 'soft' or reform communism heralded many years before in Lenin's New Economic Policy. However, neither of these bureaucratic approaches was easily able to provide small-scale services. Some, especially market and kiosk retailing, were provided by the small private sector that remained in most countries. At the same time an informal, shadow or second economy developed to provide them. This varied in form. A cashless communal economy was common, in which people provided each other with goods and services often associated with extended families. An illegal economy of unrecorded performance of services also developed and was characterised by an elaborate 'tipping' arrangement. Everybody new the rate for particular services. In Hungary, during the 1980s, the government took steps to bring this second economy into the official economy and to extend its provisions. Elsewhere these activities remained illegal.

The Communist Cultural Project

Communist modernity politicised culture. Its application in Eastern Europe was spelled out by the chief Hungarian ideologue in 1951:

Soviet culture is the model, the schoolmaster of our new socialist culture. We can absorb and use the rich experience of the Communist Party [Bolshevik] of the Soviet Union not only in state-building and in the economy, not only in the techniques of class struggle, but also in the creation of a new socialist culture.

(Quoted in Rupnik 1988: 194)

One side of this new socialist culture encouraged self-development. As I discussed (Chapter 1) Inkele (1976) anticipated Berman in recognising the traits of personal and individual development of 'modern' man. He stressed that the Soviet Union made a concerted effort to bring about these traits among its population. This was also true of regimes modelled on the Soviet system. However, individual development was encouraged within a framework of the Party as an organisation and collective community values.

Communist governments encouraged reading, fought illiteracy and had a high regard for literary and cultural values. However, the Party sought to control all aspects of cultural experience. The education system was an important transmitter of cultural norms and values. School and university curricula were brought under Party control. Recreational activity also came within the supervision of the Party. Sport, youth groups, literary and musical societies were all organised as part of the Party framework and were frequently connected to the workplace. All aspects of the mass media were strictly controlled. Newspapers, magazines and journals were all published under the auspices of Party institutions. The broadcast media, radio and, as it developed, television were directed by the Party. Not only the news and current affairs broadcasts, but also educational and entertainment programming were subjected to Party control. Radio and television were state-funded and any form of commercialism was banned.

In the arts the ruling ideology was 'socialist realism', all other aesthetic alternatives were banned. The arts and architecture were required to conform to the style and content of this doctrine. It was the duty of writers and other intellectuals and broadcasters to promote the Party and its ideas and values. All literary and intellectual output was censored. The cinema was a popular medium and from the 1970s often offered attractive conditions for artistic filmmaking. There was no need to pay attention to the commercial restrictions as subsidies were provided, but artistic licence was not censored so absolutely as earlier.

This totality of state control over culture was slowly relaxed as time went on. It occurred in different ways in different countries. Some retained their rigid and stifling control until the end. Others practised more liberal censorship régimes. Still, many intellectuals and artists were treated as criminals. They were imprisoned and their works banned. This led to the development of unofficial printing presses and networks of circulation that published banned works in typescript and often made western banned books available to their readership. In the 1980s emigrés and home-based intellectuals reinvented the idea of Central Europe. This formed a direct challenge to the imposed Russian societal and cultural system that was symbolised by the idea of Eastern Europe. Many Soviet intellectuals were shocked when confronted with the idea that Sovietisation was not seen as liberation, but rather imperial occupation. It seemed not to have occurred to them that there may be a perception of the Soviet Union other than their own (Rupnik 1988).

Conclusion

Communist modernity and the building of socialism was, perhaps, constructed on a series of fictions around Stalinism; the vision, the myth and the lie, as Z claimed. This chapter has reviewed the broad social and cultural trends in the Communist Project in the Marchlands. It has shown that the Communist Project was very different from what went before in the way that it practised and experienced modernity. It seemed at first that Stalinism, imposed as a monolithic hegemony of Russification and Sovietisation to create a new Eastern Europe, was an end game. It was not. Enlargement westwards of Soviet communism posed new challenges that would require the modernisation of communist modernity. Communist modernity, like all modernities, needed to be constantly evolving. Though ideologically based there was also a need for a measure of pragmatism in sustaining change. It failed the test of enlargement, in the lands of its farthest western extension. The key place, from the point of view of this book, that its spatiality constructed, Eastern Europe, collapsed and disappeared in that failure.

Communist modernisation can be interpreted as attempts to modernise Stalinism. These were associated with the revival of soft communism or reform communism that originated in Lenin's New Economic Policy. The struggle between reformers and conservatives in each country produced a different outcome in each of them that was associated with the gradual loosening of Soviet control. This became pragmatic in the idea of 'separate paths to socialism'. The new faith from the east could not take root in the lands of western civilisation. The idea of Eastern Europe that it created was challenged by the revival of the idea of Central Europe among the intellectual dissidents of Poland, Czechoslovakia and Hungary. Its failure to clear out all the antecedent societal and cultural ideas of an earlier modernity undermined it, and in the end contributed to its collapse. Once again contesting spatialities made the Marchlands a zone of conflict. The dominant idea of catching up with the west economically was undertaken by ideas that came from the east. These ideas were reshaped by local initiative, coloured by 'national' modification. In these processes the idea and practice of Eastern Europe reproduced the geo-historic characteristics of 'the lands between'.

8 The production of localities as an experience of communist modernity

This chapter examines the production of localities in the spatiality of communist modernity that was dominated by central élites, decision-makers and local representatives. The vital scalar resituation in the production of localities through development was the centre–local relationship of the Party-state. Local power relationships were also important in producing localities. Both were surrounded by a tension between ideological ambitions and practical difficulties. I will show that whilst localities were produced by and for central state purposes, they differentiated communist modernity as an experience of development in time and space. The chapter begins by referring to representations of power relationships, within the organisational and institutional framework, that are important for local development. It then explores the details of how these institutions produced rural and urban localities. A short description of some cities in the 1960s gives an eye-witness account of the experience of early communist modernity. Romanian settlement planning provides an example of the extreme local impact that could occur in communist modernity.

The institutional framework

Representations of central and local power

The main institutions ordering communist modernity were highly centralised in a state framework. Central decision-makers and planners were far from the places upon which their decisions had an impact. Such places only existed in their imagination or as representations on their maps. Key actors were often insensitive to local circumstances, needs and experiences. Central authorities decided on the location of various forms of investment and local agents, enterprise managers, local administrative officers and Party representatives were simply required to implement them (Enyedi 1990a). Central planning was sectoral and decisions about investments in housing, transport, industry, health and education, for example, were made independently by central ministries. Local councils and officials had to try to co-ordinate the impact on localities of central decisions made in a sectoral framework. The development of localities was forecast in the local preparation of master plans, representations of the future. Such plans incorporated economic, social and physical aspects. They were intended to guide the spatial co-ordination of investment, but local actors had no territorial decision-making power. Local plans were easily flouted by central ministries. Settlement planning was especially important in the development of localities because it furnished a direct framework for investment that specifically differentiated between settlements. However, it did so within a nationally formulated strategy. Settlements were not primarily conceptualised as individual places, but as part of a settlement system or had a standardised layout imposed upon them. Particular settlements were subordinated to the overall conception and the perceived needs of the system. Local investment could also come indirectly through regional planning, a topic discussed in detail in Chapter 9. Figure 8.1 demonstrates the principle of 'dual subordination', to the central sectoral plan and the Communist Party, upon which local governance was founded.

This 'traditional' representation of communist centre–local institutional power relations has been challenged as misleading by Pickvance (1992). Pickvance argues that it fails to reflect differences between countries, puts too much stress on formal relations, when informal networks were more important, and is unrealistic because decision-making is never concentrated in one place. Pickvance's representation sees power as more devolved and more informally practised. Local agents, especially enterprise directors, are able to influence the central ministerial allocation of investment funds. They are subject only to soft budget controls. Party officials and elected representatives, though very much under the control of superiors, will in their own career interests, promote enterprises in their own localities, from whom they derive income. Furthermore, all local agents are faced with an institutional system that does not work. They must develop 'coping strategies' to make up for short-comings in the system. Co-ordinating the impact of central decisions is an outcome of the informal networks connecting coping strategies formalised in the Executive Council of the locality. However, power within these networks is far from evenly distributed. Enterprise managers are in a strong position because of their privileged access to central funds. In addition, enterprises provide services to local residents. Social clubs, holiday programmes, even health care and housing are supported by the enterprise's social fund budget. Local authorities are weak. They are subject to central directives but officials can achieve some autonomy. Furthermore, local authorities have no territorial planning powers and they depend on taxes from enterprise turnover and profits. However, there is scope for the formation of 'spatial coalitons' or 'territorial class alliances', when in a locality the interests of enterprise managers, party officials and local authorities coincide.

In this representation, there is a degree of devolution of central power to the localities that differed between states. Within the locality, power is exercised through informal networks. Enterprise managers are the most powerful local figures. They dominate local executive councils. There is evidence of co-ordinated strategies for local development. This was an important factor because it created effective lobbying for development funds from central

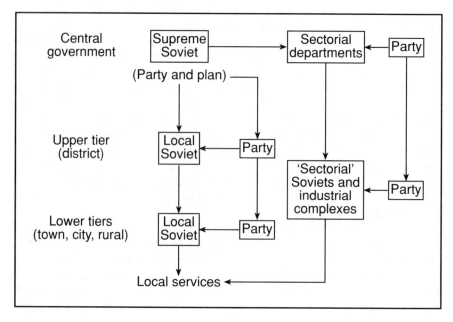

Figure 8.1 The Soviet model of local government.

Source: Bennett (1998) 'Local Government in Postsocialist Cities', in G. Enyedi (ed.) *Social Change and Urban Restructuirng in Central Europe*, 37 Akadémiai Kiadó, Budapest.

budgets. Capital cities, housing central institutions, were in a strong lobbying position, and regional and industrial cities were also potentially very strong. Rural localities were especially weak. However, it must not be forgotten that ideologically communism aimed to transform all localities, to improve material conditions for residents in every locality. The effective power distribution as described above therefore created a tension between the ideological stance and the practices of social and economic management that informed the spatiality of the élites in the production of localities.

The production of localities in communist modernity therefore depends on the relationship between ideological and practical balances of power, the economic structure of the locality and the strength of local lobbying of central authorities, worked out in the framework of a state's development strategies. The central competition for funds between ministries mirrored central priorities. Industrial branches, especially heavy industry, had more investment money to spend. Agriculture, housing, transport, health, education and the rest were less well funded. Infrastructural development was funded only as support for industry (Enyedi 1998a). Urban–rural relations were conducted under the priorities of the 'internal colonisation' model that concentrated resources in urban and industrial localities at the expense of the rural agrarian ones.

Communism was not so monolithic as to be able to 'clear out' everything that preceded it at one fell swoop, or indeed to try to do so. Local representatives had to mediate central decisions through existing societal and cultural conditions. The production of rural and urban localities by the building of socialism would not be achieved through a uniform or unproblematic transformation. It would itself differentiate localities within and between these categories and so produce new places across Eastern Europe. Towns, conceptualised as centres of industrial production, were favoured for investment. Rural localities, conceptualised as agrarian, were at first starved of investment and drained of human and physical resources by a régime of primitive accumulation under conditions of extensive development. Later the balance between urban and rural localities, between economic branches and between local and central functions shifted as a result of evolving national development strategies. These transformations all had an impact on the spatiality of the élites producing the localities and resituated localities in multi-scalar place relationships.

Rural localities

Land reform, state farms and collectivisation

Rural development produces and is an experience of rural localities. The production of socialist rural–agrarian localities began in the practice of land reform. This was politically popular because it was directed at alien owners and those who had collaborated with the Nazis during the war. Though native aristocracies and the Church were also dispossessed, large peasant proprietors were not. The land was largely redistributed to the landless and the peasant proprietors of dwarf holdings, but some remained in state hands. These actions removed the last vestiges of the traditional social and cultural control exercised by the nobility and derived from the ownership of large landed estates and weakened capitalist landowners too. The impact of land reform on localities depended, therefore, on the particularities of location and inherited socio-ethnic make-up.

What was first given was soon taken away. The communist vision of rural–agrarian localities was not one peopled by independent peasant proprietors. Development soon took the form of drives towards collectivisation. However, in the event, several types of socialised agriculture emerged and in many localities private peasant farming remained the norm. Collectivisation met resistance in localities where before the war there had been successful individual commercial farming. Where pre-war peasant co-operatives had been successful, as in Bulgaria, there was greater willingness to participate in early collectivisation. Deepest suspicion of change lingered in localities where the attachment to family, church and personal independence, the traditional strengths of the peasant, were most deeply embedded. The removal of land from wealthy peasants was not opposed by poor peasants who had sometimes been treated harshly by prosperous neighbours.

The introduction of collectivisation was achieved in several waves. Bell (1984) describes the experience of one Hungarian village, situated on the northern edge of the Great Hungarian Plain between the Danube and the Tisza, to which he gave the pseudonym Kislapos. The state's policy of land reform and then collectivisation of agriculture challenged the traditional village social hierarchy and restructured networks of reciprocity between the villagers. By the late 1940s, land reform had removed the old social hierarchy based on wealth and replaced it with a hierarchy consisting of the formerly poor, but closely connected with the Communist Party. Collectivisation came in two waves in the early 1960s. In the first wave two small co-operative farms were established with membership drawn from formerly landless labourers. A second wave witnessed the formation of two larger co-operative farms whose members came from the more successful peasantry. Collectivisation introduced a new leadership drawn from the middle peasantry whose power was enhanced. In some cases newly gained authority was used to advance personal friends and family members. The democratic bylaws of the co-operative farms conflicted with the exercise of power enforcing central directives. Particularist networking, the replication of old models of leadership by some, low levels of education and technical knowledge fuelled dissatisfaction among villagers. Thus, the reworking of the old issues of hierarchy and reciprocity remained salient in the transformation of the village towards the new values and practices of communist modernity.

In most localities of Yugoslavia and Poland the socialisation of agriculture was never completed, collectivisation was abandoned and reversed. In many of these localities the taking up by the state of vacant land, the acquisition of land from heiress deceased proprietors, the sale of land by peasants to the state were all ways that reduced the proportion of peasant holdings, changing the balance of peasant and socialised ownership in localities. These quiet, small-scale changes went unremarked.

The societal and cultural transformation of localities in regions from which aliens had been expelled or voluntarily emigrated gave special opportunities for development. The Vojvodina of Yugoslavia, the Sudetenland in Czechoslovakia and the northern and western territories of Poland are examples. The new territories of Poland exemplify how locational attributes affected the production of socialised rural–agrarian localities. The state gained large tracts of land for disposal giving the government the opportunity to create state farms in greater numbers than elsewhere in Poland. Many former German-owned, large estates were turned directly into state farms that were given preference in investment allocation and favoured with funds for machinery, land improvement schemes, farm building and housing developments. Streets of newly built detached family houses were constructed for employees of the state farms. The government encouraged new settlers to come to the area from the lands lost to the Soviet Union and other areas of Poland. The first arrivals chose farms with little war damage, fertile land and greater accessibility to transport, creating larger, consolidated farms that contrasted with smaller, fragmented farms elsewhere in Poland (Hamilton 1975).

Rural and village planning

Village planning was an important instrument for producing socialised localities, physically as well as societally. Rugg (1985) points to the construction of state and collective farms on the edge of villages as the first step in redesigning rural lifestyles. He cites examples from East Germany where theoretical designs for socialist villages were implemented in newly planted village settlements across the northern plain. Separate, but adjacent, residential and economic areas followed ideological principles for reducing the differences between towns and the countryside. The state or collective farm together with its farming and ancillary buildings became the 'factory in the village'. Housing constructed for the workers took the form of small apartment blocks or single family dwellings.

Rural localities in Eastern Europe were generally formed around village settlements. Exceptions were found in mountainous areas, in some newly settled territories of Poland and in the Tanya districts of Hungary. Localities in all these districts were perceived by central élites as posing special problems for the practices of constructing and socialising place. The Tanya localities are of particular interest. The Tanya farmstead is a unique form of rural settlement because of its particular link with the town from which the Tanya farmers originated. Tanya farmsteads became a means of recolonising the Great Hungarian Plain from the eighteenth century onwards as the Turks withdrew. As Tanyas spread across the Great Plain they extended the administrative territory of the towns from which the farmers came. At first the Tanya smallholdings were occupied only seasonally, but as time went on they became permanently inhabited. Tanya farmsteads were lined up along dirt roads intersecting at right angles. As a low density form of settlement it was expensive for the government to upgrade infrastructure and provide schools and clinics for Tanya residents. The Tanya Council was created in 1949 to provide a 'final socialist solution' to the Tanya problem by means of a radical, voluntary plan. It was established as an interministerial commission – itself an unusual procedure. The Council prepared a scheme that would invest in infrastructural facilities to build over one hundred new central villages that had high level public facilities; demolish the Tanya farmsteads; and move the population, by voluntary means, into the new villages. The Council ceased operations in 1954. Some of its central villages turned into important rural communities and developed as townships, but in other localities the public buildings did not become the nucleus of new settlements (Hajdú 1992; Duró 1992).

Rural diversification

Some localities were developed as the headquarters for agro-industrial complexes. Such poles of development (Franklin 1969) were at the centre of horizontal integration combining the outputs of several state and collective farms. They often had close links with the private sector, either through associations with private plots of farm employees or links with private proprietors. At the same time, vertical linkages of production units formed enterprises that engaged in the growing, processing and marketing of food products. Thus several localities could be linked together bringing a variety of new forms of socialised societal organisation and co-operation to the countryside.

Rural planting of new industrial capacity and infrastructural improvements contributed to the production of socialised rural localities. Primary industries such as forestry and quarrying were located in rural areas. The planned dispersal of energy plants and factories into rural localities was a feature of many national development strategies. They were primarily associated with attempts to raise the standards of living in rural areas that might stabilise rural population and achieve the ideological vision of communist modernity by equalising the social experience of rural and urban living. The provision of infrastructure for rural localities, including electricity and piped water supplies, improved roads and transport facilities, depended on the central allocation of funds and so was linked to the plans drawn up by sectoral ministries and their national conception of development.

As well as transforming societal aspects of rural localities, communist modernity envisaged an enhanced rural–cultural experience. This was expressed in the addition to the built fabric of a 'house of culture' or a club or reading room. The aim was to replace the church or the tavern as the focal point of village life. It was a venue for recreation and social contact. It served also as a conduit for Party contact with the locals. Radio and television was a common facility in many villages for those who could not afford their own receivers.

The enjoyment of rural or semi-rural recreation by urban flat dwellers was perceived as a desirable activity by the central authorities, especially when it was associated with the collective activities of factories or clubs. It gave rise to the spread of holiday homes and hostels. These were officially understood as gardening recreation. Recreation districts, many looking like allotment gardens, were found only a short distance from many medium sized towns in Eastern Europe. Lakeside, riverside or hill country sites were particularly popular. Recreation areas were laid out and plots sold to individual families or enterprises. Most were of purely local importance attracting residents from surrounding urban localities. In Czechoslovakia between 1971 and 1991, 239,400, overwhelmingly private, recreational homes were formed, by constructing new second homes, converting farmhouses and utilising apartments for recreation, raising the number in the country to 395,800 (Bicik and Dana 1997). In the vicinity of Łódź, Poland, in the early 1990s there were 2,179 ha. devoted to second homes, 25 per cent of which were in extensive recreation areas, over 50 ha. each, attached to five villages, but a further forty-five villages had between 10 and 50 ha. of recreation areas in them (Matczak 1997). In resort areas the garden nature of the recreation gave way to more typically holiday activities. The towns and villages around the shores of Lake Balaton in Hungary provide some of the best examples of the creation of socialised localities formed by this particular intermixing of individual and collective visions of holiday recreation. In the 1970s, as real incomes were rising, but foreign travel was restricted, thousands of holiday homes were built along the lake shore. Land for recreational development was sold by state farms, collective farms and local authorities eager to increase revenue by diversification. Official books of standard designs were prepared and the National Savings Bank gave loans to individuals and associations to fund building. It is a good example of how apparently ideological opposites – individual family recreation was counter to the prevailing collectivist ideology – could be combined to produce localities as distinctive places (Dingsdale 1986).

Rural depopulation

Central priorities of industrialisation and collectivisation in tandem ensured that farming activities evolved against a background of rural depopulation. Whilst natural increase remained high in some localities, out-migration could still produce a net population loss. Farm work was not attractive to many young people. They preferred to find better paid work in local industry or move to factory jobs in towns. Because of the problems of accommodation in many towns some family members lived on the farm, commuting long distances, generally weekly, to work in factories. There also developed a class of worker-farmers who combined farm work with factory work.

In these diverse ways the experience of communist modernity produced new rural places. Physical, locational, societal and cultural relationships were reshaped and localities resituated in local–centre place relations. Diversification of employment and improved infrastructure contributed to the emergence of new lifestyles in many localities. Whilst the old peasant culture remained strong in many localities, others embraced change more willingly. However, transport and communications infrastructure within and linking rural localities was very poorly developed. Over huge tracts of rural Eastern Europe the horse and cart remained the normal means of private local transport, and telephone networks were thin.

Urban localities

Representations of urban development under socialism

Urban development is the production and experience of urban places. Whether or not there has been a distinctive socialist process of urban development or urbanisation has been the font of much debate. A question posed by many scholars is: Is there a Socialist city? (French and Hamilton 1979). Here, we might ask, is there an Eastern European socialist city? Geographers and urban sociologists have adopted Marxist, neo-Marxist and neo-Weberian theories in answering the question of the distinctiveness of Eastern European socialist cities (Enyedi 1990a). There have been two principal strands to the discourse that has produced the socialist city. The first compares the ideal, imagined city with the reality of practice. In this discourse the ideal as set out by the Moscow Master Plan of 1935 is contrasted with the outcome of the practicalities encountered in transforming the cities of Eastern Europe, with their legacies of imperial, nationalist and capitalist iconographic fabric and spatial structure (Fisher 1966; French and Hamilton 1979). The second strand compares the process of urbanisation under socialism and under capitalism. In a discourse that engaged many, I concentrate on two eminent scholars whose contrasting perceptions summarise the main elements of this discourse.

Enyedi (1990a), from empirical investigation, perceived urbanisation in the East Central European socialist countries as having special features, but not representing a new model of modern urbanisation. There were many universal features including: rural–urban migration; separation of workplace and residence; and the development of functional zones in cities. He found a common basis for land use structure within capitalist and socialist cities in the concept of optimum location, the root of locational value. In capitalist cities it worked through the market, in socialist cities through building regulations, norms, physical planning and resource allocation. Despite these different mechanisms, similar sites were adopted for government building, industrial, residential, shopping and recreation areas because both systems were subject to the universal principle of optimum location. Individual urbanites in the west and east had the same concerns, aspirations and needs, but they were satisfied in different forms. Enyedi, then, perceived the universal impulse of industrial urbanisation as the key process. The societal, cultural and locational context did no more than shape the mechanism through which similarities in long-term urban development were expressed.

Szelényi (1996) argued that between capitalist and socialist urbanisation similarities were superficial, differences fundamental. He highlighted three 'uniquely socialist features'. First, industrialisation was achieved with less spatial concentration than in capitalist societies at the same stage of economic development. This 'underurbanisation' occurred because the growth of urban population fell below the growth of urban industrial and service jobs. Rural localities provided workers for the towns, but remained rural in character, whereas capitalist rural areas urbanised. Second, socialist cities were less 'urbanised'. They had less diversity, fewer shops and restaurants and were less densely built up at the centre. More attention was given to aesthetics than commerce. Socialist planners were not constrained by narrowly defined economic priorities, because land was of no value. Third, there was less marginality, fewer people did not conform. City life was less dynamic, less innovative and less tolerant. There were fewer cases of deviance, crime, homelessness or prostitution. But there was social segregation. Investment was directed to outer areas of the city and the inner city deteriorated physically and socially.

Part of the reason for the different conclusions arrived at by these scholars is that they base their view on different kinds of evidence reflecting the considerable ambiguity in the meaning of urbanisation. However, neither used the same place approach adopted here, though both make points related to it.

Planning ideas

The primary driving force behind urban development was industrialisation; communist élites imagined towns as sites for industrial production. Within this general context three other conceptions that permeated socialist thinking on planned urban development can be discerned in the production of urban localities. The first is the idea of the urban agglomeration; the second is the idea of settlement hierarchy planning, a statement of socialist egalitarian principles and a belated signal of the recognition that towns were not only industrial sites, but also service centres (Enyedi 1990a); the third is the idea of the socialist new town, a local sign of the complete ideological utopia. These ideas ensured that most categories of towns would, in practice, experience population growth; have existing quarters redesigned; add extensions of a particular form and appearance; and become invested with new cultural meaning.

Urban agglomerations and settlement hierarchy planning – ideas in action

These two ideas illustrate the way that specific towns were perceived as part of a wider grouping or network. Planners imagined an urban agglomeration as an extensive urban area of intensive development formed from separated but functionally linked, specialised and interdependent towns and cities. It could have a single polyfunctional dynamic centre surrounded by satellite dormitory or industrial towns; or be formed by a cluster of cities with similar or complementary functions in which no city is dominant. Links between centres are achieved by a highly developed public transport system. Thus, the problems of congestion and overcrowding associated with rapid capitalist growth are overcome by planned dispersal of economic activities and residential estates, each provided with retail and welfare facilities (Hamilton 1979). The agglomeration idea incorporated resonances of the Garden City concept. It was used to transform existing conurbations, redesigning the perceived imperfections of the capitalist conurbation to an improved socialist form, in Budapest and Halle-Leipzig, for example; and also as a blueprint for the development of new urban regions.

Even more ambitious was the idea of national urban hierarchy planning. Its contribution to the production of urban localities took several forms. From the locational point of view the theoretical distribution of centres of a particular status was determined, generally following Christaller's central place theory. If there were too few centres in existence more would be created, so that the required thickness of the hierarchy would be achieved. But, if there were too many, then some had to be downgraded. From the societal point of view towns allocated a certain status had to carry a certain level of infrastructure and facilities. Any that were perceived to lack these would receive investment to bring them up to the required level. From a cultural point of view the meaning of 'urban' had to be defined. Places that were designated urban and given a particular rank within the hierarchy conveyed a certain iconography of the meaning invested in urban places. In effect, the whole urban network that was being constructed was a symbol of the meaning of socialist urbanism. The process incorporated the interplay of imagined, perceived and experienced urban places. Some already existed as they needed to be, others required upgrading to the level they were perceived to need, whilst all were measured against an imagined urban ideal. The key societal and spatial practices of investment allocation relied on this scheme, whereby a vision of the future was transformed into a reality, through a process of perceived need.

The new socialist towns – icons of socialist urbanism?

Forty-two socialist new towns were established in Eastern Europe (Grime and Kovács 2001). The socialist new town was the best chance of creating the conceived urban experience of socialist modernity free from the hindrance of earlier forms. It was to be a very different kind of place from the dingy and overcrowded inner city slums from which many of its new residents came. The new town was also different from the rural poverty known to many others among its residents, who had previously lived in villages with no amenities and houses that lacked piped water or electricity. The socialist new town is the essential icon of socialist urban experience. Socialist new towns were relatively small, enabling a new community to form. They were planned to provide a balance of comfortable, if small, flats in new estates and industrial jobs in new factories on industrial estates. The residential estates were provided with kindergartens, schools and health clinics to a higher standard than elsewhere. Open green areas separated the blocks of flats and new public parks were provided for family recreation. The town centre was set out for public celebrations and festive parades to mark important events in the socialist calendar. The new jobs in industry offered a guarantee of permanent employment and a small but regular wage, benefits many workers had never previously enjoyed. New towns were industrial towns and not intended to be service centres, but many would in time provide work and shopping centres for the inhabitants of surrounding villages and so play a part in the regional urban hierarchy.

The image and the practice, however, proved different in many cases. Many 'new' towns were formed as extensions of existing towns. Some had a distinct political text. Nova Huta, Poland, a workers' town was constructed adjacent to Kraków, an ancient bourgeois city. Some were groupings of existing villages, supplied with new central public buildings such as Tatabanya in Hungary. The housing of workers on large residential estates and the construction of extensive industrial plants, usually in the heavy industrial sectors, produced drab and polluted places. Investment for maintenance and renewal was not allocated. There was little variety in entertainment or cultural opportunities. Residents came to feel trapped in a built environment of decaying standardisation without much opportunity for self-development.

The urban fabric and the experience of socialism

The layout and fabric of socialist town development was often depressing and criticised as monotonous and unimaginative. It relied on relatively few model layouts and mass produced building materials. The original 'socialist realist' architecture was modified to make greater variety as time went on, but only slightly. The zoning of residential and industrial premises also contributed to the appearance of uniformity in the town's urban fabric as large tracts of land had the same use, especially in the bigger towns. Planned housing estates, with high rise, pre-formed, system-built dwelling units, shops, schools and clinics, developed large tracts of urban land as extensions to existing small or medium sized towns. Living conditions were rendered into uniformity of floor space area per inhabitant. The square metre area became the means of facilitating exchange of flats and a measure of equality. In small towns, even some big ones, the monotony of large-scale estates was toned down and relieved by some individually constructed houses or renovations of pre-socialist housing styles. Many towns underwent the redevelopment of their central districts to provide sites for monumental public buildings and squares to accommodate communal celebrations symbolic of the heroic achievement of building socialism. New pedestrian shopping precincts were constructed in the 1970s as living standards increased and more consumer goods became available (Perényi 1978). The display of goods, however, even in the best department stores was never very exciting. Alexander Platz in Berlin was a good example on a large scale of developments that occurred in many places in more modest proportions. Apart from the period of reconstruction, on the whole investment in town centres was small as compared with outer districts. Inner city housing areas that pre-dated the socialist period deteriorated and their physical decay extended in many cities to the quality of social life as well.

In towns that escaped selection for development under socialism, large-scale industrial plants and the replication of extensive high rise housing estates were avoided, but some deteriorated for lack of investment. Certain historic cities escaped socialist industrialisation and the layout and style of the buildings in the central districts were retained, conserved and renovated by Institutes for Conservation from the Ministry of Culture. This represented the presence of a strand of national and historical consciousness in urban design. The renovation of the town centre of Gdansk in the postwar reconstruction period was a statement of Polish rebirth in a city that had for centuries been German. Monuments to historical national heroes were renovated or even newly constructed in redesigned central areas, as in the Hungarian town of Eger, famous for the heroics of its defenders during a Turkish siege. Spa and resort towns also often escaped industrial investments, yet received funds for renewal and hotel and holiday hostel developments channelled through the social funds of factories, other enterprises and institutions such as the army or universities.

The contradictions of sectoral and spatial thinking in the socialist system produced imbalances between investment in industrial job creation and the provision of dwellings and urban infrastructure. Inadequate investment in housing as compared with investment in industrial plants arose in specific localities because of the lack of central co-ordination and limited local input. One consequence of this was the development of long-distance, weekly, commuting. Another was the need for workers' hostels in many towns. Many young people were forced to remain in their parental home, even after marrying, because of the shortage of flats. But social segregation in residential districts was reduced. The central allocation of flats encouraged the mixing of all classes of inhabitants, who had little choice in where they lived. The emphasis on towns as industrial centres and the stress on collective activity made the workplace an important link in social and recreational events. This in part stemmed from the lack of alternative voluntary associations, especially in the small and medium sized towns and cities.

Thus socialist urban places were produced by a very particular, centrally supervised societal and cultural experience. However, each town or city furnished a particular, complex, relationship between ideology, principles and ideas of central management and specific socialist place forms and functions. Sailer-Fliege (1999) summarised diagrammatically (Figure 8.2) the passage of ideas and practices from ideology into the built environment, the morphology, and the functionality of the socialist city that I have described above.

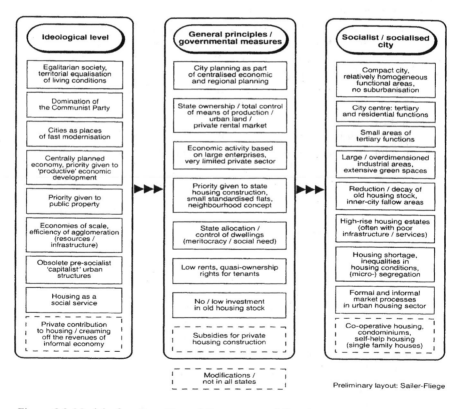

Figure 8.2 Model of an East Central European socialist city.

Source: After Sailer-Fliege 1999.

The experience of communist modernity produced new urban places. Whilst a general air of drabness and delapidation was pervasive in the old areas of some cities, there was often an enjoyable collective social life. In the new housing estates and new towns a monotonous, but at first fresh, urban fabric enclosed a limited social experience.

A western perception and representation of some Eastern European cities

There were few western visitors to Eastern Europe in the 1960s. Eyewitness accounts of conditions in Eastern Europe of the kind that travellers such as Tiltman (Chapter 5) had provided for readers in Western Europe before the war were few and far between.

A. Alvarez, who visited Eastern Europe in 1961, 1962 and 1963, provides an eyewitness account that testifies to the variety of experiences in the urban places of Eastern Europe at that time. The following extracts come from his descriptions of Warsaw, Budapest, Zagreb, Belgrade and Skopje.

In the socialist city the drabness of the urban fabric often contrasted with the vibrancy of the cultural life:

> For the casual visitor, the first glimmerings of the Polish sense of trouble came from the mere look of the place: the hopeless deep-rooted drabness of everything . . . Warsaw is one of the liveliest capitals in Europe: theatres, cinemas and art galleries teem with dissidence and life, the jazz is excellent, the dancing modish, the young people uninhibited, quarrelsome, bloody minded, the University seethes with argument, the artists clubs with gossip. Yet if this talkative city is like Paris in the twenties, physically it is like nothing so much as London in 1945. Drabness is all: in the food, the clothes, the discomforts. There are still wastelands of cleared devastation, rebuilding is slow and shoddy, there are few cars on the streets . . . the shops are poorly stocked, the restaurants awful and life generally hard.
>
> (Alvarez 1965: 24)

Of Budapest Alvarez said, revealing the power of early Cold War rhetoric, overcome by experience:

> I don't know what I'd expected of Budapest; something grim and beleaguered and restless I suppose. Such is the efficiency of our propaganda . . . Budapest is a remarkably beautiful city and a remarkably comfortable one. The Danube cuts it in two . . . On one side are the hills of Buda, dominated by a great cliff with a bad statue and a good restaurant on top. Buda itself is full of villas – whole ones for party and embassy officials, bits of ones for writers and the rest of the official élite. There are trees and flowers everywhere . . . This was not the darkness at noon I had expected. On the Pest side of the city, is the business

area, the boulevards are wide and crowded, the people well dressed and the shops full of consumer goods and of customers, apparently, to consume them.

(Alvarez 1965: 32/33)

In Yugoslavia he found:

Zagreb ... is purely Austrian, with all that implies of fussy elegance, cream cakes and the pervasive smell of coffee ... South and east from there is Belgrade, a noisy modern town, flat as a cowpat on the plain, whose ugliness is gradually being modified by some fine modern buildings after the style of Mies van der Rohe. South again from there ... is Skopje, capital of Macedonia, where the ambitious beginnings of a new industrial city jostle with an old Turkish town: mosques, minarets and bazaars, where the women wear baggy trousers, the men fezzes and the narrow streets are lined with the shops of silver-workers and sandal-makers.

(Alvarez 1965: 77)

It might be that socialism had not yet worked its homogenising spell, but then perhaps it never did except in the imagination of the Cold Warrior.

Case example: Systematazare – the building of Utopia?

Systematazare illustrates all the most important development aspects of socialist localities. It was an attempt to sweep away the physical fabric of settlement and the ideas and values that had created it. It intended to reconstruct a new 'socialist man' and a new socialist landscape. Remote central decision-making engineered a gigantuan scheme. Both rural and urban localities were caught up in the scheme. Traditional villages were to be swept away and new urban–industrial complexes built to replace them. The Bucharest agglomeration was redesigned. Settlement hierarchies were replanned. Romania as it existed was perceived to be outdated – a new vision was to remake it.

'Balkan elegy in Bucharest', 'Romania bulldozes its past aside', 'Ceausescu's clearances', 'The Balkan Tragedy' (*The Independent*, 14.6.88; *The Sunday Times*, 14.8.88; *The Guardian* 23.8.88; *The Observer Magazine*, 27.11.88).

So the headlines proclaimed to Britain the actual and planned devastation of Bucharest, capital of Romania, and the destruction of as many as 8,000 ancient villages, in a programme known as Systematazare. Bucharest, 'a beautiful and intriguing city' (was to be) 'sacrificed to the vulgar megalomania of a dictator'. The culture, society and landscape of Romanian villages inhabited by Romanians, Hungarians and Germans were to be destroyed and evidence of their former existence confined to an open air museum, a 'Madame Tussaud's' of Romanian village life, carefully sculptured and preserved'.

The Project: Systematazare, the programme to transform Romanian settlement.

The Purpose: to clear out the reminders and symbols of the past and on the tabula rasa to build the culture, society and landscape of 'new socialist man'.

The Developer: the *Conducator* with the unified vision of thought and action was Nicolae Ceauşescu, head of a communist dynasty and President of the Socialist Republic of Romania 1965–89.

Systematazare emerged quietly enough out of the collectivisation programme of the 1960s. It slowed down and was then reinvented by Ceauşescu in the 1980s. Ceauşescu's project was breathtakingly gigantuan in its vision and brutal in its practice. Virtually the whole of the settlement system of Romania was affected as a unified system was to bring urban conditions to the countryside. It involved the promotion of some villages to urban status, whilst almost 60 per cent (7,000–8,000) were to be abandoned or destroyed (Turnock 1991). The number of communes would be reduced from 2,705 to 2,000. Some 500 new advance style 'agro-industrial complexes' were to be built as places of work and of residence and villagers forcibly moved into them. Apartment blocks of three to seven storey's would be built. In practice, they were poorly built, the apartments had small rooms and communal cooking and toilet facilities. To one observer they resembled 'high rise concrete rabbit hutches'. Even the most attractive might be thought uninhabitable and entirely unsuited to villagers who had formerly inhabited their own single family house (Lambert 1989). By 2000 half of the entire rural population would have been involved in movements of one kind or another generated by the scheme. Controls were to be placed on rural migrations to existing towns. A housing programme to construct 12 million new flats was announced. The Ilfov Agricultural Sector of Bucharest had the role of pioneer in the venture and became the model upon which the rest of the country would build. Gigantuan though this was, how much was actually destroyed remains difficult to assess. However, the culture of thought and action existed to achieve it.

At the same time in central Bucharest a Haussemanesque redesigning of the city was underway (Figure 8.3). An urban fabric reflecting the historical influences of Ottoman Turks, Orthodox Christianity and the more modest houses of Balkan character was cleared out (Giurescu 1990).

In 1984, 40,000 residents were moved out of the Uranus-Antim quarter, a district with charm by the accounts of those who new it, and the district demolished. Following international protests some monuments were physically removed and reconstructed. The Boulevard of the Victory of Socialism, a grandiose civic centre and the Casa Republica arose. The Casa Republica:

> resembles several American Beaux-Arts railway stations piled on top of each other, while the buildings, along what is to be called the Boulevard of the Victory of Socialism, flats for government officials, are in the

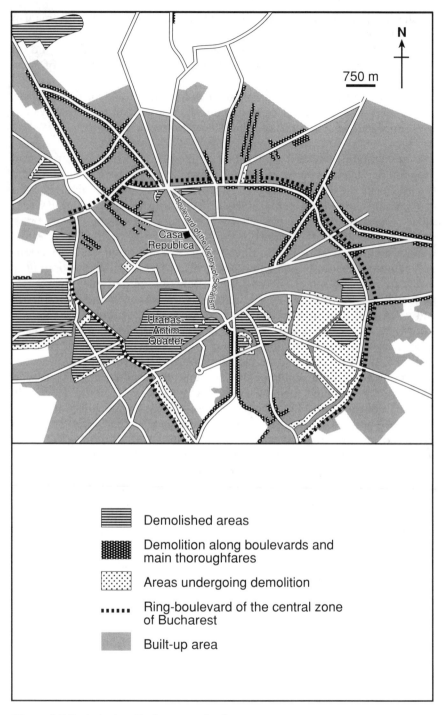

Legend:

- Demolished areas
- Demolition along boulevards and main thoroughfares
- Areas undergoing demolition
- Ring-boulevard of the central zone of Bucharest
- Built-up area

Figure 8.3 Redesigning Bucharest under Systematazare.
Source: Giurescu (1990) *The Razing of Romania's Past.*

fashionable Post-Modernism favoured by property developers around Paris and London.

(*The Independent* 14.6.88)

Ceaușescu's new presidential palace was taking shape with even more lavish bad taste under his own personal detailed supervision. In this way, arbitrary and absolute power was gratified in a supposedly collective culture.

Systematazare was an extreme, but many elements of its whole were found in isolation elsewhere in Eastern Europe.

Conclusion

Communist modernity brought new dynamics emphasising economic production to local development and transformed rural and urban localities in Eastern Europe; and local responses differentiated the experience of communist modernity to produce new places. The dominant theme was localities in a spatiality of central-élite controlled investment. Rural and urban localities were produced as part of a whole, rather than as singularities. Urban centres were often perceived by academics as denuding rural localities of physical and human resources. There was little scope for individual or local initiative. The outcome told the story of socialist modernity as a text of imagined and perceived places transformed into experienced ones by planned central action. The variety of these texts has been illustrated to show the articulation of the mosaic of localities produced as experiences of time and space in communist development. It emerges very clearly from the above discussion that, far from producing uniform places, developing what was perceived to be wrong into an envisioned ideal by the application of limited standard practices, actual existing socialism was a variegated modernity at the local level. Although central élites, decision-makers and planners were often insensitive to local needs they were also often pragmatic in responding to local circumstances. As they encountered a variety of subtlely different societal and cultural antecedent conditions and responses to their project they had some capacity to revise their thoughts and actions. The experience of communist modernity in the localities was a set of minor variations on a clearly ordered, dominant theme. However, as demonstrated by the example of Systematazare, in which physical destruction was allied with the elimination of societal and cultural values, communism in Eastern Europe could reproduce in both rural and urban settings the worst nightmares of Stalin the Developer.

9   The production of the Party-state and its regions

Place, modernity and the Communist Project have their fullest expression in the idea and practices of the Party-state. The experience of communist modernity was most sharply defined by the building of socialism in the production of the state as a socialist place. This chapter first discusses state boundary changes and migratory movements following the Second World War. It then follows national development strategies, assessing the possibilities of the Stalinist model of development for the states of Eastern Europe. The movement away from the Stalinist model is presented as the dynamism of development in communist modernity. This dynamism articulated the experience of building socialism in the separate states producing distinct places. Attention is then focused on regional development.

Delimiting state boundaries and migration

The redrawing of state boundaries in Eastern Europe following the Second World War was contrived in Great Power meetings at Yalta (1943) and Potsdam (1945), and was dominated by the westward enlargement of the Soviet Union. The eastern boundary of Poland was shifted westwards to follow the Curzon Line, proposed as the boundary in 1920, but ignored because of Polish military successes. One-third of inter-war Poland's territory was taken away. In compensation, as it were, Poland's western and northern boundaries were changed to incorporate former German territory, bringing in almost half of East Prussia and the lands as far west as the Oder-Neisse rivers. Czechoslovakia lost its eastern province, Ruthenia, giving the Soviet Union a boundary with Hungary. Romania lost Bessarabia, which it had annexed from the Soviet Union in 1920. Yugoslavia gained the Istria Peninsula and lands to the north from Italy. The German Democratic Republic (GDR), founded in 1955 with the boundaries of the Soviet Occupation Zone, was the only new state to come into existence. Apart from these changes, pre-war boundaries were confirmed, and boundaries redrawn in the approach to and during the Second World War were abolished.

The defeat of Nazi Germany induced voluntary and forced movements of ethnic Germans from across Eastern Europe. The boundary changes induced a generally westward shifting of ethnic populations. Russians, Ukrainians, Belarusians moved in to colonise the newly acquired Soviet territories. These territories were vacated by Finns, Poles, Ruthenes, Hungarians and Romanians. Poles migrated into the newly acquired northern and western territories as Germans moved out. These migrations had a significant impact on the ethnic make-up of the territories that had been transferred between states, but the population change was by no means total. Ethnic identity was also subjected to more subtle transformation in these territories as individuals adjusted to the attitudes and values brought by the new authorities (Kocsis and Kocsis-Hodosi 1998). Ethnicity was subordinated in official policies in the effort to create 'socialist man'. However, in most countries ethnic national identity remained strong and where state boundaries encompassed clusters of ethnic minority settlement tensions simmered. In Yugoslavia, a state of many nations, the nationalities question was never satisfactorily solved by communist governments. The republic boundaries that were intended to provide territories for national identity were never able to do so, because of the complexity of ethnic settlement. Emigration was an ever present theme in the development of all the states.

The national development strategies

Once firmly established in political power, communist governments set about the task of building socialism, their guiding principle of modernisation, creating the experience of communist modernity. The building of socialism was to be achieved through a national development strategy. To do this they imposed the Stalinist model of development, partly because they wished to imitate its success, partly because it was the only model of building socialism that they had to follow and partly because it was required by Stalin's political control.

The Stalinist system promised rapid economic and social transformation. Its institutional organisation, public ownership of enterprise and property, central decision-making and planning meant that it was capable of marshalling resources and directing them to specific development objectives. The objectives of industrialisation, urbanisation and the collectivisation of agriculture as the general aims of the development strategy were positive, tangible goals for achieving socialist economic transformation. The goals of equal opportunity of access for all citizens to education, health care and social well-being, and the equalisation of standards of living between urban and rural areas, promised a fulfilment of social and cultural transformation. The concentration on national self-sufficiency allowed anticipated success to be heralded as a triumph of national achievement, victories for socialism, blows against capitalism in an enthusiastic political rhetoric.

Because it was honed as communism in one country, Stalinism created the new states as containers for a particular development model. The Stalinist model was based on combined and uneven development, a variant of the internal colonisation model (Smith 1989). It drained the financial and human resources of the countryside to supply funds and labour for industrialisation and urbanisation. It concentrated resources on specific industrial branches and neglected agriculture. It was an extensive form of development, applying extra resources, especially labour, to raise production. But it used resources very wastefully. Stalin succeeded because of his terror tactics, the huge resource endowment of the Soviet Union and the population increase as the Soviet Union passed through the demographic transition. The latter two factors were not present in the countries of Eastern Europe.

The modernisation of Stalin's modernity (development model) in the building of socialism in the states of Eastern Europe

Berman showed that the dynamism of development was central to the image of socialism. Taylor pointed out that whilst the idea of modernising modernity might seem odd, in fact it was not. Because modernity means incessant change, 'it follows that what was modern, must become unmodern' (Taylor 1999: 25). Hence the idea of modernising Stalin's modernity carries the meaning of the development of different ways of building socialism, in particular changing the Stalinist model of development.

Communist governments were driven by the ideological conviction that socialism was more effective than capitalism, in generating and equitably distributing material wealth and prosperity. Socialism was more rational and therefore more modern than capitalism and would supersede it. Socialism would be able to achieve the objective of catching up with the west and even surpassing it in levels of development. However, because they practised a Soviet imposed system, communist rulers often found themselves faced by a hostile public. There was a strong sense of 'them' and 'us' and communist rulers were forced by popular disturbances to undertake changes to the development model. In addition the system itself generated pressures for change, but it lacked an effective mechanism for achieving this smoothly.

Stalinist ideas and practices produced a particular place – the Soviet Union in the 1930s. The states of Eastern Europe were different from each other in their location and territorial extent, their population size and dynamics, their resource endowment and their level of socio-economic practices. They were all very different from the Soviet Union in the late 1920s when Stalin embarked on 'the building of socialism in one country'. It was soon realised that trying to reproduce the Soviet Union of the 1930s was not an effective path of development. The attempts of communist governments to modernise Stalin's modernity produced states in Eastern Europe as different places through the building of socialism in different ways; different places of socialist modernity.

Governments undertook modernising reforms to strengthen socialism and maintain their political power against a frequently hostile public. New development models contained four main elements. One was a transfer from extensive to intensive and then selective development. A second was to end autarchy and increase foreign contacts. A third was the relaxation of the institutional rigidities of the system and the introduction of 'market mechanisms'. A fourth was to find ways of streamlining the system to make it more effective. Each country produced development models that uniquely combined these elements. All adopted the first and second elements, but the latter two were vital and contrasting dynamics to modernising Stalin's economic development model and the socialist modernity it represented.

Krushchev's denunciation of Stalin provided the first opportunity to introduce economic reforms. However, the extent to which local Communist Party élites could break free from Soviet domination was strictly controlled at first, as a fine balance between political and economic reform. Not until the 1960s were the political conditions right for some local élites to essay radical new socialist models of economic development. Even then reforms had to be kept within the economic sphere. Popular discontent played a specific role in some countries as élites, at least temporarily, responded to public demands for change. Modernising Stalinism, by adaptation, by a search for new socialist models of development, and to meet practical problems arising from the development process itself, differentiated the experience of communist modernity in the countries of Eastern Europe. In none of the countries, however, was it a story of linear progression, but rather one of contested ideas of communist development. The introduction and implementation of innovations reflected a struggle between reformers and conservatives within the cadres of the Communist Party. Yugoslavia and Hungary evolved 'market socialist' models of development that, in different ways, pushed out ideological limits, elaborating new forms of socialist economic development, but not without periods of retrenchment as conservatives gained the upper hand. Czechoslovakia and Poland reflected the changing strengths of conservatives and reformers by changing models, trying to introduce some measure of market socialist principles, but on the whole retaining the essential elements of the command system. The GDR, Bulgaria, Romania and Albania formed a 'conservative' group that, for different reasons, tried to make the Stalinist model of development more efficient, whilst remaining within strict ideological tenets of faith.

The market socialists

Yugoslavia and Hungary modernised communism by devolving decision-making, introducing constrained market influences and encouraging the emergence of a significant private sector. Neither abandoned formal Five Year economic planning, but it lost much of its importance. In the 1950s Yugoslavia abandoned collectivisation and introduced 'self-management' of enterprises by their employees. In the 1960s fixed pricing and production targets were dropped; banking and investment policy devolved to republics and communities; and local communities were made responsible for providing and financing local services. New ideas of the 1970s weakened market influences by creating 'self-management agreements'. Decisions on investment and production were vested in officially recognised groups of workers under a compact supervised by the Yugoslav League of Communists (Dawson 1987). In the 1980s, following the death of Tito, Yugoslavia began a descent into political and economic chaos that ended with the collapse of the state.

Hungary's modernisation made it the most celebrated example of 'market socialism'. Early signs of independent thinking about the nature of socialist development ended with the Soviet invasion of 1956. In 1968 the government introduced the New Economic Mechanism (NEM). NEM was intended to create 'market socialism' that would combine macro-economic efficiency, indirect government regulation and socialist ethical principles. It removed obligatory plan targets, incentives based on quantity of production and central control of resources. It replaced them with economic measures (market forces), price and wage flexibility, incentives based on quality, and greater responsibility for company managers in business decision-making and investment. In 1971 some economic and political powers were devolved to local governments. However, despite the much heralded reforms, central, state ownership of enterprise was retained and minimised the effects of the NEM.

The most radical developments came in the 1980s. New laws further decentralised decision-making; abandoned the guarantee of employment for workers; and removed from enterprises protection against bankruptcy. At the same time, government restructured state-owned enterprise, reducing subsidies to heavy industries and drawing up plans to dismantle multi-enterprise trusts. Furthermore, it encouraged the formation of new kinds of small businesses. It chose to regard these as new forms of socialist co-operatives, but they were privately financed and profit orientated (Dingsdale 1991a). The operations of the 'second economy' previously unofficial, but thriving, were legitimised and taxed. Personal income tax and Value Added Tax for businesses were introduced. Joint ventures with western firms were agreed. Hungarian television broadcasting and cinema were also far more open to imported programmes and Hungarians had greater freedom than existed elsewhere to personally import technical items such as western videos and computers. By these means the government modernised the experience of building socialism in Hungary, searching for new forms of socialist development.

The undecided

The Czechoslovak experience was summarised by Carter (1987a: 104): 'the Czechoslovaks have had the occasional flirtation with market socialism since 1948, but the Soviet style Command Economy . . . has been the usual form of management'. The vital time for the Czechoslovak experience of the modernisation of communist modernity came in the late 1960s with the 'Prague Spring'. The reformers of the Prague Spring aimed to introduce a 'market socialism' that retained collective ownership, but required enterprises to compete in a market situation and not merely meet plan targets. Underlying the idea, however, was a link between economic decentralisation and political decentralisation. The declaration of the reformers also hinted that they regarded Czechoslovakia as a more modern state than its Soviet counterpart and so maintaining Soviet practices was a step backwards. Whilst the Soviet leadership and many members of the Czechoslovak Communist Party could accept economic reform, they could not accept the political implications. It challenged their privileged position. Consequently, the Czechoslovak flirtation with 'market socialism' was quashed by military intervention, because of its political content. The Husak government that then came to power 'normalised' Czechoslovakia into 'traditional', command methods of economic and political management.

The first scheme to modernise the Stalinist model by introducing market elements was proposed in Poland in 1956. Stimulated by Krushchev's denunciation of Stalin and as rioters filled the streets of Poznan, a 'socialist market' was proposed. The intention was to retain central planning objectives but operationalise them through market practices. An alternative scheme was also advanced which was to follow Yugoslavia by introducing workers' councils. By this means 'direct industrial democracy was to be a counterweight to the Party's economic bureaucracy' (Rupnik 1988: 181). A new leader, Gomulka, began by introducing reforms emphasising the production of consumer goods and improvements in standards of living, but soon backtracked into the old ways. Popular disturbances ended the rule of Gomulka in 1970. He was replaced by E. Gierek and efforts were made to introduce 'intensive growth'. To achieve this rapidly the government turned to western sources of capital, technology and expertise. However, this did no more than graft advanced western practices onto an otherwise unchanged command economy. The economic collapse of 1980 demonstrated that this did not work. It did, however, pile up foreign debts; and signalled that economic change must be accompanied by social and political change (Rupnik 1988). The success of the Solidarity workers movement in opposition to the government in 1980 led to the imposition of martial law, an expedient not used elsewhere under communism as a means of controlling development.

The conservatives

The conservatives were the GDR, Bulgaria, Romania and Albania. The modernisation of Stalinism within the framework of Soviet principles was the prime path of the conservative states.

Their motivation and circumstances were different. The GDR, the creation of Stalin's system, could exist only within it; no radical departure was conceivable. So, with German thoroughness the Kombinat system was operated to streamline the rigidities and ease political control. Grouping industries together in relatively few horizontally and vertically integrated organisations improved efficiency and facilitated political supervision. The GDR was severely handicapped by autarky and extended links with socialist countries breaking free from this restriction, but with mixed effect (Mellor 1987). Contact with the Federal Republic of Germany (FRG) was always a special feature of modernising the GDR. Under a protocol of the Treaty of Rome, establishing the EEC trade between the GDR and the FRG was treated as 'intra-German' giving the GDR some special benefits (Marer 1984). Even so, because the GDR was a product of Stalinism, it was the only one of the more developed northern tier of countries of Eastern Europe that did not contemplate development towards market socialism.

The Stalinist system served the needs of Bulgaria and Romania by rapidly accelerating their economic development. Bulgaria's path to development was strongly influenced and prompted by that of the Soviet Union, primarily motivated by official friendship that assured Bulgaria of Soviet technical assistance and generous credit facilities. '[O]f the post-war states of Eastern Europe, Bulgaria has been one of the least affected by the winds of change and self-assertiveness which has touched her neighbours' (Carter 1987b: 67). Bulgarian development is full of the rhetoric of reform – the Great Leap Forward of 1958–60, the New Economic System of the mid-1960s and the New Economic Mechanism all proclaimed radical change, but none delivered it (Moore 1984). Bulgaria developed the principles of technical development and organisation in agricultures to shift from extensive to intensive production. Collectivisation was completed by 1958. Between 1959 and 1970 collective farms and co-operative farms were merged to accumulate funds to advance technical and managerial expertise and purchase equipment. Finally, from 1971 there was the development of agro-industrial complexes replacing co-operative and state farms that ceased to exist. 'Bulgaria now boasts one of the most concentrated agricultures in the world, vested with stupendous possibilities for further development, for combining the advantages of socialism with the revolution of science and technology' (Grozev 1981: 291).

After 1965 Romania developed under the stern rule of N. Ceauşescu. Development maintained Stalinist priorities but asserted a distinctive national independence from Soviet influence. A cult of personality was constructed around Ceauşescu leading to a streak of erratic decision-making and personal arrogance-promoted dynastic ambitions. Centralisation was intensified, discrimination against minorities increased and standards of living deteriorated. In the 1980s food and energy shortages were routine for Romanians, causing persistent hardship whilst the schemes of Systematazare (Chapter 8) brutalised normal existence amidst lavish presidential spending on personal comfort. President Ceauşescu and the Party élite became ever more remote from the population and national development strategy became subject to meglomaniacal power.

Albanian national development lurched from one model of communism to another (Hall 1987), but was always characterised by strong central control, imitative of external ideas and dependent on outside assistance, despite being cacooned in a myth of self-sufficiency. Thus, at first closely linked with Yugoslavia, Tito's break with Stalin prompted a shift to Soviet dominance. In the late 1950s, Soviet pronouncements on Albania's best way forward met with official disapproval and Albania's leaders increasingly supported the Chinese in the Sino-Soviet ideological conflict. In the 1960s Chinese influence and aid replaced the Soviet Union. In the late 1970s China was adjudged to be adopting revisionist policies and so links were broken off. Albania became increasingly isolated and strove for the impossible goal of 'self-sufficiency' leading to austerity and stagnation. These shifts in allegiance brought substantial outside expert and financial aid, but also interrupted fulfilment of planned progress. Projects were slowed or left incomplete and standards of living regressed.

Crucially, none of these modes of modernising Stalinism invested much in transport or communications technology for general purposes. The Soviet Union developed complex military technology and communications systems but these were not shared with Warsaw Treaty Organisation states and not transferred to public usage. The process of development was contained within an institutional establishment that feared technologically advanced communications and contact because they were difficult to control. Interpersonal communication would become too open, the no-alternative Party line too easily challenged. Even in the Party controlled media individual initiative and inventiveness were discouraged.

Regional development: regional identity – the identity of the region

Three domains in the production of regions

Development is a process that produces regions as places constructed as contested spatialities. Regional development under communism was produced in practice and discourse within three principal, separate, but weakly interactive domains. First, there was the official domain, made up of the ideas and practices of the Party-state. Second, there was the academic domain, made up of the discourses among geographers, economists and sociologists. Third, there was the popular domain, made up of understandings of regional development issues that were often invested with a strong sense of cultural and social meaning. In the relationship between these three domains the practices of the Party-state as the arena of public policy was pre-eminent in articulating the meaning and experience of regional development under communist modernity – as the building of socialism. Regions were transposed from national space in a variety of ways that were often locational mismatches, but the idea of the region is best understood as an outcome of communist social relationships, particularly the exercise of élitist centralised power.

The official domain

All communist governments had broad spatial aims as part of their project of building socialism. The warp of the fabric of communist modernity was industrialisation, urbanisation and the collectivisation of agriculture. The weft was a concern to equalise standards of living across the whole of the country, between urban and rural areas; and to harness the physical and human resources of every region to maximise national development potential. Turnock (1987: 230) writing about Romania states that: 'The communist government deplored the spatial contrasts which it had inherited in levels of development and set out as one of its goals the more equitable distribution of economic activity.' The sentiment was true in all communist countries. The settlement and regional development policies communist governments introduced were intended to achieve this equalising goal, though sectoral economic policies often counteracted them and were, furthermore, dominant in development thinking.

> Nobody would deny the time dimension or historical character of social development, yet it is all the more difficult to convince theoretical econ- omists or economic decision-makers about the spatial (*regional*) dimension of the economy. It occurs not infrequently that regional development is reduced to the problem of the local council budget or state support to some region.
>
> (Enyedi 1979: 113)

The official domain encompassed the thinking in three official areas of development:

(a)   the national development strategy;
(b)   planning ideas and practices as the framework of action;
(c)   the territory of the state that shaped the locational, societal and cultural nature and intensity of regional development issues, perceived ideo- logically and practically.

Regions as places are produced by development that is either sectorally or spatially structured, or by a combination of the two. Sectorally structured development produces regions mediated through locational principles and perceptions of current distributional attributes applicable separately to each sectoral category. Spatially structured development produces regions by directly differentiating between them as places for the purpose of locating investments. Within national development strategies, sectoral thinking and spatial thinking competed for priority in strategic approaches to regional development.

The authorities thought of the strategy for development as a national task, not a regional one. They measured success by national indicators of growth, particularly economic growth. They conceived development as first and foremost divided into economic sectors for the allocation of investment funds. Spatial division of the country for the purposes of investment and development was only a leit-motif against this background. The sectoral mode of thinking dominated national development strategies. Throughout the period of communist modernity, the authorities with the key decision- making powers favoured sectoral modes of development over spatial modes, especially at the regional scale. This was reflected in a variety of ways across the Eastern European countries. Whilst each country is different there are examples of: the general absence of clearly thought out spatial strategies; the late introduction of regional strategies, for example in Bulgaria (Carter 1987b) and Hungary (Enyedi 1979); the prioritising of national efficiency over regional equalisation, for example Romania (Turnock 1987); and the rejection of carefully prepared schemes of national regional development as in Poland (Dawson 1987) and Czechoslovakia (Carter 1987a).

In so far as spatial thinking played a part in the perceptions of the central authorities, as expressed in the national development strategy, it did so primarily through administrative policy and settlement policy. All the countries engaged in redrawing administrative boundaries at various scales. Administrative divisions were significant in the production of regions because administrative boundaries were used in economic and social development strategies and most regional strategies. In only a few cases were spatial planning projects undertaken without regard for administrative boundaries. There are examples from Hungary in the 1960s, such as the Lake Balaton Plan (Ministry of Building 1977) and Czechoslovakia under the 1959 Town and Country Planning Act (Carter 1987a). Settlement development policies contributed to the production of regions through local development and were also an important dimension in national development strategies (Chapter 8).

Planning was the framework of action. Issues of scale, practices, perceptions and conceptions of place, among planners, produced the delimitation and character of regions as places. A rich vein is tapped in examining the production of regions in the realm of planning ideas and practices, even though the outcomes were largely ignored by the authorities. In each country, planners devised a whole array of regional plans. Representations of regions and their physical, societal and cultural futures were carefully drawn and articulated in detail on thousands of maps. Progress was made in establishing the relationship between territorial planning, national economic planning and settlement planning. This permitted plans to be drawn up to promote regional policy objectives and have a direct impact on the production of regions by physical and economic development. Within approaches to regional planning in some countries, Poland and Hungary, for example, there is a general evolution from conceptualising regions as separate areas, with particular socio-economic problems or special potential for development, to conceptualising regions as integral to country-wide systems of socio-economic planning (Dingsdale 1991b). This was associated with the switch from extensive to intensive and selective development. It built on the earlier surveys of resources. Thus spatial planners increasingly perceived regions as complex areas, whose internal and external relations were crucial to the process of national development. It is difficult to say to what extent this shift was influenced by the academic debate on theoretical economic regions referred to below. It is clear though that its acceptance by central decision-makers was limited.

Within the territory of the state the perceived nature and intensity of problems and possibilities of regional development are defined. In Hungary the main perception was of regional development polarised around the status of the capital, Budapest, on the one hand, and the rest of the country on the other. This resulted in policies, introduced in the 1960s, designed to reduce the dominance of Budapest and stimulate economic activity in the regions. These policies were underpinned by 'growth pole' theories of regional development. In Czechoslovakia the main issue was perceived as the persistent contrast in levels of development between the Czech lands and Slovakia. Slovakia always lagged behind the Czech lands in economic development. Strong policies to redress the balance were pursued by governments. The planting of new industrial premises in Slovakia began early and

continued throughout the communist period. In Yugoslavia the contrasts
in development lay between the more prosperous northern republics and
the less developed southern republics. In Poland an early task was the integration of the new northern and western territories into the new territorial
state. Subsequently the contrast in development between the east and the
west was perceived as a vital problem. In all countries border regions were
restricted areas, patrolled by border guards, denied investment funds and
left blank on maps. A lack of international co-operation produced border
regions as highly polluted areas. The Polish, Czech and GDR border region
suffered especially badly from pollution for this reason.

The academic domain

For geographers, economists and sociologists within Eastern Europe and
from outside, regional development under communism was a topic of considerable interest. One strand of debate concerned regions as spatial patterns
of uneven development. It analysed the nature and causes of regional problems and contrasts in regional levels of development (Demko 1984;
Mihailovic 1972; Robinson 1969). In this debate regions were represented
as 'dynamic', 'backward', 'intermediate', 'peripheral' and so on. This
terminology described regional characteristics in a way that tended to equate
industrialisation and urbanisation with development. This was not unreason-
able given the official emphasis on these processes as at the root of development thinking, but it certainly implied value-laden assessments of change.
A second strand was engaged with the description and evaluation of regional
development policies and planning (Dawson 1987; Hamilton 1975; Lackó
1985; Osborne 1967). These debates often pointed to the ineffectiveness of
regional equalisation policies. A more theoretical debate on economic regionalisation was also vigorously pursued. Individual academics and academic
institutions introduced theoretical schemes of economic regions following
detailed empirical and theoretical research. The idea of 'economic regions'
was a prime concern, because academics wanted to suggest how the state's
territory might be most efficiently organised for the planned maximising of
resource utilisation.

The popular domain

Whilst spatial thinking at the regional scale in the official and academic
domains was primarily concerned with economic and social development,
it was underscored in some countries by an awareness of an ethnic cultural
dimension. The production of regions in popular consciousness reversed
this conceptualisation, owing much more to ethnic national identity that
was sometimes thought to be the reason for contrasts in the pace and level
of socio-economic transformation. This was particularly so in the two Federal
Republics, Czechoslovakia and Yugoslavia. In Czechoslovakia the persistent
weakness of Slovakia in socio-economic and cultural development underlay
the practical and popular issue of regional identity and attitudes between

Czechs and Slovaks. In Yugoslavia the contrast was between the northern republics as compared with southern republics. Popular regional conceptions, identities and frictions were fuelled by persistent socio-economic contrasts in development and ethnic consciousness.

In Romania, a unitary state, ethnic tension was particularly associated with Transylvania, which was settled and historically disputed by Hungarians and Romanians. For a time, in the communist period, Transylvania enjoyed autonomous status and its economic performance was relatively strong. The withdrawal of regional autonomy by the Ceauşescu government antagonised ethnic Hungarians. In the late 1980s, the Hungarian media, and popular demonstrations in Hungary, emphasised the ethnic content of Systematazare in Romania. Even the Hungarian government was forced to protest to the Romanian government. The ethnic content of Systematazare is contested (Chapter 8), but there is no doubt that Transylvania was an emblem for both Romanian and Hungarian nationalisms. The discrimination against Hungarians practised by the Romanian authorities at numerous administrative levels was one cause of a stream of Hungarian emigration from Romania in the late 1980s (Dingsdale 1996).

The construction of regions in communist development in Eastern Europe was contested, but conditioned by an overwhelming desire to draw lines and divide up state territory. The production of regions was a tool of bureaucratic power and control. In official and academic domains there was a concern to define regions in terms of distributional attributes. The spatial distribution of economic activity, urbanisation, and social and cultural development resulted primarily from decisions made by central sectoral ministries. The thinking underlying the delimitation of administrative divisions was often obscure but, it seems, professional and intellectual regionalisation of the state territory did little to influence those decisions. In practice, regions were produced within countries as administrative divisions differentiated on the basis of locational, societal and cultural attributes as a layering of different locational decisions of sectoral ministries. Thus separately structured locational, societal and cultural spaces were transposed into regions by administrative bounding. Such boundaries did not match popular regional consciousness. To some extent they were designed to change this. Neither did they match socio-economic distributions, but they became the basis for the allocation of a portion of investment funds. Because of this they conditioned the process of development.

Case example: The production of the Great Hungarian Plain as a backward and peripheral region in communist development

The Great Plain is a physical region, lying east of the Danube, that occupies almost half of Hungary. During the communist period the Great Plain was produced as a backward and peripheral region, materially and in a discourse of dominance within Hungarian regional relations. Budapest was the dominant place in Hungarian regional relations and, although its relation-ship with the Great Plain was changed during the communist period, this only served to change the form of its dominance.

In 1945 Hungary had a relatively simple regional development profile. Budapest had an articulated economic, social and cultural structure that localised all the country's powerful government, business and cultural insti-tutions. The north east was industrialised by coal mining and engineering, mainly around the cities of Miscolc and Diosgyör. The north west was industrialised by mining, and heavy and light engineering centred on the city of Győr. Southern Transdanubia was agricultural with industry confined to the city of Pécs and the nearby Komló coalfield. The Great Plain was almost exclusively agricultural; industry was located in some towns. The standard of agricultural practices was low throughout Hungary, but was especially poor in the Great Plain (Enyedi 1979) (Figure 9.1).

The First Five Year Plan heralded a period of accelerated industrialisa-tion in Hungary. It included references to industrialisation in the Great Plain. However, regional development was determined by the priority of investments in heavy industry, and agriculture was neglected (Table 9.1). The plan unfavourably resituated the Great Plain in national development as investment went into developing the existing industrialised regions and their mineral resources. Divergence in regional development adversely affected the Great Plain. The 1950s, however, held out some signs of a brighter future. The country was divided into nine theoretical regions and surveys were undertaken in a search to discover new resources. Even so, hopes for the development of the Great Plain were dashed, when it was decided to transfer oil and gas resources, discovered in the Great Plain during the course of these surveys, to existing plants in the north for processing. A series of special regional plans were drawn up, including one for the Tanya farmstead districts of the Great Plain, but this turned out to be a plan for their destruction (Chapter 8).

After 1958, government priorities changed and agriculture was collec-tivised and the first comprehensive regional development policies formulated. The broad aim of the regional strategy was to reduce the dominance of Budapest and industrialise agricultural areas, especially the Great Plain. What emerged was a triumph of élitist manipulation that ensured that the Great Plain became industrialised, but its industrial profile was formed by low

Table 9.1 Regional distribution of industrial investment in Hungary, 1945–85 (%)

	1956–60	1961–65	1966–70	1971–75	1976–80	1981–85
Central	34.4	31.8	29.6	29.6	24.2	25.7
North west	25.6	22.9	24.8	23.6	21.9	22.1
North	17.7	19.7	20.8	21.0	18.8	12.6
N. Great Plain	4.2	7.0	7.1	7.3	10.7	9.3
S. Great Plain	3.3	7.2	9.6	9.1	9.0	10.1
S. Transdanubia	14.8	10.6	8.1	9.4	15.4	20.1

Source: *Regional Statistical Year Books.*

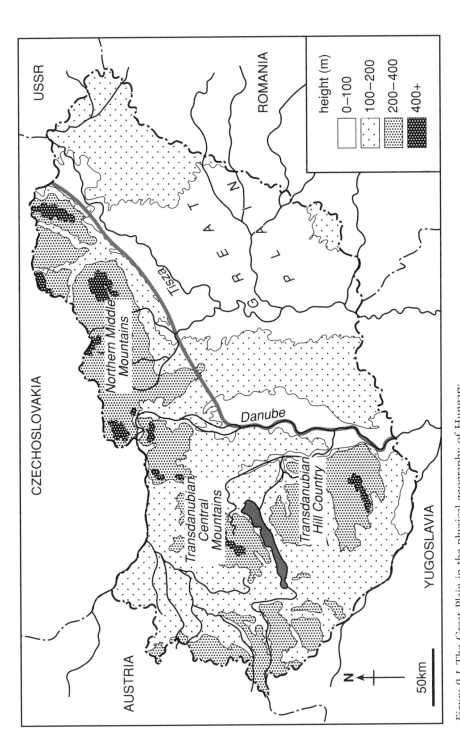

Figure 9.1 The Great Plain in the physical geography of Hungary.

Source: Adapted from Dingsdale (1991a) 'Socialist Industrialisation in Hungary', in F. Slater (ed.) *Societies, Choices and Environments*, 441.

technology, low labour skill branches. What happened was that Budapest's unwanted branches were dispersed to bring low grade industrial jobs to the Great Plain.

The dispersal scheme was devised by the central government in co-operation with the Budapest city authorities. It offered substantial benefits to Budapest, where political power lay. The capital's industries were sorted into three categories: those that were to be relocated; those that could stay but not expand; and those that could stay and expand. Budapest retained those branches that had growth potential, but expelled those that had not. Between 1965 and 1970, sixty-seven plants were relocated to the Great Plain (Tatai 1978).

The Great Plain's development was not, however, only promoted as the place of reception for Budapest's cast-offs. Under the terms of the New Economic Mechanism (1968) managers had more power in decisions about plant location. In the late 1960s there began, on the initiative of enterprise managers, the spread of branch plants into the Great Plain (Figure 9.2). Behind these moves lay two important factors. First was the transformation of Hungarian enterprises from single plant to multi-plant operation (Barta and Dingsdale 1988). Second was the shortage of labour in Budapest. The industrial investment, however, was concentrated in the regional cities and towns under the terms of the National Settlement Development Strategy that was dominated by the idea of 'growth poles' (Compton 1987). The Great Plain also developed as a result of innovation in agricultural organisations. Several highly efficient agro-industrial organisations were established. The foundation of two extensive national parks introduced the potential to attract tourists. However, the rapid expansion of the private or second economy in Hungary during the 1980s that was associated with service activities, including tourism, was only weakly represented in the Great Plain.

In summarising this process Barta (1984) states: 'Despite the reduction in spatial differentiation of production the most dynamic control elements of industry remain in Budapest.' Nemes-Nagy and Ruttkay (1985) comment: 'In spite of decentralisation in the Hungarian economy . . . in the 1960s and 1970s, no profound change took place in terms of the regional distribution of sources of innovation. What is more the . . . trend of decentralisation in the sphere of production occurred with a growing level of concentration of economic management, control and research functions in the hands of company headquarters in the capital (Budapest).'

Thus, the Great Plain developed materially, as increasing levels of industrialisation, urbanisation, agricultural organisation and infrastructure testify. However, at the same time it became more firmly under the economic control of distant enterprise headquarters located mainly in Budapest. Neither did its level of development keep pace with that of Budapest, where the dynamic branches of the economy remained prominent.

The discourse on development confirmed the idea of the Great Plain as backward and peripheral. When describing the spatial patterns of development of 1945, Enyedi (1979) referred to Budapest as 'the only dynamic centre' and the Great Plain as 'backward'. This terminology echoed that of

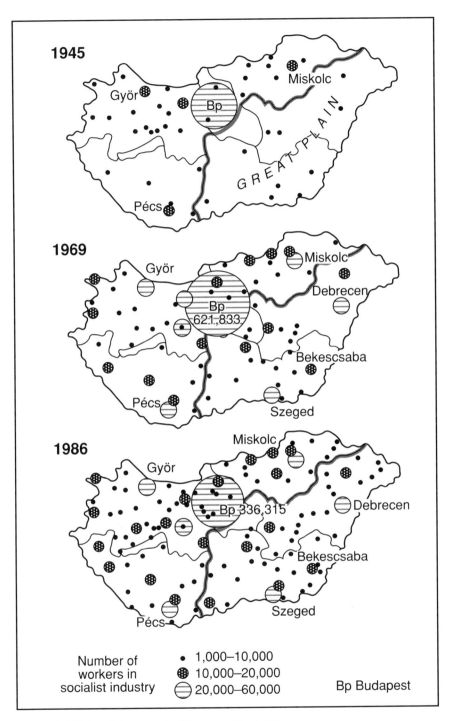

Figure 9.2 Industrial centres in Hungary, 1945–86.

Source: Adapted from Dingsdale (1991a) 'Socialist Industrialisation in Hungary', in F. Slater (ed.) *Societies, Choices and Environments*, 475.

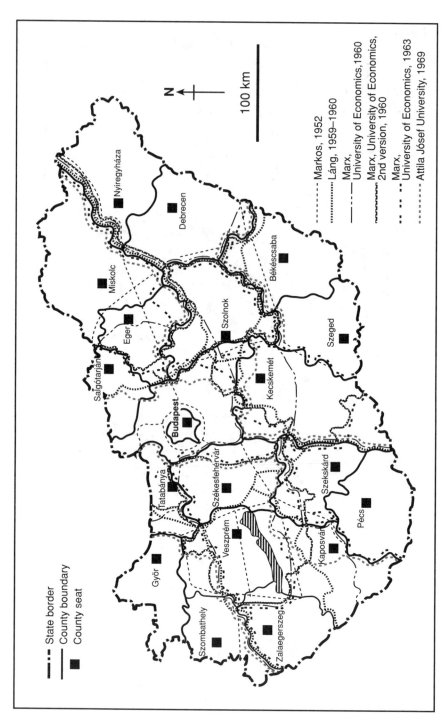

Key:
- ----- State border
- ——— County boundary
- ■ County seat

- --------- Markos, 1952
- ·········· Láng, 1959–1960
- —·—·— Marx, University of Economics,1960
- ∿∿∿∿ Marx, University of Economics, 2nd version, 1960
- ··—··— Marx, University of Economics, 1963
- ········ Attila Jósef University, 1969

Figure 9.3 Proposals for economic regions in Hungary.

Source: Hajdú, Z. (1984) Geography and reforms of administrative areas in Hungary' in G. Enyedi and M. Pécsi (eds) *Geographical Essays in Hungary*, 1984, 67.

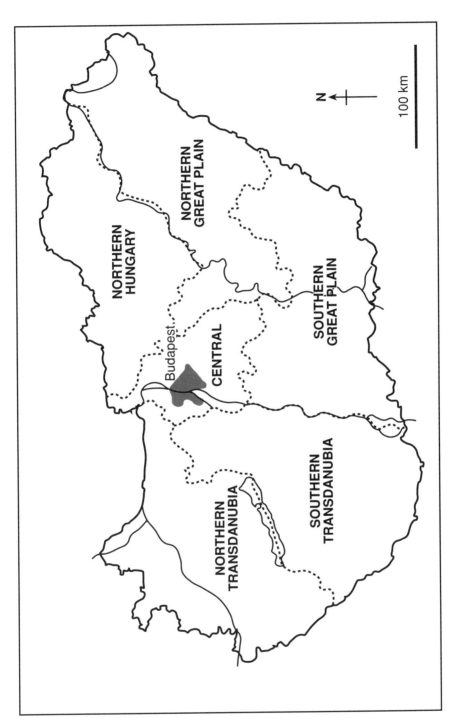

Figure 9.4 Economic planning regions in Hungary, 1967

earlier researchers such as Köszegi (1969). There was a less pejorative strain of academic research engaged in proposing schemes of economic regions. No fewer than eight complex schemes were suggested for Hungary (Figure 9.3). The government accepted none of these but rather forced them into administrative units to create six economic planning regions (Figure 9.4). The Great Plain was divided into two such regions.

Popular regional thinking also polarised around a dominant image of Budapest and an undesirable countryside. Young graduates were unwilling to look for jobs in the rest of the country and clung to the lifestyles of Budapest. In the popular mind, the Great Plain was still often imagined as the romantic place of Hungarian literature learned at school and recaptured for visitors in the idea of the Hortobag National Park. In practice though, the material conditions of the settlements of the Great Plain did not attract young people. Romantic as the image of the life of the Plain's horsemen might be, it could not dislodge from people's minds the dominant image of Budapest as the place to be.

Conclusion

Communist modernity brought new material practices and dynamics of discourse to the production of states and regions. The Marchlands in communist modernity can be conceptualised as a group of separate states, adopting and then adapting the Stalinist model of socialist development. Communist élites were inward-looking, they closed their borders and shunned international contact, except of the formal institutional and diplomatic kind. The Party-state was produced through the building of socialism as a modernisation of Stalinist modernity, but largely retained the model of 'socialism in one country'. The vision of a communist future to be achieved through the building of socialism legitimised their dominant spatiality within the state's territory. Particular paths in the building of socialism differentiated the states one from another and produced them as particular syntheses and contrasting places. The production of regions was an integral part of national development, combining sectoral and spatial thinking among central decision-makers and planners, tinted by academic and popular ideas and practices. The experience of communist modernity always had the hallmark of shortages and, except in the market socialist countries, a lack of consumer goods. It was very much a make-do-and-mend modernity; a colourless, flattened experience.

10 The production of Eastern Europe in the European and global spaces of competing modernities

This chapter explores the production of Eastern Europe in the European and global spaces of competing modernities. It employs the concept of simultaneous contrast, that is comparing different perspectives at the same time, to elaborate five spatialities producing Eastern Europe as a place in practice and discourse. These spatialities are: the communist world system, interstate relations, the idea of development, the capitalist world economy, and cultural relations. These spatialities are melded in the construction of European space by the primacy of the East–West ideological division. The global space of competing modernities is constructed in multiple discourses. European space is dwarfed by the global competition of modernities in the enlarged East–West division of ideological and security conflict, around the two 'superpowers', the USA and the Soviet Union, articulated in the idea of the Cold War. The idea of the First, Second and Third Worlds constructed global space in terms of the ideological divisions and non-alignment. This became transformed into the development debate in the form of the West–East–South construction. The contacts and superficial similarities of the West and East in development characteristics extended this idea to a North–South development division. Meanwhile in the predominant projection of Western global modernity – the capitalist world economy – a tripartite core–periphery conceptualisation that characterised the early period became overlain by the first signs of globalisation and the production of global space as a single place. The communications technology and media associations of the globalisation concept extended the discourse to the idea of cultural imperialism.

Eastern Europe in a European space of competing modernities

Between 1950 and 1990 the predominant conceptualisation in continental Europe was of an East–West divide. Generations of Western Europeans conceptualised Europe as bounded on the East by the Iron Curtain. The appellation 'Europe' signified contemporary representations of Western European modernity (Chapter 1). In popular Western perceptions Eastern Europe was an unknown neighbour, sharing continental space, and sometimes, curiously, associated with some aspects of European culture, but at the same time not Europe. School curricula contained nothing about Eastern

Europe, newspapers and the broadcast media were equally silent and their, occasional, reports were tinged by Cold War assumptions and images – Budapest 1956, Prague 1968, for example. Visits were even rarer. Within Europe international travellers circulated around the two separate halves, but there was negligible contact between them. Emigrant communities living in the West painted, on the whole, an unattractive picture of lands that they had fled. The East became the 'Other Europe'. In the East this was perceived, and often resented as, 'Yalta Europe'.

Eastern Europe in the European communist world system

Relations with the Soviet Union and the perceptions and practices of the Soviet leadership endowed Eastern Europe with multiple identities. Aspaturian (1984: 16) described these as simultaneously:

> a defence *glacis*; a springboard for possible expansion westward; an ideological legitimization of her universal power; a laboratory for the application of the Soviet model of development; a reservoir of human, natural and economic resources for Soviet recovery; a collection of diplomatic pawns and surrogates to be used in international politics; and a source of psychological and even quantitative comfort in international organisations and conferences, where the Soviet Union might otherwise be isolated or alone.

This succinctly summarises a Stalinist perception that was reflected in material practices some of which persisted under later leaders. However, events in Eastern Europe, related in Chapter 9, shaped new perceptions and Eastern Europe became perceived as a collection of unruly and increasingly independent-minded extensions of Soviet modernity, sometimes needing corrective attention. Yugoslavia and Albania soon rejected Soviet domination. Hungary in 1956, Czechoslovakia in 1968 and Poland in 1980 were perceived to be challenging Soviet hegemony. Romania had strong independent tendencies. The 'Brezniev Doctrine' conceptualised an Eastern Europe that was fragmented by varied national styles of independence, but remained within the Soviet socialist political sphere, and undertook action to enforce this in practice. The accession to power of M. Gorbachev transformed the perception of Eastern Europe within the Soviet leadership. The modernisation of thought and action inside the Soviet Union foreshadowed by the policies of *glasnost* and *perestroika*, in part forced by the strain of the global role played by the Soviet Union, transformed the perception of Eastern Europe. Still conceptualised as socialist, Eastern Europe was perceived as draining Soviet material and psychological resources. The 'Sinatra Doctrine' perceived Eastern Europe as no longer vital to Soviet interests. Eastern Europe and the Soviet Union would both benefit from *perestroika* and a relaxation of previous levels of Soviet intervention. Each Eastern European state could be free to be socialist in its own way and the consequences must not be allowed to damage the evolution of the global

US–USSR relations. The Soviet leadership now perceived Eastern Europe as a place in which the costs of intervention were greater than the cost of tolerance (Linz and Stepan 1996). Gorbachev encompassed this in a vision of a 'common European home' that united Europe from the Atlantic to the Urals. Thus, as time went by, Eastern Europe became resituated in the European and global spatiality of the Soviet leadership.

This sketch of transforming perceptions should not be taken to imply a simple or easy progression. As the next paragraphs will show, Eastern Europe was perceived as a place of prestige as well as challenge for the Soviet Union.

The material and spatial practices of the Council for Mutual Economic Assistance (CMEA) known popularly in the West as Comecon, were significant in changing Soviet perceptions of Eastern Europe. Formed in 1949, the CMEA did not play an active part in Soviet–East European relations until after 1971. The 1971 Complex Programme was intended to provide an opportunity for Eastern European states to throw off autarchy and through a variety of new contacts, strengthen each other economically, under the direction of the Soviet Union, within the communist world system. Despite its cumbersome, inefficient, bureaucratic practices the CMEA built a series of bilateral exchanges in which the Soviet Union provided raw materials and technical assistance to Eastern Europe and received manufactures in return. The Soviet leadership soon realised that this polycentric model was a burden on the Soviet economy (Bozyk 1988). On the one hand, the Soviet Union faced rising costs of raw material production, as the most accessible sources became exhausted. On the other hand, imported manufactures were of inadequate quality, coupled with unreliable delivery. Changes to a more balanced exchange flow of manufactures and technical co-operation were attempted. Bozyk calls this the 'repartitional-exogenic model'. However, the Soviet Union still carried the substantial burden of keeping economic relations confined to the communist world system, compounded by the lost opportunity of selling raw materials, at higher prices, on the wider world market.

The practical burden of supporting Eastern Europe through the CMEA was counterbalanced, in the minds of the Soviet leadership, when forming its perception of Eastern Europe, by important non-material considerations. The Soviet Union gained prestige and ideological credibility among the leaders of other communist states, and those in the West through its particular relationship with Eastern Europe.

> Eastern Europe as a bloc of communist states modelled on the Soviet Union ... continues to validate the Soviet Union's credentials as an ideological and revolutionary power as well as the residual centre of a world revolutionary movement ... it is also an extension of the Soviet social system; and as the first and most important extension of the Soviet system ... it represents the residual of Moscow's former ecumenical pretensions ... the first step in universalising the Soviet system.
>
> (Aspaturian 1984: 46)

Eastern Europe and the capitalist world economy in Europe

Western Europe forms one of the three cores of the global geography of the capitalist world economy. As a result of multiple forms of contact, Eastern Europe took shape as a minor peripheral place in the European sphere of the capitalist world economy. It was transformed only slightly and slowly, and retained a different mode of production, but a new Western modernity made its first invasive steps. Contacts between the European core and Eastern Europe occurred through trends in production, trade and capital investment, and the processes producing them. Until the mid-1960s contacts were few, but from then on a process of reciprocal benefit, or so it seemed at first, promoted contacts. Western European companies, responding to global challenges, especially the turbulence of the global scene in the 1970s, began a 'European shift' in production, trade and capital investment. Eastern European states, modernising Stalinist practices, as shown in Chapter 9, sought to open their economies in order to acquire, quickly, higher technology equipment, improved management practices, and Western sources of capital. A variety of forms of technology transfer and industrial agreements were adopted. These included the supply of complete industrial plants or production lines; licensed production; co-production and specialisation; and joint ventures. However, much of the production technology that was transferred was already redundant in the West, for example the FIAT production lines transferred to Poland and the BAC 111 aircraft and Renault 12 production lines that went to Romania. Hamilton and Linge (1981) emphasise several aspects of the trade relationship that were detrimental to Eastern Europe: East–West trade expansion was mainly financed by Western supplier and bank credits, causing huge indebtedness in the CMEA; East–West trade was largely complementary, Western manufactures were traded for raw materials, fuel and agricultural products from the East, but some Eastern exports of lower order manufactures were restricted; the East depended more on the West than vice versa. Western companies and banks perceived in Eastern Europe a politically stable, financially reliable place to do profitable business.

Investment funds also passed from East to West. Brun and Hersh (1990) noted 331 instances of investments in Western Europe from Eastern European countries by the end of 1983. Investment was primarily aimed at companies offering commercial services and often associated with import–export activities. The investment seems to have been motivated by the need to facilitate trade and understand Western business practices. West Germany (72), the UK (53) and Austria (40) were the principal countries to which investment was directed. Hungary (95) and Poland (76) were the most active countries. Hungarian investment was especially prominent in West Germany (28) and Austria (23). This reference gives no indication of the monetary value of these transactions, but serves to illustrate that the interconnection of West and East in the European capitalist world economy was not entirely a one-way process.

Eastern Europe in European international relations

In international relations the perception of Eastern Europe was produced and given its identity, in Western Europe in the framework of the Cold

War and the protracted attempts at economic, social and political integration around the EEC idea. Bearing this in mind, it is important to remember that in the post-World War Two period the states of Western Europe faced a crisis of identity. As they lost their global imperial roles, they turned to the idea of Europe to find a new sense of place identity and purpose. Furthermore, the Treaty of Paris that established the European Coal and Steel Community (ECSC) and the Treaty of Rome that established the European Economic Community (EEC) declared as an aim the ever closer union of the peoples of Europe. Although this declaration was made in the name of the heads of state of the six founding members it clearly was meant to apply to a larger territory. The Treaty of Rome proclaimed that any European state may join, but did not specify what was meant by 'European'. It is clear that the appellation 'Europe' was not denied to the states of the East of Europe; it was the states of the East that were forced, by the Soviet Union, to stay out of the early process of integration that began in the West (Van Ham 1993). Neither the ECSC nor the EEC included all the states even of Western Europe, but the aspirations they proclaimed came to be a focal point in reshaping Western identity.

One of the motivating forces of Western European co-operation was the perception of Eastern Europe as a springboard of Soviet military advance westward. In the years immediately following the war it was not clear which states would be part of the integration process; which would be Western and which Eastern. Czechoslovakia and Poland had wanted to participate in the Marshall Plan. Even as late as 1956, Hungary hoped to break free from the Soviet sphere, but Western European states, still engaged in a wider global role, and the USA did not move to assist this. The events of the 1950s drew the bipolar lines and Western European co-operation and integration began in the mutual antagonisms of West and East. Eastern Europe was constructed in a fiercely anti-communist atmosphere in the West. Meanwhile, the Soviet Union and the Eastern European states under its influence always opposed Western European integration for ideological, economic and political reasons. As Cold War rhetoric ebbed and flowed, the Western perception of Eastern Europe was adapted in response, but the conception of a divided Europe persisted; the East was a distinct place, ideologically different from the West.

The perception of Eastern Europe in the West was constructed from commercial and political practices of Western European states. The EEC never produced effective commercial or political policies for dealing with the East. Member states acted in their own commercial and political interests. Commercially EEC and CMEA contact shifted from early neglect, to meaningless negotiation, and then to hasty mutual recognition in the maelstrom of *perestroika* (Van Ham 1993). Politically, the major states followed their own agenda, German Ostpolitik, French l'Europe des Patries, British Commonwealth preference and US Special Relationship guided relations with the East. Official commercial and political contact between the two halves of Europe were fitful and the East remained a place of deep ambiguity for the West.

Thus Eastern Europe was produced in a Western Europe that was uncertain and seeking a new identity in global, European and national space, through new practical relationships, perceptions and conceptions of itself and Europe. It projected its internal tensions onto Eastern Europe as an opposite mirror of its own modernity. The construction of Eastern Europe in Western European international relations was permeated by fragmented ideological, commercial and political structures, most particularly: a wider global bipolarity; the actions, attitudes and perceptions of the Soviet Union and the East European states; wider visions of European unity; and the practices of national interests by member states in relation to the EEC process.

The prominence of this East–West geo-political division put into the shadow a third European zone. The belt of states, from Finland in the North to Yugoslavia in the South, and including Sweden and Switzerland, the traditional neutrals, and Austria, whose international role was subject to Soviet agreement, were all but ignored in the formation of the dominant conceptualisation. Yet, these states played an important part in beginning the non-aligned movement that gave original meaning to the idea of the 'Third World'. Some of them also played an important European role in the European Free Trade Area (EFTA).

Eastern Europe and Soviet cultural imperialism

Aspaturian (1984) detected a geo-cultural cleavage in the foreign policy stances of East European states that divided the North and centre from the southeast. He identified in the behaviour of Albania, Yugoslavia and Romania, a Balkan opposition to the Soviet Union, interrupted only by Bulgaria's pro-Soviet positions. He recognised an episodic intensity in Balkan attitudes formed through their perception of the interventionist intentions of the Soviet Union. Enyedi (1990b) noted that the 1980s saw a strengthening of wider Balkan identity represented by the intensified contacts of the southern socialist states with Greece and Turkey. The Balkans were also separated from other areas of Eastern Europe in Western perceptions too (Todorova 1994). Enyedi noted also a refusal to accept, or more precisely a protest against, the Yalta division of Europe. His reference here was to a revival in the idea of Zentraleuropa (Chapter 2); M. Kundera, V. Haval and G. Konrád were most notably associated with this. The cultural roots of this revival were most distinctly revealed in Kundera's representation of Zentraleuropa as having been 'kidnapped' by the Soviet Union and alien, Eastern, values imposed. There was a strong anti-Russian sentiment in the idea. The idea of Zentraleuropa was associated with Western European civilisation and civic society, in particular the Habsburg, Roman Catholic formulation of that tradition. This was perceived by these intellectuals as producing a golden age of Zentraleuropa culture. For Kundera, Haval, Konrád and others Zentraleuropa was a symbolic place. It stood opposed to the idea of Eastern Europe that was constructed from Eastern cultural and historic forms and imposed by Soviet, Russian hegemonic power. Zentraleuropa was in Garton-Ash's (1986) phrase 'a kingdom of the spirit'. Konrád acknowledged that Zentraleuropa was a symbolic place.

It was never meant to convey a place with precisely drawn boundaries that could be represented on a map. It signified opposition to Russia, the Soviet Union and Eastern norms and values. It was an affirmation of Western European cultural values and a challenge to the symbolism of Eastern Europe as it was constructed in geo-political experience after 1945. The idea of Zentraleuropa became a place critique of Eastern, Soviet cultural imperialism.

Enyedi (1990b) rejected, as an insider, this imagined place, Zentraleuropa, because it seemed to him to imply an inferiority complex, a desperate need to feel European that was unnecessary when surrounded by a rich cultural environment. He perceived in East–Central Europe a cultural life that was 'more complex and colourful, and hence more European, than some great Western European cultures which are compassed within a single language and a single national culture' (Enyedi 1990b: 143). For Enyedi the idea of Zentraleuropa was nostalgic for a fin-de-siècle image of a place that never existed. What is more, he thought that the idea of Zentraleuropa turned towards the past when the search for a democratic East–Central Europe should look to the future. Finally he believed that the Balkans should not be excluded as Czechoslovakia, Hungary and perhaps Poland escaped and joined the West. Enyedi asserted that there was no 'emergency exit' from a geo-political region. Whilst I do not share Enyedi's implied conceptual stance of permanent regional boundaries, this does not invalidate the strength of his main points.

The idea of Zentraleuropa was a protest against the Russian, Eastern, cultural imperialism of Soviet communist modernity. Other examples can also be cited. The protracted and unsuccessful efforts to establish an effective nationalities policy in Yugoslavia can be understood as a struggle between communist attempts to create a modern socialist citizen and forces championing a maintenance of national cultural identities. The local and Western outcry against Systematazare in Romania (Chapter 8) also stemmed from the perceived conflict between communist modernity and national cultural identity.

Eastern Europe in the global space of competing modernities

Smith (1993) pointed to four main geo-political conceptual models of post-1945 global space. Traditional geo-politics, building on Mackinder, portrayed the Soviet Union as an aggressor, striking out from the world core into its Rimland and contained by the US and its allies as they defended the 'free world'. Balance of power models, common in the realist international relations literature, emphasised interstate relations in regional and global hierarchies. World Systems Theory represented state relations as taking shape within the capitalist world system. The post-Stalinist/Atlanticism duality stressed the intersystemic character of global space, where intersystemic relations were perceived, by some as competitive, and by others as complementary. The Cold War was unique as a geo-historical period because of the 'systemic exceptionalism' of these two systems. Whilst I recognise these insights, spatial modernity understands this global spatial setting differently.

It approaches global space as a conjunction of the communist world system and the capitalist world system each of which is conceptualised as a formation of modernity. This implies a complex global development space of First, Second and Third Worlds together with a North–South divide. It is within this conjunction that Eastern Europe is produced.

In global space, by 1985, the communist world system had a member state on every continent except North America. In the 1960s some fourteen states were self-defined as socialist states; by the mid-1980s this number had risen to thirty-four (Figure 10.1). This was the Second World of global space. With enlargement came diversification. Smith (1989) pointed to three sources of diversity. First, the context within which socialism was established; second, the form of socialism adopted; and third, the level of development. Smith conceptualised context as the socio-economic structuring of development. This included the elements of cultural heritage; natural resource endowment; social structure, agrarian or industrial; ethnic make-up; contacts with the capitalist world economy; and political relations with other states, especially neighbours. By form Smith meant the 'different brands' of socialism, mentioning Maoism, Titoism and Leninism, each a distinct ideology affecting social organisation in different ways. In level of development Smith argued that the communist world had its own 'First' and 'Third' Worlds that paralleled Western global divisions.

Within this communist global space, these sources of diversity produced different places. Eastern Europe was produced as a distinct and particular place. It was distinctive in its trajectories of national development strategies. Smith identified two forms of socialist urban centred development: uneven development with urban industrialisation and intensive development with slower urbanisation. Eastern Europe began with the first of these and progressed to the second (Chapter 9). This experience was in marked contrast to the rural-orientated strategies of balanced development with limited urbanisation, most clearly exemplified in China between 1958 and 1976; and uneven development with de-urbanisation, at its most extreme in Kampuchea 1975–78 (Smith 1989).

Eastern Europe was also distinctively part of a 'Soviet bloc' in this communist system (Aspaturian 1984). The associational profiles – membership of international organisations and associations – of Eastern European states coincided with that of the Soviet Union and they had the strongest treaty relationship with the Soviet Union. Furthermore, Eastern Europe had a role, as one of the most industrial areas of the socialist world, in supporting the Soviet Union in the framework of the CMEA's provision of development aid. As Smith (1989) pointed out, the socialist world had its own 'geography of dependent development' that was reproduced as the Soviet Union and China promoted communist modernity in competition with the capitalist West and sometimes with each other.

Eastern Europe, development aid and interworld co-operation

The CMEA considered trade a key component of development assistance. Levigne (1992) represented the communist world as divided into four

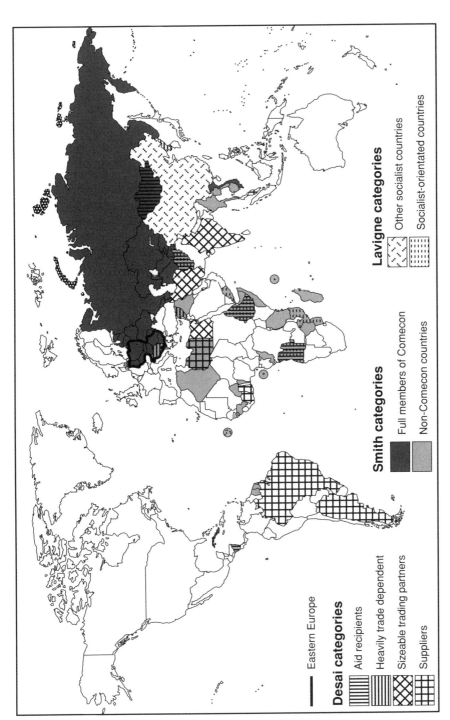

Desai categories

— Eastern Europe

▦ Aid recipients

▥ Heavily trade dependent

▨ Sizeable trading partners

▤ Suppliers

Smith categories

◼ Full members of Comecon

◻ Non-Comecon countries

Lavigne categories

▨ Other socialist countries

▦ Socialist-orientated countries

Figure 10.1 The global communist world system.

categories of state. These were CMEA donor countries; developing countries that were members of the CMEA; other socialist countries; and socialist-orientated countries (Figure 10.1). Most aid, about 70 per cent, went to the developing countries that were members of the CMEA, the rest to socialist-orientated countries. The composition of trade encouraged by the East was intended to be part of a wider socialist strategy of collective development within developing countries. It sought to provide outlets for the traditional exports of developing countries and at the same time to contribute to the technological and technical assistance they needed to achieve rapid industrialisation. A system of preferences and credits promoted socialist development strategies in client states. Michalak (1995) argued that this was arranged to the benefit of CMEA countries and the Soviet Union in a way similar to that practised by the West in its dominant relations with developing countries. The CMEA and the Soviet Union sold uncompetitive machinery and secured in return raw materials and foodstuffs whilst keeping their own economies closed to industrial manufactures from 'socialist Third World' countries. Levigne also showed that CMEA aid was only a small percentage of its GNP and that the Soviet Union supplied as much as 85 per cent. East Germany contributed 30 per cent of the rest. Whilst Czechoslovakia and Bulgaria contributed mainly technical assistance, Hungary, Poland and Romania contributed very little. Aspaturian (1984), on the other hand, thought that the Eastern European contribution to socialist assistance was 'more than a modest share'. He noted that the efforts of the GDR, Czechoslovakia and Bulgaria tended to be co-ordinated with Soviet interests and policy. Yugoslavia and Romania in contrast pursued their own interests. Poland and Hungary were only very marginally involved. At the same time he observed that the distribution of CMEA aid was rather variable in amount at different times and uneven in the number of countries receiving assistance. Eastern Europe also complemented the Soviet Union in providing military assistance and general economic, educational and training aid.

Trade directed aid and assistance also promoted communism beyond the socialist 'Third World'. Desai (1992) pointed out that the trade–aid contacts between the East and the South were founded on two distinctive elements that do not always go together – bilateral balancing and the planning of trade. Contacts also ranged from developing countries closely associated with the CMEA to developing countries whose contacts with the East were no different from those they had with Western states. From this perspective he constructed East–South connectivity in terms of four types of developing country: first, those countries in receipt of Eastern aid; second, those countries that were heavily dependent on trade with the East; third, those that had sizeable trade links with the East; and fourth, those that primarily supplied the East (Figure 10.1). All the states in the first category, except Mongolia, were involved in local or regional conflicts in pursuance of which they received assistance, whether or not they were substantial trading partners. Cuba and Syria, in the second category, were also front line states in the conflict with Western modernity. The states in the third and fourth categories were not generally communist states.

Together Levinge's and Desai's analyses show that whilst Eastern aid and trade with the South favoured communist-orientated governments it was not confined to them. Direct East–South contacts were paralleled by direct West–South contacts and so became represented as North–South contacts. The global North–South development divide was the global structure that underpinned the Brandt Report that shaped development thinking in the 1980s. However, West–East–South contacts were more complex than this. 'Triangular commerce' was practised principally by the Soviet Union which sold manufactures, mainly arms, to the South in exchange for raw materials that were then, in a separate operation, sold to the West in exchange for technology and equipment. West–East–South relations were further extended during the 1970s by 'tripartite industrial co-operation' (Gutman and Arkwright 1981). This pattern involved Western firms co-operating with central foreign trade agencies in socialist countries acting on behalf of socialist enterprises, who joined together to construct industrial plants requested by governments in semi-industrialised southern states. Gutman and Arkwright recorded 188 cases of this kind of co-operation between 1965 and 1978.

Eastern Europe and the United Nations

All the Eastern European states had independent membership of the UN. Some of the aid they offered to the developing world was channelled through UN agencies. Semi-official recognition was given to Eastern Europe as a distinctive region when it was decided, informally, that at least one state should be elected on a rotating basis as a non-permanent member of the Security Council (Aspaturian 1984). During Stalin's time Eastern Europe was completely subservient to the Soviet Union in the forum of the UN. However, as time went on the states individually and collectively played a bigger part in UN organisations. By 1983 Yugoslavia was affiliated to seventeen organisations and Romania to sixteen, whilst the GDR and Albania were involved in only nine and eight respectively. The remaining East European states ranged between these figures. Activities within the UN were used by some states to give leverage in relations with the Soviet Union. Thus, UN membership helped the states of Eastern Europe to find their role and identity across the bipolar global divide.

Eastern Europe, global shift and cultural imperialism in the global capitalist world economy

Dicken (1992) perceived a transformation in the material and spatial practices of the capitalist world economy, as it intensified and enlarged its global reach, in the period after the Second World War. He conceptualised this evolutionary transformation as 'global shift'. The Second World War all but destroyed the map of the pre-1939 global economy. The capitalist world economy was practically begun anew in the bipolar geo-ideological global structure after 1945. At Bretton Woods in 1944 a new institutional global space was conceived with the intention of avoiding the extreme economic conditions that had been produced by the market dependent economy of the 1930s. It included the World Bank, the International Monetary Fund (IMF) and the General Agreement on Tariffs and Trade (GATT). Within this regulatory framework, until 1971 using fixed exchange rates, thereafter with floating rates, the capitalist world economy evolved. It was driven by three dynamic processes: rapid technological progress in all sectors, but especially transport and communications that 'shrank' global space; the development of transnational corporations that evolved from the US model devised by President Roosevelt to ensure stability and prosperity in the American economy; and an intensification and diversification in the number of international financial transactions and flows. These processes produced a tripolar geography of the global economy. North America, Western Europe and Japan were core zones surrounded by peripheral areas. As the relationship between cores and peripheries evolved, driven by the dynamic processes, it restructured economic activity within the different zones. The speedier growth in trade than production shifted the emphasis in core zones to business and financial services, whilst manufacturing and assembly shifted globally to the peripheral zones, especially the Third World. The cores became the focus of corporate and financial control, the peripheries the location of production and foreign investment.

Eastern Europe was at first completely separate from this evolution. However, as I discussed earlier, partly from its own development needs, from the late 1960s it became more closely connected with the European sphere. It did not, however, have much contact with the US core, except for its television and film exports, or with the Japanese core. Nonetheless, some East European countries joined the World Bank, IMF and GATT and by the late 1980s Eastern Europe was being drawn more closely to this dynamic capitalist world economy as Western media products, high technology equipment and management ideas were imported. Eastern Europe, however, was still a very marginal partner.

The transformation Dicken describes also gave rise to several cultural discourses, in which the idea of 'cultural imperialism' was prominent. Tomlinson (1991) pointed to the ambiguities of the idea of 'cultural imperialism'. He treated it as a critical concept in the related discourses of 'media imperialism', 'national cultural identity', the critique of 'multinational corporate capitalism' and 'cultural modernity', all of which bear on my discussion. Tomlinson suggested that these critical discourses of cultural imperialism were best understood as protests against the rise of global modernity. The discourses, as Tomlinson readily admits, are themselves culturally bound. Like the discourses themselves, Tomlinson's discussion omitted reference to communist modernity as I have conceptualised it, though Marxist ideas permeated the critiques. Consequently, Eastern Europe as a place in the European and global space was not referred to. Yet Eastern Europe, as I touched on above, was not immune from the varieties of Western cultural imperialism Tomlinson discusses. The early sense of catching up with the West implied a fertile ground for Western ideas, despite the claim of communism's socio-economic superiority. Western media products, especially television programmes and films, portrayed an ambiguous, but often glamorous, image of the West and its lifestyles. Young people attempted to produce their own versions of Western popular musical styles. The greater efficiency of Western corporate organisation and higher levels of technology and quality of Western products were tacitly admitted. However, the far more virulent force of modernity at work in Eastern Europe, as I noted above, was Soviet communism, opposition to which reflected similar themes to those introduced in the numerous discourses of Tomlinson's discussion.

Conclusion

Eastern Europe was produced within a lattice of several competing spatialities of European and global space. When simultaneously contrasted with other places in these different spatialities, Eastern Europe took on different identities. As the European and global spaces within which Eastern Europe was produced changed, so did the experience and meaning of Eastern Europe. Although I have delimited Eastern Europe in terms of the fixed territory of eight states, it was many different places within the European and global space of competing modernities. The layering of these different partial perceptions produced the complex identity of Eastern Europe in global spatial modernity.

Part 4

Spatial modernity and the Neo-liberalist Project

11 The Neo-liberalist Project

The assertion of self-development and from geo-politics to geo-economics in global modernity?

The First World War and the Second World War were epochal events that ushered in new global configurations of spatial modernity that in the Marchlands of Modernities I have suggested could be understood in the ideas of the Nationalist Project and the Communist Project. The collapse of communist governments in Eastern Europe and the Soviet Union between 1989 and 1991 was equally epochal. It ushered in a new configuration in spatial modernity. Neo-liberalism sought to oust communism and rid the Marchlands of its 'evil' legacy. However, this time there was no sharp break of the kind the wars had created. There had been no widespread destruction of the built environment, or decimation of the populations. This time a modernity had collapsed as a dominant hegemony, but its artefacts, practices, perceptions and conceptions had not been totally discredited or destroyed by several years of military conflict. In Yugoslavia, however, war was a consequence of the crisis of communism. There was, however, a sense of a new beginning across the Marchlands.

The new configuration of spatial modernity after a decade is unclear. However, I will suggest that it is being produced by a multifaceted Neo-liberalist Project. This Project, within capitalist and more widely western modernity, can be understood in terms of a new spatial modernity. It is a reassertion of self-development at the expense of collective institutional development. It is also an intense renegotiation of multi-scalar place relations. This guides a set of current processes that are interacting with the inherited communist antecedent conditions to produce a spatially diversified and directed experience of development. The triumph of neo-liberalism is that its practices and discourses, though they have not been unchallenged, have rapidly become hegemonic in the Marchlands.

The production of the Marchlands in the 1990s

To trace and interpret the production of the Marchlands after 1989 we need first to consider the causes of the crisis and collapse of communism. The events of 1989–91 took everybody by surprise. Subsequent analysis of the 1980s has pointed to the main factors that led to the collapse. The increasing openness of Eastern Europe, during the 1980s, to the capitalist world system, freshly driven by the dynamic of neo-liberalism, was showing up the failings of the communist system. Neo-liberalism challenged the state on the one hand and the large-scale organisation of production on the other. These were the two main props of communism. Communist governments faced pressures for change from all quarters. Individual, intellectual, dissident challengers to the régimes had kept alive the spirit of opposition. The general public had never given the communist governments true legitimacy through the ballot box. The political system and the economic system were manifestly failing to provide the outcomes that they had promised. A sense of nationality had been supressed in supposed fraternal international relations, despite the retention of the 'national' territory as the framework for action in the Party-state. A strong sense of nationality persisted. The Churches, though they had reached accommodations with communist élites, in most countries, had assured the presence of an alternative world view. The Churches retained international contacts and had also kept in being a practical organisation across each country. A wide range of interest groups, including environmentalist, began to assert particular forms of social contact separate from the state. These developed into broad political movements that challenged the idea of the Party-state in asserting a civil society separate from the state (Elander and Gustafsson 1993). The introduction of *glasnost* and *perestroika* reforms in the Soviet Union with the accession to power of M. Gorbachev in 1985 led to the abandonment of the threat of military intervention from the Soviet Union as a measure to retain communist governments in power. With this the governments, by then assailed from opponents who were gaining greater confidence, lost the will to retain control and began the process of voting themselves out of power. It is certain that many communist politicians and officials were already assessing the opportunities of a new situation. When it came, the end was negotiated and with the exception of Romania and Yugoslavia relatively bloodless. The disintegration of the Soviet Union soon followed.

Where and what are the Marchlands in the 1990s?

In Chapter 2, I discussed the difficulties that historians, geographers and writers have had in conceptualising where and what are the lands of which this book speaks. To some extent this problem was eased for westerners during the communist period because of the perceived, but as I have shown misleading, distinctive wholeness of identity implied by the Communist Project and the Soviet Russian hegemony. With the collapse of communism the difficulty has returned. Following 1989/90 the Marchlands are being produced in a newly intensified maelstrom of people, events and ideas, and a variety of competing spatialities.

Three competing representations that rely on the former situation have been prominent in discourses reproducing the areal extent of the Marchlands. These are 'post-socialist', 'post-communist' and 'post-Soviet' (Dingsdale 1999a). None of these is satisfactory because each emphasises a past modernity, which though its legacies remain in material and discursive modes of production has been more or less cleared out. The term 'Eastern Europe' is

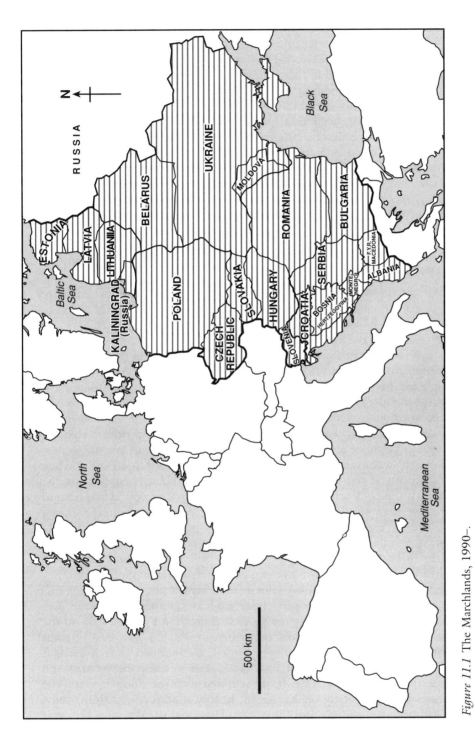

Figure 11.1 The Marchlands, 1990–.

Source: Adapted from Dingsdale (1999b) 'Redefining Eastern Europe', *Geography*, vol. 84(3), 205.

sometimes used to include Eastern Europe as I conceptualised it in Part 2 plus Russia and the former Soviet Republics that are now independent states. There has also been a tendency to perceive the area as three separate units: the GDR, now integrated into the Federal Republic of Germany, is usually, but not always, discussed in the context of a reunified Germany; Russia and the former Soviet Republics; the old Eastern Europe. These conceptualisations emphasise the communist legacy as a primary factor of spatial order. An alternative conceptualisation has been to refer to Central and Eastern Europe, but not to elaborate either a distinctive territorial extent, or a separate meaning of these two terms. I will redefine the territory to follow the traditional conceptualisation of 'the lands between' Germans and Russians, Germany and Russia (Figure 11.1). This most clearly encapsulates the idea of the Marchlands of Modernities in the new configuration, materially and discursively, though as I will show local contacts transgress this bounding.

The return of the West: a return to Europe, but not to the past

Communist modernity was the high water mark of eastern influence imposing a particular spatiality built on a foundation of ideological and military control, but never enjoying popular legitimacy. The collapse of the system laid open the way to a return of the West.

So, a broad way of thinking about what is happening in the Marchlands is through the idea of a return of the West, or as some in Eastern Europe saw it a return *to* Europe. The distinction is an important one because it informed popular responses to transition and transformation. The West and Europe are not, however, the same place. In both cases the appeal of the West and of Europe was its modernity, perceived as individual freedom, openness, dynamism and material prosperity. The hidden personal insecurity inherent within this western modernity was not fully appreciated. When in practice it was realised, it came as a shock and the early euphoria at the collapse of the communist system gave way to a more sceptical response to the new western invasion. Furthermore not everyone, even from the outset, welcomed a return of the West. Short-term personal interests and long historical cultural attitudes and values are at play.

Many in Eastern Europe did welcome the collapse of communism. They saw it as opening the way for a return of western ideas and influence. They looked forward to the establishment of democracy, the market economy, civil society and much more as western ideas returned to prominence. Jeszenszky (1998) recalls imagining a New European Frontier, akin to the American frontier of the nineteenth century, in which a Wild East of lands devastated by communism would be tamed, transformed and civilised. A feeling of a return to the West was widespread as many intellectuals and ordinary folk saw themselves as being reunited with the European, and the emphasis is very strongly on European, historical antecedents from which they had been cut off for almost fifty years. However, the West that returned to the East in 1990 was not the place that left it in 1945. So, those who thought of *returning to the West,* imagined as a historical, cultural Europe, as distinct from *the return of the West* in its 1980s garb, may not have anticipated the outcome. The locational, societal and cultural traits of the West, as Western European civilisation, that were so admired by some were now transformed by four decades of development and were at that moment grappling with the dynamism of neo-liberal thinking and the consolidation of the European Community.

Neo-liberalism and western capitalist modernity

Neo-liberalism began to transform western capitalism in the 1970s and was rampant in the 1980s, especially in the USA and the UK. Its agenda was to destroy the architecture of state-regulated, market mechanisms of growth that had been constructed in the western world system after the Second World War. It was multifaceted. It was founded on the reassertion of free-market economic theories championed by such academic economists as Milton Friedman. It made its assault on corporate America in the activities of Micheal Melkin, Sir James Goldsmith and others, who as 'corporate raiders' claimed to be recapturing control of multi-national corporations for shareholders, from managers and, in the process, reinvigorating them through individual entrepreneurialism. However, these buccaneers of free enterprise were soon displaced, often for corrupt practices, and control of the great corporations passed to financial institutions and the caprice of the financial markets. Such was the global strength of US business that its new world view permeated the global regulatory institutions. The managers of the World Bank, and especially those of the International Monetary Fund, succumbed to policies formulated on the propositions of the gurus of neo-liberalism.

Right-wing politicians such as President Reagan and Mrs Thatcher embraced the ideas of the free market as the most efficient engine of growth. They pursued policies of deregulation of markets, privatisation of state-owned assets and entrepreneurialism in public administration. They tried to bring everybody into the business world and imbue them with its sense of priorities, by selling shares in privatised companies. They espoused social policies that made individuals responsible for themselves and reduced the welfare underpinning of the state. The ideology of the private and individual replaced a sense of the public and communal good as the cement of societal relations. These changes were accompanied by an intensified sense, even assertion, of feminine identity. There was a feminisation of the labour force as brain replaced brawn in the tertiarised, de-industrialising economy. There was a recorded increase in single person households. Social relationships were restructured, new groupings created, and a new social terminology invented to describe them. Urban–rural relationships were transformed, as personal mobility and personal communications became much easier. Furthermore, neo-liberalism was not only a right-wing political agenda. So-called socialist governments from Spain to New Zealand adopted measures flavoured by neo-liberal thinking (Allen, Massey and Cochraine 1998).

Neo-liberalism encouraged consumption liberated by personal taste and encouraged new societal relationships. Hundreds of 'niche' markets became sufficiently large to support new patterns of production, to meet rapidly changing fashion. New forms of commercial organisation generated new types of employment in the service sectors. Advanced technology shortened product life cycles, transforming the productive sectors. Leisure, entertainment and intensified personal communications permeated lifestyles.

Neo-liberalism transformed societal relations, but it also transformed cultural experience. It restructured the vocabulary of meaning. It promoted hyperbole, an accelerated pace of change and a preference for image over substance. It promoted competition and confrontation. It created an information culture. The quantity of information available multiplied dramatically, but not its quality. It created an instant reward culture. What was worth doing was specified by what was measurable or graded. Perhaps most important of all it commodified place and brought it into the realm of marketing and competition in a way that had not happened before. It created a sense that the culture of business was the predominant form of meaning.

A returning Western Europe and the production of a New Europe

The return of the west, its societal and cultural norms and values has created a sense of a New Europe. Smith (1993) envisaged three geo-political New Europe scenarios. The first of these he called the common European home. This idea originated with General de Gaulle and was promoted by President Gorbachev. The second Smith called not quite all Europe. This scenario centred on the European Union Project. The third Smith termed a return to the past. This conjured up the flawed national solutions of 1918–22. Miall (1994) envisaged the New Europe in terms of new patterns of security, conflict and co-operation. The maelstrom of transformation in the Marchlands since 1990 has been coloured by all of these envisioned scenarios. Each of them was present in particular places to a greater or lesser degree.

Perhaps the most important of these scenarios has been the European Union Project. The most successful attempt to unite Europe began with the ECSC and evolved to the EEC, established by the Treaty of Rome in 1957 (Chapter 10). The most rapid steps to 'the ever closer union' of the peoples of Europe envisaged by the founding fathers occurred between 1986 and 1992. In 1986 the Single European Act and the Single Market Programme envisaged a strong economic union. Events in Eastern Europe contributed to an acceleration in thinking of the EEC as a political entity, always a part of the Project's agenda. This culminated in the European Union (EU), established by the Treaty of Maastricht in 1992.

Integration and disintegration in the two halves of Cold War Europe between 1985 and 1992 laid a key foundation for a New Europe, in which the west and the east could develop closer links. It would soon lead to the idea of the eastward enlargement of the EU. But this has been a highly problematic and contested process.

Globalisation and globality in neo-liberal western modernity

For some authors, such as Swain and Hardy (1998), the return of the West is primarily the reintegration of the Marchlands into the capitalist world economy that is being transformed by globalising and internationalising processes. Altvater (1998) argues that three 'ideal' forms of capitalism, extractive, productive and arbitrage, combine to determine the development path of any particular place in global space. Arbitrage capitalism is predominant because it most sharply exploits differences between time and space as productive forces are restructured. The Marchlands are not uniquely different from the rest of Europe or any other major world region but part of the same process of global change. Globalisation postulates a single global arena of complex connectivity. Neo-liberal societal relations inform business organisations, institutions and electronic communications that together are constructing a borderless globe as time/space is collapsed. Globality postulates a sense of global wholeness through notions of earth community, environmental stewardship and a global village or neighbourhood. In spatial modernity the root of the globalisation–globality discourse is the core dichotomy of spatial modernity writ large. The discourse is underscored by individual development versus institutional, environmental development. The conflict is mediated by the enabling/disabling force of technology. Technology promotes the neo-liberal project, but also assists challenges to it. Globalisation and globality, each with a global/local spatiality, are contesting the spatiality of World Systems Theory.

The policies of governments in the new states of the Marchlands

Locally modified versions of the neo-liberal model of capitalist, western modernity influenced the general philosophy of many of the first governments to be elected after the communist abdication of power. This was a vital, political, component of transformation. They introduced ambitious policies that were designed to replace all the principal features of the apparatus of communist power. First, they proclaimed a commitment to democratic institutions to operate at national and local level. They transferred selected administrative responsibilities to county and district level and introduced elections for local representation. Particularly important was a commitment to respect human rights and freedom of association. Second, they announced measures to introduce market economies; privatisation of state assets and encouragement to new private enterprises, especially small and medium sized businesses, providing opportunities for individual entrepreneurship. They enacted laws that permitted the repatriation of business profits and the constitution of corporate enterprises. In short, they transformed the technical and legal constraints that had inhibited a wide participation in the capitalist world economy in the previous decades. They also sought to reconstruct their political, economic, military and cultural international relations. Where they were not already members, they sought membership of the World Bank, the International Monetary Fund, the Organisation for Economic Co-operation and Development. They also sought membership of the North Atlantic Treaty Organisation and the Council of Europe. Perhaps most prominent of all, they wanted to join the European Union.

These western, neo-liberal influenced current processes did not flood onto a tabula rasa, but encountered antecedent conditions inherited from the stock of locational, societal and cultural attributes of communist modernity. Amongst these inherited conditions were particular bureaucratic, economic and administrative structures; unbalanced manufacturing profiles characterised by large-scale, state-owned 'smokestack' industries and a poorly developed tertiary and financial sector. There was poor infrastructure, inadequate communications networks and a horrendous environmental legacy. There was a lack of multi-party political structures and a highly centralised state government with very weak local institutions. There was no civil society and traditional expressions of ethnic-national and religious identity were supressed.

Democratisation and marketisation

Unlike communist modernity that, when it was first introduced, brought a unified economic and political system in the form of the single Soviet model, the new western modernity is diverse and fragmented. The response of 'the west' to the collapse of communism in Eastern Europe can be divided into two major categories: (a) commercial – characterised by corporations seeking to extend their markets and invest for profitable returns, and (b) public – characterised by governmental and non-commercial organisations seeking to extend 'western' ways of thinking, promote specific attitudes and values and assist in restructuring politics, economies and the security agenda (Kramer 1993). Both are ideologically based and seek to expand the theory and practice of democracy and the market economy in a neo-liberal mode. Western European governments were also afraid there might be a huge in-migration from former communist countries and this too motivated their actions.

Democratisation and marketisation have had their own separate pathways and their own internal conflicts, but their interaction in reshaping geo-political and geo-economic relations is a vital dynamic of neo-liberalist modernity in the Marchlands. Democratisation and marketisation are separating and decentralising institutional power structures inherited from communism, but raise questions of their own. Democratisation has raised the question of where should political power lie, by whom should it be exercised, and how should it be shared? It has introduced a conflict between state, regional and local interests, as well as between ethnic-national and citizenship identities, and all overshadowed by European Union integration and senses of Europeanness. It has changed the parameters of geo-political relations. Marketisation has raised questions of ownership, organisation and operation of enterprise and, through commodification, changed the nature, patterns and intensity of geo-economic relationships. All these new issues are negotiated by the contrasting physical, societal, locational and cultural legacies from communist development and varied responses to current processes from individuals, organisations and institutions. This recasting of place particularities is resituating them in multi-scalar place relationships.

A new cultural vortex

The social, political and economic events described above, and underpinned by the Neo-liberal Project, have undoubtedly had a big influence on the production of the Marchlands. They have reshaped the societal, material and spatial practices, producing places, and they have transformed the perception and conception of the Marchlands from the point of view of outsiders and insiders. They have thrown the inhabitants of the Marchlands into a new vortex of competing cultural influences. Released from the enforced communist collective identity that few accepted, they are faced with a revival of local and national identity and new European and global cultural forces as well. The revival of ethnic national consciousness is rooted in feelings of home and community attached to particular places. Gender consciousness is now a more prominent dimension in personal identity and issues of equity (Meurs 1998; Regulska 1998). For many, the disorientation of this whirlpool is heightened by the context of local territorial disputes (Chapman 1996). The return to Europe is a culturally derived idea, reconstructing the past from a particular, uncertain present. Global images are now beamed directly into homes and are reshaping cultural meaning. Culture, it is argued, is becoming globalised and deterritorialised, that is separated from localities as global modernity (Tomlinson 1999). The return of the west has brought a heightened sense of contested meanings from which to make sense of place and space.

The assertion of self-development and the renegotiation of multi-scalar place

Neo-liberalism is a concept that has been employed primarily in socio-economic discourse where it has been used to emphasise commercial freedom, the market and deregulation. Its principal hallmarks here were individualism and entrepreneurialism. As the above discussion has suggested, neo-liberalism is only one dimension of an intensified sense of individual self-awareness. The heightened sense of self-awareness has been expressed in demands for equity, challenges to authority and increased individual confidence and assertiveness. Intensified self-awareness has led to an outburst of myriad contradictory and contested conceptions, perceptions and actions that has been labelled post-modernity. Even Berman himself used this kind of metaphor when he saw modernisation extending practically to the whole world, before shattering into a multiplicity of fragments and losing its meaning for people's lives.

In the terms of my understanding of modernity, this is the dynamic of self-development. It is asserted that personal ideas and interest, operating through a free market, are the basis of economic efficiency; and personal motivation is the only valid, naturally legitimated, platform for action. Communal institutions and the physical environment are objectivised and separated from individual development in thought and action. The crucial new dimension is the dramatic increase in the potential for interpersonal communication and with it the sense of personal freedom from an asserted, supposed, collective restraint imposed by organisational or institutional rules. However, this ideology is tempered by the need to work through organisational team effort within institutional rules that generally change only slowly, but in the Marchlands have been put under pressure to change rapidly.

This had a particular resonance in the Marchlands because of the restrictions that communist modernity had on individual identity, inventiveness, innovation, mobility and action. Personal self-development uniting thought worlds and action worlds was liberated by the return of the west, which in its neo-liberal guise promoted, encouraged and applauded the very qualities that had previously been most supressed, frowned upon and even criminalised. In releasing these energies in the commercial arena, it also lifted the lid from the pressure cooker of communist repression that had supressed all forms of individualism within its vision of collective, societal and cultural, meanings and actions. The people of the Marchlands became a part of the explosion of personal communications, intensified self-awareness and dynamic, media imagery that propelled societal and cultural development in the 1990s. Much of this has become focused upon and translated into a newly complex sense of place. Individuals, organisations and institutions are opened to new ways of thinking and are renegotiating their relationship with multi-scalar place. Spatial modernity traces the place particularities and multi-scalar space/place transpositions that this implies.

Development as transition and transformation

To what extent the collapse of communism in Eastern Europe was internally generated and to what extent the result of external contacts is a matter of debate. The debate is not without significance for the conceptualisation of what is going on in the Marchlands of Modernities. It is generally taken for granted that socialism is giving way to capitalism, but this is a very skeletal idea that needs to be clothed. The most prominent way of conceptualising this process has been in the ideas of transition and/or transformation. Smith (1997) distinguished between the conceptualisations of 'transition' and 'transformation' to describe development since 1989/90. He suggested that an early conceptualisation of 'transition' imagined that the collapse of state socialism had left a tabula rasa on which to construct capitalism. A later conceptualisation 'transformation' contends that the processes of change are 'complex … and "embedded" in the nature of the state socialist system, the struggles that arose out of that system and its current transformation' (Smith 1997: 331). Hamilton (1999) sees ideas of transition and transformation as a controversy within scholarly discourse. Economists use 'transition to a market economy' to conceptualise change. Other social scientists see a more diversified transformation.

Pickvance (1996) problematised the significance of the events of 1989/90 in the process of social development, suggesting continuity rather than discontinuity. Potter and Unwin (1995) saw the speedy penetration of capitalism as a result of the failure of the socialist system itself. Harloe (1996) stressed that transition was *from* socialism to emphasise the uncertainty of the future and the importance of the contrast between countries in the socialist period in conditioning specific pathways of change.

Turnock (1996b) suggested that to interpret the 'momentous experience' of disintegration and reconstruction in Eastern Europe and the former Soviet Union, attention should be directed to three processes of change: in the political, economic and social arenas; in the balance of continuity and change, creating new structures; and in the spatial contrasts both between countries and within countries. In other words, in current societal contacts and relationships; the relationship between the formerly dominant and the currently dominant ideas and values; and in multi-scalar geographical space.

Hamilton puts this same framework a little more directly, claiming that three components are interacting to reconstruct Central and Eastern Europe and the former Soviet Union. These are:

1 progress from a predominantly, centrally-managed, state-owned . . . economic system . . . within a single-party communist political system *towards* a more privatised market-type economy . . . a myriad of decision-makers within a more democratic, civil society and new regulatory frameworks;

2 legacies and impacts of structures, institutions, networks, behaviours and mind-sets . . . inherited from the . . . socialist era [and] . . . those more deeply 'embedded' from pre-socialist times;

3 human, organizational and institutional responses at various geographical scales to opportunities or constraints and pressures imposed by the transformation process itself.

(Hamilton 1999: 136)

The ideas of transition and transformation emphasise the introduction of new current processes. There is a basic contrast, however, in approach between the idea of a transition *to* capitalism as against a transformation *from* socialism. Whilst transition tends to be focused almost exclusively on the evolution of necessary economic policy measures, transformation gives greater prominence to other societal and cultural dimensions. Transformation's wider perspective is to some extent summarised by the idea of a re-emergence of civil society. At the same time, transition tends to forget that the territory existed before 1989, whilst transformation stresses the importance of earlier periods in shaping individual and institutional processes of development. Is it the dynamism of transition, from socialism to capitalism, that is driving and directing transformation, or is it transformation that is driving and directing transition?

Development in the Marchlands has become universally represented as transition and transformation. However, transition is taking many forms.

The challenge is to negotiate ways in which we can understand the *diversity of forms* of transition . . . Transition is not a one-way process from one hegemonic system to another. Rather transition constitutes a complex reworking of old social relationships in the light of processes distinct to one of the boldest projects in contemporary history – the attempt to construct a form of capitalism on and with the ruins of the communist system.

(Pickles and Smith 1998: 4)

But this new capitalism is dominated in practice and discourse by the Neo-liberalist Project that, like other modernities, seeks to clear out what preceded it. Pickles and Smith attest the force of neo-liberalism, as they mount a critique of its prominence in models of transition:

There simply was no third way – a closure produced in part by the growing public imaginary of a West offering wealth, freedom and opportunity . . . and in part by the direct and indirect effects of western propaganda throughout the 1980s.

(Pickles and Smith 1998: 4)

Conclusion

The variety of conceptualisations of the processes producing the Marchlands reflects two situations. First, that there is an intensified material experience of maelstrom for the peoples, events and ideas within the area. Second, that understanding events in a hitherto neglected area of western interest and one in which local academic research was hindered by a variety of impediments (Hamilton 1999) is demanding a widening of the theoretical underpinnings of geographical discourses. The Neo-liberalist Project, and the particular form of capitalist modernity that it symbolises, is central to both of these situations. It is transforming material and spatial practices; their representation; and conceptualisation of the Marchlands, and so reconfiguring meaning as place/space transposition. It is restructuring place as a lived experience for the inhabitants of the Marchlands, and is reshaping the perception of insiders and outsiders about the Marchlands. It is informing discourse as a critical and interpretative concept (Hamilton 1999) as it changes the agenda for the production of place.

The theory of spatial modernity brings a unifying order to these disparate material and discursive modes of development that are producing the Marchlands in the 1990s, because it directs attention to the particularities of place as relational and multi-scalar, and neo-liberalism as a dominant ordering category of modernity. The multi-scalar renegotiation of place, driven by individual development, is the vital dynamic of the impact of neo-liberal modernity in the Marchlands. Whether this can be represented as a shift from geo-politics to geo-economics in a setting of global cultural modernity is an important question. What is certain is that in the 1990s there was, more than ever, a direct conflict of modernities, as ways of thinking and acting, in the production of places within the Marchlands as new vibrant current processes of western neo-liberal capitalist modernity clashed with legacies of antecedent communist modernity. As an official of the city of Łódź, Poland, told Stefan Klosowski:

> The fight with the communists is still ahead of us. We have a hard time changing that former mentality . . . The basic task is to change the mentality . . . For Americans, liberty and commerce is a natural situation whereas the Polish officials are used to having someone tell them what to do.
>
> (Klosowski 1995: 37)

Klosowski then commented: 'The struggle that we have here in Poland is not just one of a new market economy, but of a mind set that should accompany these changes.' At the same time the attitude adopted to the indigenous population of the Marchlands, by most western academic and professional contributions to the debate, is well summed up in the title given by Hardy and Rainnie (1996) to their study of Kraków: 'Desperately Seeking Capitalism'. This view pervades the debate in which neo-liberal material and discursive models of transition are hegemonic.

12 The production of localities in transition

This chapter looks at the production of localities in neo-liberal modernity. The conjunction of democratisation and marketisation created localities in new forms of individual and environmental development. Localities were produced by myriad individual decision-makers with contrasting interests and purposes in a multi-scalar renegotiation of places of complex connectivities. The chapter begins with a brief comment about the discourse on local experience, followed by reference to the important structural features of democratisation and marketisation relevant to localities. Attention then turns to a discussion of the experience of localities, with particular cases illustrating the main lines of the practices and discourses of their production.

Staddon (1999) has pointed out that after an early concentration on the national scale, considerable attention has been paid to local experiences of transition and transformation. However, he argued that there was a gap between theoretical and empirical approaches that left the scholarly discourse:

> unable to contribute to the development of empirically-informed and theoretically pragmatic 'alternative development futures' as opposed to either purely idiographic case studies or nomothetic high theory ... a new way of understanding the socio-economic dimensions of transition as a function of myriad local transitions the key to which is a renegotiation of the relations between different geographical scales and between localities and their natural environments is needed ... within a new politics of peripheralisation.
>
> (Staddon 1999: 200)

The production of localities within the concept of spatial modernity of neo-liberalism contributes to answering Staddon's call.

Democratisation and marketisation

Democratisation and marketisation are separating and decentralising institutional power structures inherited from communism, but localities have contrasting physical, societal, locational and cultural legacies from communist development and varied responses to current processes from individuals, organisations and institutions. This recasting of local particularities is resituating them in multi-scalar place relationships.

Across the Marchlands the production of localities was most usually ordered by the peaceful reworking of social, cultural and locational relationships by democratisation and marketisation. In the localities of Yugoslavia the early experience was one dominated by interethnic violence as war erupted when the constraints of communism were released. The practices and perspectives of multi-scalar place relationships were conditioned by the war. Television pictures and newspaper reports thrust the experiences, particularly of Bosnia's localities, into the living rooms of the west. Although the local experience of the war was different in the different localities of Yugoslavia, the war itself marked out all Yugoslavia localities as different from those in other countries. My discussion of localities will not make reference to the war, but this tragic circumstance must be remembered. The question of the war in Yugoslavia will be discussed in Chapter 14.

Democratisation and the (re-)emergence of local self-government and civil society

The establishment of subnational, autonomous, representative institutions is a vital element in transforming hegemonic regimes into democratic ones (Elander and Gustafsson 1993) and so the (re-)emergence of local self-government has been an important aspect for the production of new localities. However, central reforms introduced local self-government in the context of liberalising policies that shifted responsibilities from public to private action, following the western neo-liberal models of governance. Thus not only was the relationship between local and central government affected, but also the relationship between local governments and quasi-market, market and non-governmental organisations as well. At the same time central constitutional reform applied the principle of self-government to the commune, the very lowest level of public administration (Bennett 1998). This was accompanied by a sense of 'local utopianism' in some quarters that resulted in the splitting up of communes and unrealistic expectations of local government action.

Democratisation has devolved local administration to very small areas that face great difficulty in acquiring financial, human or technical resources for effective action and provision of local services. Local governments given wide-ranging responsibilities have few resources. To play an important part in local development, new self-governments must develop strategies either of co-operation or competition. The formulation of such strategies is hampered by the inexperience of new local government representatives and in some localities the persistence of nomenclatura networks. They are often reliant on officials who were in office during the communist period and who did not welcome change. In many cities previously in the Soviet Union, there is a complex ethnic dimension as Russians are the majority and this colours local socio-cultural relations, and also affects local urban–rural relations. The weakness of local government has led to many of the old tensions continuing and to suspicions in some quarters of the general public that real change is not occurring (Bennett 1998). Even so, as local people they have a strong local interest but cannot, for reasons related above, engage in the practices of localism.

The formation of permanent and transient private interest groups is an important aspect of informal political processes of the transformation to democracy and civil society. In western societies there is a vast array of such groups including trade associations, chambers of commerce, professional associations and environmentalist and local protest groups, but they were absent from communist societies because of the leading role of the Communist Party. In 1997, for example, there were as many as 3,020 local environmentalist groups in the Marchland countries. They existed generally on annual budgets of less than $500 and mainly found a role in educating the public, fighting apathy, raising awareness and monitoring changes, rather than in lobbying government or mounting protest action (Dingsdale and Lóczy 2001). Individual development through a heightened awareness of informal political action stemming from democratisation is becoming an important element of local societal and cultural structures and the spatiality of transformation.

Marketisation: commodification, wild entrepreneurialism and the pulse of capital

Marketisation implemented through indirect means of privatisation, including land restitution, price liberalisation, the encouragement of private enterprise and a wide range of other measures, embodied a process of commodification that had a radical effect on the thinking of all kinds of decision-makers (Hamilton 1995). Marketisation through commodification produces localities as particular economies within a complex connectivity of enterprise organisation, factor endowments, labour markets and communications media.

Hamilton (1999) suggests that an approach derived from regulation theory specifies three ways that large-scale enterprises are adjusting to new economic realities: *de-industrialising enterprises* are state owned or labour managed. They operate in capital intensive sectors, are technically backward and depend on CIS markets. They are overmanned and deep in debt, but still behave in a socialist mode, despite the withdrawal of state financial aid. *Paternalistic enterprises,* also state owned, have the problems of de-industrialising enterprises, but receive state subsidies or other financial support. They may be national flagships, market-niche defenders or be in strong worker-supported sectors. Some have used merchant capital to create conglomerates in which cross-subsidies support inefficient core activities. *Globalising enterprises* are usually multinationals headquartered in North America, the European Union or East Asia. Direct foreign investment has gone into privatised state enterprises, forming joint ventures, or making complete acquisitions. In other cases new, wholly-owned subsidiaries have been formed. They are deploying the newest technology, the latest management organisation and selecting sites for new development as part of their European and global competitive business strategies. This last category of enterprise is the most keenly engaged in the assessment of location-specific factor endowments, including local labour markets and national and international competitive advantage.

Hamilton's categorisation stresses large-scale enterprise that was typical of the socialist period. At the same time small and medium sized companies (SMEs) have been formed. Typically, they have a single owner or are family owned and operate in the broad services sector, but are also found in agriculture and manufacturing. Many were formerly state owned but some are new start-ups. The productivity of these companies is often much lower than the big companies, and they find it difficult to obtain capital. Formation and bankruptcy rates among them are high, but they make an important contribution to employment and in providing family incomes. There are literally millions of these enterprises across the Marchlands. Though SMEs face many difficulties, they can have, even individually, an important impact on the experience of their home localities. The EU PHARE programme promotes SME development. There are conflicting perceptions of the empirical evidence of operational contacts among SMEs in the current processes of local and regional development. Smith (1997) points out that some linkages among SMEs have been interpreted as forming western style 'industrial districts', but he is sceptical of this interpretation.

Alongside these large-enterprise and small-enterprise industrial sectors, commodification in genuine and pseudo-marketisation has spawned a wild entrepreneurialism (Smith 1997). Wild entrepreneurialism is embedded in neo-liberalism and takes the form of a variety of fast money economic relations. Each is characterised by the rapid recycling of money by legal, semi-legal and illegal means and each has its own dynamic and relationship to de-industrialisation, economic collapse and socialist antecedent conditions. Motley entrepreneurs get rich quick by providing a range of agency services whilst other individuals are engaged in a desperate struggle for survival. The mobile phone is an essential item of equipment, widely used to maintain contacts with associates in fast moving markets that frequently brush against illegality. Smith (1997) thought of these as counter to the neo-liberalist project, but I perceive them as its very essence and triumph, because they reflect an extreme of individualism and self-development. Smith drew attention to four principal types of activity. First, formalised capital and money markets operate through official stock exchanges, but these are deformed because the national currencies cannot perform all the necessary functions of money. Second, political and merchant capital represent new relationships between old political power and new command over money. Old élite networks are used to secure legitimate capital accumulation in industry and commerce. Third, the kiosk economy is a small-scale merchanting sector flourishing because of the privatisation of retailing and low overhead costs of entry. Fourth, mafias operate illegally to acquire shares and engage in extortion, racketeering and illegal speculation. Hamilton (1999) points to barter as integral to much of this kind of activity, and he believed that it constrained commodification. The share of GDP generated by these activities is declining, but remains high in some countries.

Thus, commodification is creating new local business environments. These environments are also shaped by the transformation of spatial relationships by the commodification of connectivity. Localities are situated in multiscalar place relations in which connectivity has a complex cost structure. Transport, communications, information and transactions linking local enterprises with suppliers and customers now attract costs related to time–distance, infrastructural characteristics, labour productivity and skill. At the same time, the variety, intensity and complexity of connectivity has multiplied rapidly with advances in personal and commercial transport and communications technology.

It is through this complex connectivity of formal, informal and illegal business activity that the pulse of capital is felt in the production of localities. Investment or disinvestment, embedded in commodification, creates new social and spatial practices that permeate inherited local networks and resituate localities in multi-scalar place relations. Localities are produced in the practice of rigorous comparative and competitive assessments of economic potential, by diverse local, national and foreign entrepreneurs and companies. Foreign perceptions may be influenced by specific local promotional activities and by broader channels of information transfer, such as the promotional efforts of regional and national development agencies. The local impact of western assistance programmes such as the British government's Know-How Scheme or EU programmes such as 'The Poland and Hungary Assistance for Economic Reconstruction' (PHARE) and the Technical Assistance to the Commonwealth of Independent States (TACIS) also work through complex procedures in having an ultimately local effect.

Rural and urban localities

The pulse of neo-liberal capitalist modernity has coursed through the localities of the Marchlands, producing *winners* and *losers*, by re-evaluating and reconstructing the rural and urban fabric, revising the text of representation and re-imagining the identity of villages, town and cities. *Winners* implies the notion of localities that have gained from the changes and *losers* implies localities that have been adversely affected. This differentiation is used by Horváth (1995), by Nagy and Turnock (1998) and in some official publications (Hörcher 1998). Each local story is unique, but its particularity lies in its synthesis as place from a neo-liberalist perspective. Similarities derive from democratisation and marketisation, differences from specifics of state transformation policies, inherited societal, cultural, legal and infrastructural conditions, and the nuances of local collective and individual responses to the new circumstances.

Discernible strands in the material and discursive order imposed on the maelstrom of ideas, events and actions that are producing localities in the neo-liberal vein are: the siting of foreign direct investment (FDI), primarily dependent on investor motives; inherited local societal structures and infrastructural conditions; the vigour of local entrepreneurialism including the energy of local self-governments; the situation in the national urban hierarchies; the pace of home state transformation; and location in the Marchlands as a whole. The synthesis of these elements in local, national and international competitiveness has produced localities as winners and losers in transition.

FDI is generally regarded as crucial for successful transition. Localities are produced in geographical discourse as representations of the motives and perceptions of foreign investors. However, approaches to understanding the role of FDI have been narrowly focused on economic decision-making by investors and been particularly associated with the automobile industry (Sjöberg and Josefson 1998). Too little attention has been given to socio-cultural factors and host activities. Approaches underpinned by the principle of investors seeking market access, endorsed by ERBD as a prime motivation, give prominence to capitals as core locations. However, the regulationist and new international division of labour approaches stressing investors seeking production cost advantages, give prominence to non-core locations. Sjöberg and Josefson propose several extensions to these approaches, but still focus on investor economic decision-making as the prime factor. This work offers theoretically underpinned pointers to the localities that have been favoured by foreign investors, but needs balancing by reference to locally promoted initiative and refining from a place perspective.

A model of local FDI investments across the Marchlands (Figure 12.1), developed from Sjöberg and Josefson's model on the basis of country year-books and World Bank Development Reports, stresses several key features. Most investment goes into Central Europe and comes from Western Europe. Investments across the area are most usually directed to core or capital localities. At the same time it is clear that the level of investments declines eastwards. Investments by small or medium sized enterprises are usually concentrated in the immediate neighbour country. However, large-scale investors are the most prominent.

Harvey (1989) conceptualised a shift from local, urban managerialism to entrepreneurialism as a strategy in cities competing for investment in a global arena. Growth coalition theory proposes that local individual and institutional co-operation proactively promotes local economic development (LED). Ideas and actions such as urban regeneration, city marketing and re-imaging are combined in place promotion strategies for local development that serve the interests of dominant groups and are undertaken for internal reasons as well as international competition (Short 1996). The vigour of local entrepreneurialism has not been much explored in the literature on the Marchlands, but is often referred to in the business press, where short reports update economic development in particular cities and generally stress the importance of local initiative. Several dimensions are seen as important including: the rate and capitalisation of new start-ups; the number and type of de-industrialising and paternalistic enterprises; and the dynamism of new management in locally owned, privatised, state

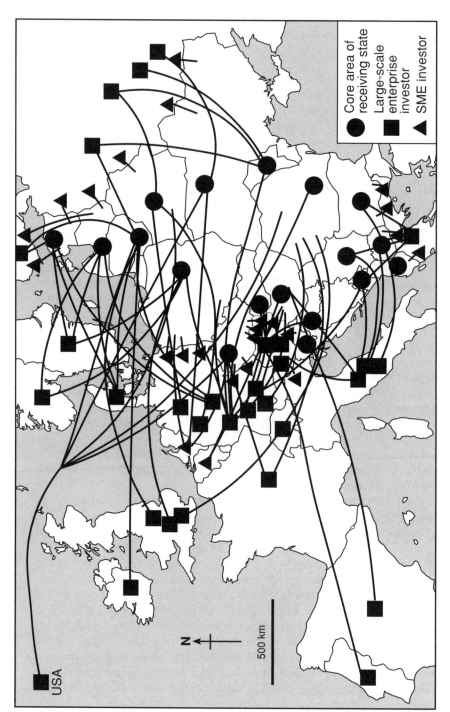

Figure 12.1 A model of FDI flows into the Marchlands.

enterprises. A less specific factor is the presence of an entrepreneurial culture remaining from pre-capitalist days or invoked by a 'second economy'. It is also complemented by the energy and entrepreneurialism of local self-governments. Place promotion is becoming a vital component of local development strategies, but has received little attention (Young and Kaczmarek 1999). In some localities there has developed a close association between the local business community and the local self-government with businessmen holding prominent positions in local government. This suggests that régime theory and growth coalitions may help in understanding success and failure. However, as Ashworth and Voogt (1988) pointed out, place promotion is not a simple task. It is clear that local place entrepreneurialism is not practised everywhere but that it has been a part of managing transition in some localities. City promotion is a 'major instrument of the transition to market economies, but also a symbolic reflection of progress in that direction' (Ashworth 1994: 347).

Theories of FDI location confirm empirical studies that point to the importance of status in existing national urban hierarchies in achieving 'success' in transition. All capital cities have localised high proportions of FDI, concentrated local entrepreneurial activity and been reinforced as the control centres of large enterprises and financial institutions, as well as centres of political power and cultural experience. Regional cities have had more mixed fortunes. Those founded on regional profiles of heavy industries have had more difficulty than those serving more mixed regional economies. Ports have generally done well because of increased and shifting trade patterns. Small towns and rural areas have faced the most difficult problems of structural adjustment (Nagy and Turnock 1998; Turnock 1998). They have the widest spectrum of 'success' or 'failure', because their fortunes can rest on a single major investment or disinvestment. Many that inherited huge environmental problems have experienced the greatest difficulties.

The significance of status in a national settlement hierarchy is elaborated within the wider setting of Marchlands space. Proximity to external borders, primarily the western border, has been seen to confer locational advantages on all types of localities. Sjöberg and Gustafsson (1998) pointed to the local nature of some foreign investment activities. However, Döry (1999) found that local cross-border communication and co-operation between Hungarian and Austrian enterprises was quite low. Yet there is no doubt that local cross-border contacts have increased. New crossing points, additional transport services and local government co-operation have all been recorded as intensifying local cross-border contacts (Turnock 1997b). Increased shopping trips promoted growth in some Polish border towns at the expense of others, often in the framework of Euroregion co-operation (Potrykowski 1998). Towns split by the Polish–German border have the opportunity to grow together again (Kowalczyk 1997; Matykowski and Schaefer 1998). Across the Marchlands as a whole, opportunities of 'success' seem to be enhanced in the west and deteriorate the further eastwards localities of particular hierarchical status are located.

Rural localities

The practice of privatising collective and state farms in many localities of Lithuania and Estonia was regarded as symbolic of a return to independence and nation building (Maciulyte and Maurel 1998; Unwin 1998). In some Bulgarian rural localities privatisation was not welcomed at all (Carter and Kaneff 1998). In Albania, it was surrounded by incompetence, corruption and deep suspicions (Hall 1998). In Romania state farms remained intact (Turnock 1996a). In many rural localities, programmes of land privatisation through the practices of land restitution to former owners and distribution among formerly landless collective and state farm workers were highly controversial. Rural, agrarian localities have been produced as a mosaic of private family farms that are too small and fragmented for commercial activity. They play an important part in survival strategies for individual families, especially where rural services, generally poor, are deteriorating. However, villages often have individuals with entrepreneurial energy who can give a lead in private enterprise and local government. Younger, better-educated farmers have shown most initiative in some Polish localities (Grykien 1998).

Many new farm proprietors are attempting commercial operations. In some localities the agricultural sector has seen the transformation of state and collective farms into commercial private businesses that have remained in local ownership. Many small and larger-scale farm enterprises are now contracting with western retail chains who are engaged in fierce competition to enlarge and intensify their networks. Whilst this practice is not widespread, powerful retailers are putting farmers under pressure to produce higher quality produce at lower cost. Big food processors are also demanding improved quality. These processes have increased differentiation between rural localities that now rely on local initiative and environmental resources.

Rural transition as a myriad local experiences is well illustrated in localities facing serious problems of structural adjustment. However, insiders and outsiders often perceive local attributes of place in different ways. Key themes of geographical discourse are the development strategies of economic diversification and environmentally sustainable practices. Within this discourse sustainable rural tourism is given special significance in multi-scalar renegotiation of local practices that hinge on contrasting insider and outsider perceptions of local environmental development. The discourse selects and produces some of the most dysfunctional localities from the neo-liberal perspective. They synthesise locational peripherality, economic, agricultural, marginality and socio-political weakness. Development strategies usually call for public funding. There is, however, an ambiguity because their perceived assets of cultural distinctiveness and landscape attractiveness depend on cultural images of rurality, commercially constructed for the consumption of urban dwellers. Some localities can be commodified and represented as offering a sunshine packaged holiday, but with equally uncertain outcomes for the indigenous population. As such, they can be represented as providing business opportunities. In other localities the environmentalists' construction of nature threatens the livelihood of local residents by restricting access to vital informally reaped resources.

Case examples of rural localities in transition

The Apuseni district of Romania

Many localities in Romania's Carpathian Mountains provide specific examples of the points made above (Turnock 1999). The Apuseni district is one such. Concepts and practical schemes for the development of the Apuseni district reflect a tension between local and outsider perceptions of how to utilise environmental resources, to meet local needs, when faced with the structural problems of transition. Abruden and Turnock (1998) discuss a combination of private farming and the expansion of tourism as a development strategy that could protect the natural landscape of the Apuseni district, if capital is provided to improve local services. But the process is not without conflict. Conservation groups propose the establishment of a national park. Whilst this will promote tourism it would restrict grazing and wood cutting for further processing, two important aspects of the local Apuseni economy. The idea of a 'natural park' has emerged from negotiations between local authorities and conservation groups. A regional scale dimension is introduced by the suggestion that cross-border co-operation with Hungarian communes will strengthen the tourist potential of the Apuseni district, reflecting the presence of an ethnic Hungarian population and the potential value of its participation in the Carpathian Euroregion (Chapter 13). The national scale factor is present through the National Association of Rural Ecological and Cultural Tourism (ANTREC) founded in 1994, that has provided promotional material and represents rural tourism at exhibitions and international fairs. An international dimension arises through ANTERC's membership of the European Federation of Rural Ecological and Cultural Tourism; this secured international assistance, especially from Gites de France (Turnock 1998). International assistance in the development programme has also been provided through the Belgian organisation 'Operation Villages Roumaine', originally established as a response to Systematazare. This programme arranges the transfer of experience by a village twinning scheme with the support of the EU's PHARE. Development in the Apuseni district is thus at the centre of a network of organisational debate that is resituating it in multi-scalar place relations.

Developing the Albanian coast

Foreign capital could help in rural development, but inappropriate foreign involvement may threaten sustainable tourism and create conflicts between local, regional and national interests. The tourism development strategy along the Albanian coast in the 1990s is a good example of this (Hall 1998). The Albanian government and western consultants envisaged a future for Albanian coastal localities as tourist resorts. They advised that tourism would form a suitable development strategy. Albanians, seeing the wealth created by international tourism in Greek, Italian and Croatian coastal resorts, assumed tourism to be a panacea for poverty, bringing easy money. Foreign commercial developers took advantage of the weak government and the country's fragile economic condition to propose schemes for the development of coastal resorts catering for mass tourism. As a result, permission was given for forty tourist development projects. National and regional perceptions were satisfied, but the economic and environmental outcome of these schemes for local interests is far from certain. Hall (1998) argues that whilst there is knowledge of the technical controls needed to avoid environmental deterioration, the confident government, stable society and buoyant economy that underpin them are not present in Albania. He also highlights dangers for localities that must avoid leakage of economic gains. The development of local wealth depends on the use of local produce, visitors staying in locally owned accommodation, the local service sector being in local ownership and local manufacturers responding to the tourist demand for craft and other products. None of these are certain consequences of outsider promoted mass tourism schemes in localities along the Albanian coast. As it is, the local contrasts in prosperity are very dramatic, with sumptuous new resort developments standing adjacent to dilapidated villages.

The Bialowieza Forest of northeast Poland

The Bialowieza Forest is on the border of Poland and Belarus, where a Belarusian minority live among the Poles. In the decade of transition the forest became a place of conflict central to a multi-scalar contest between the 40,000 local residents and a handful of Polish biologists. The forest was produced from conflicting social constructions of place and nature. It was constructed in the experience and perception of the local insiders, and in the imagination of the outsider biologists. The contest was thrust into the national political arena when the competing constructions of the forest were taken up by opposing political parties. It entered the global scene when environmentalist groups from all over the world became involved. The story is told by Franklin (2001).

For the residents, the forest was a place of economic resource, integral to and constructed in their daily experience and practice. They worked in the formal forest industrial economy, whilst informally they gathered mushrooms, berries and firewood. The informal forest activities were a key survival strategy for some families. The biologists constructed the forest as an imagined place of primeval nature in need of protection from exploitation by foresters who were destroying it. The imagined forest of the biologists threatened the experienced forest of the residents and their material welfare that depended on it.

Franklin reports that the biologists, supported by the right-wing Freedom Union Party, used fabricated data and spurious arguments to mount a 'heritage campaign' producing an image of a primeval forest that was disappearing. It was this image that they promoted nationally and globally. The biologists, through their US and EU contact networks, recruited the assistance of international environmental groups, trading on the assumed legitimacy of crisis environmentalism. On the basis of false assertions they almost convinced the national government and media that the whole of Bialowieza Forest should be designated an IUCN Category II National Park. This would have given it strict protection as a scientific laboratory and a spectacle for restricted tourism, denying the local residents free access.

In the knick of time a counter-campaign succeeded in exposing the false-hoods of the biologists' constructed image of the forest. A grassroots protest was organised by the Belarusian minority that gained the support of the left-wing SLD Party, Poland's leading newspaper (*Gazeta Wyborcza*) as well as the Polish National Parks Board and the Polish State Forestry Board. Polish 'green lobbyism' was on the way to becoming 'unmasked as a form of oligarchy by the back door'.

In telling this story of transition in the Bialowieza Forest, Franklin (2001) shows that the particular local experience was embedded in the social construction of nature. Furthermore that such constructions of nature were used as tools in the strategies in the normal struggle for political power at local and national scales. He also showed that in its construction as 'the "last fragment of primeval forest" in lowland Europe', it survived the impact of 'crisis environmentalism' and challenged Castells' assertion that most environmentalist groups were founded on grass roots democracy.

The North Rila district of Bulgaria

The tension between outsider and insider claims to develop local environmental resources has a strong resonance in the political economy of transition. In 1994/95 Sofia was short of potable water. Central government proposed to divert water between the rivers of North Rila. This was supported by Sofia's residents, but met with local opposition and resulted in confrontation at Sapareva-Bania between the national authorities and local protest groups. The episode is related by Staddon (1998) who interprets it as an illustration of urban–rural and centre–peripheral cleavage in the reconfiguration of political imagination in transition. He points out that the antagonists claimed legitimisation for their diametrically opposed actions in the western discourse of democracy. This discourse fused three strands of experience: the peripheralisation of North Rila, in 'fast capitalism'; the relationship between political identity and local social structures; and the place foundations of local cultural identity.

What all these examples emphasise, in different ways, is the need to develop close local networks of contact to strengthen local influence and locally owned enterprise and initiative in development futures. None are free from insider/outsider conflict that must be negotiated if rural localities are to achieve a prosperous future. They all show how localities whose particularities are changing are being resituated and reinterpreted in multi-scalar place relations of neo-liberal spatial modernity.

Urban localities

Capitals: the localities leading transition

Capitals have been produced as leading localities of multi-scalar negotiated transition. They concentrate national social, political and cultural power and dominate foreign and domestic investments in financial and business services. They have the best developed international links. They are self-proclaimed national gateways for east–west business and are often represented as competing with each other in meso-regional competition. New investment has brought the symbols of the west's neo-liberal capitalism to the urban fabric. New office blocks, hotels and shopping centres are prominent in all capital cities. This dynamism has in turn attracted an influx of well-educated young people seeking high paid jobs with foreign and domestic firms. But economic benefits have been unevenly distributed among the inhabitants. Whilst 'yuppies' drive round in Mercedes, beggars ply the streets and public transport.

Budapest, Prague and Warsaw

The built environments of Budapest, Prague and Warsaw have been reconstructed and the cities produced in a discourse of success and failure, as the leading centres of transition. Sýkora (1994: 1,149) represented Prague in the 1990s as becoming 'a standard Western city, considerably shaped by the forces of global capitalism . . . giving an impression of a restless urban landscape'. Kovács (1994) used a crossroads metaphor to characterise rapid change in Budapest. All three cities have developed the ambiguities of neo-liberal modernity and so have become very different places, formed by contrasting outsider/insider perceptions. For insiders the experience of the cities has depended on individual, personal, careers through transformation. Outsider perceptions have been influenced by the nature of contact, the motivation behind their interest in the cities, international media representations of the cities, and government and local promotional activities.

Transition processes have produced the new urban experience around six key socio-cultural spatial dynamics: landscape change from production to consumption; fragmented administration; housing privatisation; social inequality; economic restructuring; and the commercial property market (Sýkora 1998). The most striking changes have occurred in central areas, where conversion to commercial functions is forming business, service and tourist quarters, pushing out manufacturing and residential uses. Gentrification and suburbanisation have made substantial contributions to the advancement of spatial restructuring. The cities have much in common, but they also represent contrasting local experiences that I will illustrate by reference to particular aspects of Budapest and Prague.

Case example: Budapest: city of business and consumerism?

Budapest is an uneasy city, exuding a hectic, anarchic sense of business and comsumerist energy fuelled by rampant individualism. There is an irritable tension about the inhabitants of the central areas. Mercifully, this quickly dissipates in the outer suburban districts of Pest. The city is deluged by cars, and appalling driving standards are a most dangerous and widespread sign of individual aggressiveness. The arrogant affluence of the few is marked in the stylistic, usually tasteless, eccentricity of new individually commissioned houses that are swarming rapidly over the Buda hills. Less flamboyant gentrification and new 'housing parks' also symbolise new private wealth, whilst the poverty of the many is seen in the continued dilapidation of old apartment buildings. In a landscape of fragmented details, individual, smart, renovated frontages sit next door to crumbling building facades.

The construction of major new hotels, office blocks and shopping centres has become a prominent feature of the Budapest landscape. The process began before 1989, but the change of régime was the precursor of a boom in the demand for new commercial premises as $10 bn of foreign capital was invested in Budapest. The twenty-three, newly-empowered, district councils strapped for money have competed for, but have also been partners in, development projects. Meanwhile tensions between the municipal authorities and district councils have hampered planning for the city's future development, which had also to accommodate new central government premises such as the new Interior Ministry and National Police Headquarters. Personal observation suggests that several hundred major new commercial building projects, each exceeding a milliard HUF, have been undertaken in Budapest since 1989. The 1990s have produced Budapest, second only to Berlin, as a giant building site.

The rush to build has marked out Budapest as the most likely gateway city for western business into the Marchlands. The lack of modern, well-equipped commercial office accommodation was a serious problem that developers were quick to realise. However, the city agglomeration with a population of over 2 million inhabitants is one of the biggest and wealthiest retail markets in Central Europe. The general level of economic activity has kept unemployment in Budapest relatively low. The boost in demand is associated with economic restructuring and an accelerated trend to better paid, white-collar occupations, fuelled by western levels of pay for many. The final legitimisation of entrepreneurial activity from the 'second economy' with much stored-up wealth has revealed a class ready to spend on expensive consumer items. Yet Budapest remains a cash economy for many transactions ranging from the purchase of a hawker's paprika to the purchase of flats, allowing unscrupulous buyers and sellers to cheat the tax regime and mafias to cheat owners and purchasers. In Central Europe's self-styled primary centre of business, many banks will not cash travellers cheques!

The dominant image of the new fabric of Budapest is the shopping mall. Malls are the focus of popular controversy and serious political discussion. As a speaker in a conference on Budapest's shopping malls held in the city in 1997, I witnessed an interesting exercise in the new democracy. The shopping mall idea raised issues of local empowerment, environmental quality and international capital, contested between local elected representatives, planners, private entrepreneurs and the general public. At the turn of the millennium, it is not difficult to understand why shopping malls are important in the local debate about Budapest's development. They are vital to the iconography of contested change in Budapest. In January 2000 there were forty-seven shopping malls operating, under construction or planned, in Budapest and only two pre-dated 1989 (Figure 12.2). All but eight had over 10,000 sq. m. of retail floor space. The West End City Centre claimed to be the biggest shopping mall in Central Europe with 91,000 sq. m. of retail area and parking for 1,700 cars. However, the, as yet, only imagined Transelectro City dwarfs this with a planned 220,000 sq. m.

Many of these cathedrals of consumption are specific transformations of sites of production. An industrial boiler factory was demolished to make way for the Duna Plaza Mall. The Interspar and Eurocenter Malls, under construction at the turn of the millennium, redevelop former brickworks sites. The Pólus Center replaced that other symbol of Soviet hegemony, a Russian military base (Dingsdale 1999c). The malls are gleaming symbols of a new prosperity that many of Budapest's inhabitants seem to enjoy. The very biggest of them are full of international boutiques, famous name stores, and have multiplex cinemas. Yet, there is still uncertainty about their long-term financial success. The crowds who frequent the malls seem to window shop rather than buy. But are they uniform and tasteless non-places? It would be harsh to answer yes. Built for a functionally similar purpose, they are architecturally diverse.

On the edge of the city there are no shopping malls. There are retail parks and superstores. They cluster on the city's western gateway where the M1, M7 and M0 intersect in the 'Motorway Triangle'. Approaching Budapest along the joint M1/M7, the eye is assaulted by an iconography of brash commercialism with all the usual international logos prominently displayed and culminating in a wall of giant adverts that adorn the rooftops of communist era housing blocks.

Figure 12.2 Budapest shopping malls, 2000.

Key to Budapest shopping malls map

No. on map	Date of opening	Name	Floor area	Investment Mrd Forint	Car parking	Proportion let
1	1998	Hattyúház	16,000	2.7	100	40
2	1994	Budagyöngye	10,000	2.7	150	100
3	1998	Rózsakert	17,500	2.2	250	100
4	1998	Mammut	27,000	8.0	750	100
5	2001	Mammut 2	20,400		600	
6	2000. IV.	Rózsadomb	8,000	2.5		
7	Planned	Pagony Plaza	20,000	2.5	350	60
8	1977	Flórián	20 000	—	—	95
9	1996	Süba	10,400	1.6	242	99
10	1998	Új Udvar	17,500	4.0	400	84
11	Planned	Új Udvar 2	4,000	1.0		
12	2000. IV.	Interspar	23,000	5.0	890	
13	2000. IV.	Eurocenter	25,000	10.0	1,000	
14	1998	Árpád	10,718	1.0	m.500	
15	1997	Millennium Center	10,000	5.5	300	100
16	1994	Oktogon	2,200	0.25	Nincs	80
17	1997	Teréz Udvar	700	0.065	Nincs	50
18	2000	Király Plaza	40,000			
19	Planned	Hunyadi téri Csarnok		1.9	250	
20	1998	Orczy Plaza	12,000	110 m US$	100	100
21	Planned	Orczy 2	31,000	8.0		
22	Planned	Work Park Center	33,000	7.0		
23	1998	Lurdy Ház	42,000	7.0	2,000	60–70
24	1996	Ferencvárosi Bevárkóközpont	3,600	1.0	1,200	0
25	1997	Duna Ház	6,000	2.8		
26	1998	Eleven Center	10,000	2.8	340	99
27	Planned	Eleven Center 2	13,000	4.0		
28	Planned	Albertfalva Bevk.	16,000	3.4		
29	2001. III.	Mom Park	40,000	6.0	2,400	
30	1996	Duna Plaza	42,000	83 m DM	1,500	100
31	1999	West End City Centre	91,000	40.0	1,700	100
32	Planned	Árpád Híd		10.0		
33	2002	Transelektro City	220,000	35.0		
34	1980	Sugár	30 000		253	100
35	Planned	Örs Business Center	50,000	8.5	1,000	
36	2000	Veritas	25,000	1.0		
37	1996	Pólus Center	56,000	8.6	2,500	89
38	2000. XII.	ITC	43,000	9.0	350	
39	2001	Pólus 2	70,000			
40	2000	Füzér	5,000	1.0		
41	2000. XII	Flamingó	20,000			
42	1996	Lörinc Center	8,000	1.8	250	
43	1997	Europark	30,000	3.8	1,000	100
44	2000	Europark 2	13,000	5.5		
45	Planned	Megapark	125,000	22.0		
46	1997	Csepel Plaza	20,000	3.0		96
47	1999	Campona	38,000	28.0	2,800	80

Source: Data from Ingatlan és Befektetés 16.12.1999.

Note: Mrd Forint = Milliard Forint (one million Hungarian Forint); Nincs = there is none.

Other extensive, but visually less horrendous, out-of-town retailing centres are dominated by foreign international chain stores. The French superstore chains Aucean and Cora and the Austrian Metro cash and carry warehouse chain are located at Budakalász in the north west and Fót on the north-eastern approaches of the city.

Case example: Prague: mass tourism and the consumption of the historic centre

Marketisation is bringing new pressures from tourism to Europe's historic cities. The management of the conflict between tourist consumption and the conservation of historic cores of European cities is a serious problem (Ashworth and Tunbridge 1990). It is nowhere better illustrated than in Prague's city centre. The meaning of central Prague has become contested around the question: to whom does central Prague belong? The unique beauty of central Prague lies in its symbiosis of Gothic and Baroque architecture, complemented by the encircling neo-Renaissance palaces and monumental buildings of nineteenth century historicism (Sármány-Parsons 1998). It is filled with local societal, political and cultural meaning as a symbol of Czech identity in a rapidly globalising world (Simpson 1999). After 1990 this rich historical architectural fabric was preserved by a restitution programme that returned property to former owners and so rescued it for traditional, small-scale retail entrepreneurialism and residential use. At the same time, the authorities are promoting the image of a city of cultural heritage with strategies reflecting Austrian experience, such as in Salzburg. Music and the visual arts are used to enhance the quality of the city's built environment in a strategy regarded in scholarly discourse as alien (Sármány-Parsons 1998). Consequently, Prague was overwhelmed by tourists, and the city centre residents awakened to an experience completely absent earlier.

Tourist consumption of the city centre conflicts with residents' conception and identification with it. The reception of visitors is ambiguous; they bring economic benefits, but also intrude. Simpson (1999) found that Prague's residents and visitors attribute great importance to the ambience of the streets and their particular pattern and form in creating the identity of the city centre. Residents perceived an erosion in the quality of the built environment, but rejected planning intervention as a solution. Tourist over-crowding was disliked by residents and the tourists themselves and was especially associated with guided tours on foot. This was a major factor in some residents considering quitting the area. Variations in retailing provision were associated with a separation of residential and tourist functions of the city centre, but tourist outlets were increasing, whilst those selling everyday goods were declining. Residents were unhappy with this trend. Whilst motivation was complex there was a 'substantial erosion of the sense of place and identity of the historic core'. Tourism is not the only cause of this, but is probably the main one.

Prague is becoming a more cosmopolitan place; some 24,000 Americans, for example, now live there and for many there is a bohemian sense of 1920s

Paris (Sármány-Parsons 1998) that strikes me as ironic in the capital of Bohemia. International tourism is contributing to a wider experience for residents, even if this experience is ambiguous. Mass tourism has affected only the central areas of Prague, but the city's authorities have set their stall out to attract more tourists. Conflicting perceptions of the city among visitors and residents have resulted from the visitor-induced changes in the built environment, the local services that no longer fully cater for the needs of residents, and the lack of regulation of visitor numbers and activities. So, then, neo-liberalism has raised a new question: to whom does central Prague belong? The residents for whom it is a daily experience? The city entrepreneurs, who promote it internationally, for whom it is a business opportunity? The foreign tourists who consume it as a symbol of European heritage neatly packaged for them? There is certainly competition for its ownership. Such is the essential contradiction of its meaning in neo-liberalist modernity.

But what about the provinces?

Provincial cities have attracted investment, particularly where local initiative has been proactive. Renewal of the urban fabric and changing urban lifestyles are mirrors of neo-liberal consumerism, though less intensely than in the capitals. Maribor in Slovenia and Brno in the Czech Republic, for example, have experienced strong growth. Brno's main industries have attracted foreign investment and a new business centre has been developed and a technology park has been promoted by a coalition of local government, local businessmen and the Brno Technical University. Odessa, the Ukrainian Black Sea port, has attracted foreign investment away from Kiev, the capital, because of the openness and flexibility of the local mayor in promoting the city. Lvov, also in Ukraine, has capitalised on its historical heritage to encourage tourism. Varna, the Bulgarian Black Sea port, has prospered through local initiative, despite obstruction from the central government. Klaipeda, Lithuania, promotes itself as 'your seaport in the Baltics'.

Not all regional centres have been able, or have wanted, to utilise their potential assets for local development as perceived in the new neo-liberal world. The ambiguity of Tartu, Estonia's second city, is its bohemian unworldliness. This is, at once, its greatest asset and a barrier to its economic development. Tartu city's slogan is: 'the city with good ideas'. However, soviet era high technology industries and its prestigious university have not stimulated much innovation or investment in the city. The local government is seen as pro-business, but business has not developed (Birzulis 1999). Perhaps Tartuans really just want their city to remain the intellectual and cultural heart of Estonia. The city has aspired to become the European City of Sculpture. Consequently the Tartuans have not sought to compete with Tallin, the capital and dynamic business centre of Estonia (Kuus and Raagmaa 1998).

Case example: A brief tale of two cities

Székesfehérvár and Nyiregyháza are two Hungarian regional centres, each with a population of around 115,000, that have had contrasting experiences in transition. The contrast owes most to local initiative and enterprise, but is also a reflection of their different location. Székesfehérvár is located in western Hungary, about 60 miles from Budapest. Nyiregyháza is located in eastern Hungary, about 200 miles from Budapest. The national regulatory system and the national regional policy have also affected the experience of the two cities.

The economies of both cities collapsed in 1990 when each lost its major industrial sectors. Very quickly the Székesfehérvár local council and local entrepreneurs co-operated to begin rebuilding the economy by attracting western investors, taking advantage of its situation in western Hungary and good motorway access to Budapest. Infrastructure was improved, a training centre was established and the local specialist skills base promoted. Soon Siemens, Philips, IBM, Canon, Akai, Emerson Electric, Shell-Gas Hungary, Alcoa and Ford had been persuaded to set up on the city's new industrial estates and invest $2 bn. Western fast food corporations have come to the city and the Alba Plaza Mall, opened in 1999, added 15,000 sq. m. and eighty-five shops to the city's retail scene.

By contrast, Nyiregyháza had no early, major investment to recreate its economy. Infrastructure was not improved despite a Hungarian government scheme to assist the chronic economic situation in the region. Links with Budapest are difficult and the preferred foreign bidder designated to extend the M3 motorway into the region withdrew fearing a lack of profitability (Dingsdale 1997). It was not until early 1996 and the election of a new socialist mayor that a revitalising programme was begun in association with the Regional Development Agency. In 1998, Nyiregyháza benefited from Flemish government aid that financed the construction of a business park on a former Russian army base. The Nyirplaza Mall is scheduled to open in mid-2000. Even so, Nyiregyháza will hope to claim its share of the HUF 916 m allocated to its county in the Hungarian government's Szécheny regional assistance programme announced in March 2000.

Case example: Zaporizhzhia, a city in Ukraine

Zaporizhzhia is a city of 900,000 inhabitants in southeastern Ukraine. Its particular local experience of transition was conditioned by the retention of power among the old élites. This affected the patterns of integration of Zaporizhzhia, for the first time, into the global economy, shaping the way the city was resituated in multi-scalar place relations.

Zaporizhzhia was founded in 1921. Its production as a locality was embedded in the Soviet vision and practices of urban places as centres of heavy industry. In Soviet times the city evolved in almost total isolation from the international economy; its place relations being almost exclusively within the Soviet Union (Van Zon 1998). This has proved to be a very restrictive legacy permeating the thinking and practices of some of the ruling élites who held back innovative change after 1991.

The experience of transition in Zaporizhzhia, after 1991, was marked by rapid industrial decline, the disappearance of the dominant Soviet suppliers and the disruption of local supply chains by western imports. New place relations, formed with the west, interrupted local linkages, and encouraged local enterprises to make, if they were able to, direct contact with new western partners, abandoning their local associates. China, a relatively little commodified country, became a new market for basic metal products. Very little foreign direct investment was attracted. When foreign corporations showed an interest in investing, negotiations often became complex, confused and long drawn out. Such was the story of the predominantly state-owned car manufacturer Avtozaz. Negotiations for joint ventures were undertaken with Daewoo, General Motors, Peugeot and Rover that lasted for three years before an agreement was reached between Daewoo and Avtozaz. Van Zon (1998) explains this experience as the outcome of institutional mismanagement. Whilst local authorities tried to encourage economic development, establishing a Regional Development Agency, they lacked both a clear vision of the future and practical organisational skills. With western assistance a Business Communications Centre was set up, and financial assistance was given to the Zaporizhzhia Chamber of Commerce. Despite this help, the local authorities imbued with the lethargy of the past failed to use these institutions in a proactive or effective manner to promote the city. A progress from 'plan control' to 'clan control' signalled the emergence of unsuitable informal networks.

The experience was also influenced by the divisions in the élites created by political developments that were also conditioned by changing place relations. The political response in Zaporizhzhia to the collapse of communism was the formation of a plethora of political parties, but real power still lay with the old Russian nomenclatura and industrial élites (Jackson 1998). Enterprises continued to provide social services for employees, but not at the same level. However, the industrial élites were not simply pro-Moscow as Ukraine struggled with the issue of independence from Russia and membership of the CIS. Their political position began to diversify depending on the new economic orientation of their particular plant, in the newly emerging multi-scalar place relations. Several plants had forged new links beyond Russia and their directors, like their plants, were not Moscow orientated. Others of a military industrial nature had lost markets and suppliers and failed to gain new ones as the CMEA collapsed. The political interests of their managements still lay in old networks.

The experience of Zaporizhzhia in transition shows how inappropriate particular responses and practices combined with unfavourable outsider perceptions and new place relations can produce a locality that is unable to synthesise the effective mechanisms of neo-liberal growth.

Case example: Łódź: the experience of a loser?

Łódź is an industrial city of 850,000 inhabitants in central Poland. For Horváth (1995) representing the city as 'a formerly well supported citadel of the working class', Łódź was one of the 'obvious losers' of the change of regime and the restructuring that it heralded. The socio-economic development of Łódź in the 1990s offers an insight into the production of an urban locality in neo-liberal spatial modernity for which the change of regime might

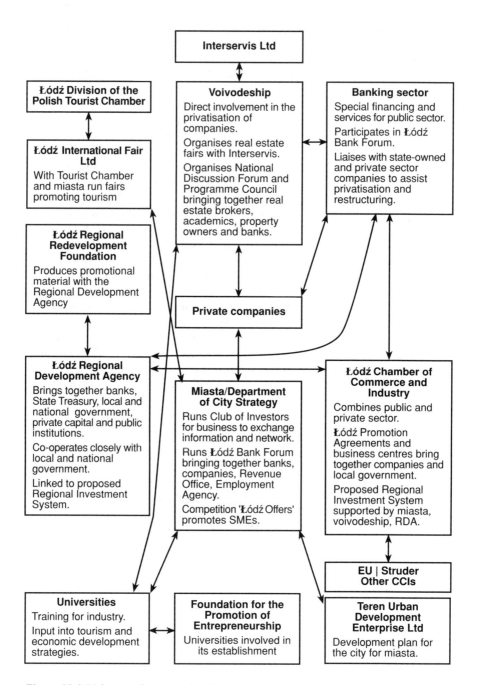

Figure 12.3 Urban coalitions in Łódź, Poland.

Source: Young, C. (1997) 'Urban Governance and Local Economic Development in Łódź, Poland: the emergence of an "entrepreneurial" local state', in A. Dingsdale (ed.) *Urban Regeneration and Development in Post-Socialist Towns and Cities*, Trent Geographical Papers, No. 2, 32–51.

have been catastrophic. Geographical discourse has produced Łódź as an experienced, perceived and conceived outcome of structural and institutional change in which both insiders – ordinary citizens and development-orientated groups – and outsiders – domestic and foreign investors – have been actors.

In 1990 Łódź was perceived as an ecological disaster, despite having living conditions comparable to those elsewhere in Poland; two-thirds of flats built since 1950, one-third since 1970 (Liszewski 1995). Housing and environmental conditions were worst in the old centre of the city (Kaczmarek, J. 1995) but weekend recreation in second homes and on garden plots within 30 km of the city was a widely enjoyed experience (Matczak 1997). The collapse of communism resulted in the emergence of an elected local self-government and the establishment of numerous agencies interested in city development (Young 1997) as well as private businesses (Figure 12.3).

Kaczmarek, S. (1995) represented Łódź as constructed in the perception of its inhabitants as bipolar. Their sense of place was highly localised in their experience of emotional ties, community relations, personification of landscape and the practicality of restrictions on housing choice. Contrasting with this 'almost punctiform' perception of place, Juianow, a district of single family houses in a garden setting, was universally idealised as a living place. Between this 'mythical synonym' of high living standards and their present residence was unperceived, empty space.

This representation related only to residential perceptions and not other aspects of urban experience, and suggests that the citizens' construction of Łódź is highly fragmented with little sense of the city as a place. This could have important implications for the activities of those groups seeking to promote Łódź as a distinctive place for the purposes of local economic development.

Riley (1999) represented the experience of Łódź as a transition in manufacturing, shifting the balance between 'top-down' and 'bottom-up' initiatives. He perceived local authorities as 'urban managers' who together with local entrepreneurs, especially those offering producer services, or opening new manufacturing firms, became powerful agents of development as transition. Working within wider structures, the actions of these local agents improved the agglomeration advantages of Łódź, to the detriment of smaller settlements that caused local place inequalities. Local agents were now so powerful that the broader transition processes that had set them in motion could be easily forgotten.

There are many groups who have a new image of Łódź and are working together to realise it. The image is derived from their perception of the city they experience in the light of the new institutional situation. They have sought to guide the transition of Łódź by regenerating its socio-economic fabric and built environment, reinventing its image and promoting its new iconography internationally and within the city. In the new institutional framework, land use and the built fabric of central Łódź were re-evaluated and redeveloped. Manufacturing has practically disappeared as industrial premises were converted to new retailing, wholesaling and office uses. Residential courts have been converted for business and financial services (Riley 1997). Newly constructed buildings, often part of prestige projects, symbolised for Liszewski (1995) the new 'cash and carry' society. These various detailed processes have transformed land use and the urban landscape of central Łódź around what Riley calls the 'office and retail spine' of Piotrkowska Street.

Klosowski (1995) perceived the transition process as a tentative progress from a consciousness of city identity to city marketing in the representation of the city in plans for socio-economic and physical development. Young and Kaczmarek (1999) found place promotion and imagery in Łódź largely ineffective. Re-evaluation and representation of place was important for Łódź in the new competitive environment. The numerous organisations active in re-imaging the city for external consumption had also recognised the need for efforts to change the attitudes of the city's inhabitants and engage their support. However, the strategies adopted, the images used and the co-ordination between agencies were weak. A crude strategy had formed in which city marketing was too dependent on isolated and 'flagship' regeneration projects in the city centre. Łódź' place promotion also 'snipped out' the communist period and returned to a mythical entrepreneurial image that reconstructed its nineteenth century industrial experience (Young and Kaczmarek 2001). It seemed, however, that place promotion had been unimportant in attracting investment. Prestigious investors had come to the city, including Shell, Gillette, Coca Cola and Wrangler, Sara Lee and ABB (Klosowski 1995), but had not been attracted by marketing efforts. The investors had selected Łódź because of its central location, skilled workforce, positive local-authority attitude and future motorway construction projects. Thus, investment had not been located in Łódź primarily on the basis of the city's promoted image, but on investors perceiving the strength of the city in traditional factors of production. The investor's Łódź was a different place from the promoter's Łódź. Though the outcome was welcome for the city, it raised questions about the provenance of re-imaging.

Łódź is unusual in the extent of its promotional and re-imaging efforts but it offers a warning for the localities of the Marchlands. Much of the problem seems to arise from a failure to properly evaluate promotional activities and their effectiveness. This may be no more than a result of inexperience, though Łódź did seek advice from foreign consultants. Recent western promotional strategies for local economic development cannot be adopted without caution. The situation in Łódź to some extent reflects the contrast between western and eastern experience recorded in the comparative study of Groningen in the Netherlands and Debrecen in Hungary in which the obvious contrast between the two was that city promotion in Groningen was much better organised than in Debrecen (Ashworth 1994).

Case example: A small town in the Czech Republic: Mlada Boleslav – a very special case?

Some smaller towns have had special advantages. Mlada Boleslav in the Czech Republic is one of them. The experience of transition in the town has been better than most. In 1991 the Czech government sold 70 per cent of the town's principal industrial company Skoda Automobilova to Volkswagen

(VW) for \$900 m. It was the second biggest FDI investment in the Czech Republic. VW agreed to invest DM9 bn, increase production to 400,000 cars annually, retain the 21,000 workers and use Czech suppliers. In return, the firm received tax concessions, the writing-off of Skoda's debt and protection for the Skoda monopoly in the Czech market (Pávlinek 1998). In 1998 Skoda Auto was the third biggest company in Central Europe, employing 20,400 staff, 17,000 in Mlada Boleslav and its surroundings – one in three of the town's population. It accounted for 8 per cent of Czech exports. This success has attracted other foreign investors. French, German and American parts suppliers operate in the town that is at the centre of a network of linkages across the Czech Republic and Slovakia. The town has also become connected in new place relations to the network of places encompassed by the multi-national linkages of VW organisation.

Success has come to Mlada Boleslav without the need for the local government to provide investment incentives. The town council gives little assistance to would-be investors offering no more than advice on factory sites. Unemployment is around 2 per cent and Skoda had to hire 3000 Poles and Slovaks to meet its workforce needs. All the town's employers face labour shortages, but workers don't seem keen to move in. A scheme to offer new housing for sale failed, only fifty of the 137 flats were sold and Skoda bought the rest to rent out. These difficulties could constrain future growth based on VW's second and third phase investments. These include a huge warehouse for Skoda–VW–Audi parts opening on a new industrial park and a further commitment of up to \$1.5 m investment by 2002 (Harris 1998).

The presence of the Skoda works was the key attraction to VW, the town in which it was located was incidental to VW. Whilst VW saved Skoda and so the town, Skoda workers have adopted practices, such as strategic in-sourcing, that were rejected by the unions in VW's German plants, and are only used in the Brazilian and Argentine plants (Pávlinek 1998). Furthermore, the Skoda operation has had some problems and the town's future in the mid-1990s depended on an acrimonious dispute between VW and the Czech government over investment, in which VW threatened to leave. This danger notwithstanding, the town council see the town's success as putting infrastructural capacity under stress, but have been unable to persuade the Czech government to provide investment for improvements. Similarly, an early plan to build an international airport has been abandoned (Harris 1998).

The experience of Mlada Boleslav is a special case in two respects. First, the extent of the embeddedness of the VW investment implies a long-term commitment. Other investments in the Czech Republic do not have the same degree of embeddedness (Pávlinek 1998). Second, there was no need for local initiatives or incentives to attract the investment. The town illustrates the effect of local agglomeration advantages and local circular and cumulative causation following an initial advantage.

Conclusion

Transition and transformation have been represented and illustrated as a myriad of local experiences, perceptions and conceptions of democratisation and marketisation. Localities are being produced in newly negotiated multi-scalar, socio-cultural and economic development relationships that constrain localism, but demand local initiative. Outsiders have often dominated these relationships, but individual local entrepreneurialism and local government activity, even local protest, have been significant factors in shaping the experience of specific local transitions. Some localities have achieved 'success' in the form of economic development, but this has been accompanied by social and spatial fragmentation, especially in the big cities. Great wealth has been quickly accumulated by some individuals, whilst others have fallen into poverty. Beggars are a common sight on the streets of the most prosperous capital cities. At Christmas 2000, news reports stated that some 8,000 individuals relied on charitable food gifts in Budapest. Many of the poor are old. Brought up in the communist world their attitudes and values are out of place in the new situation and they lack the mental and practical flexibility that is now at a premium. Many of the wealthy are young, they are enthusiastic for change and readily adapt to the new situation.

Neo-liberal capitalist modernity has produced in the Marchlands a mosaic of localities in commodified space, some as commodified places. Some localities synthesise locational centrality, financial control and socio-political power and prestige. They are experienced as vibrant and dynamic and represented as 'winners' in the neo-liberal spatiality. They have become places in the networks of international business. Others synthesise locational peripherality, economic marginality and socio-political weakness. They are experienced as slow changing and represented as 'losers', relatively isolated from the dynamics of international neo-liberal development. The text is written almost exclusively in a genre of material and discursive vocabularies of economic competition. Between the extremes is a fine-grained gradation of contrasting localities, synthesised in the spatiality of western outsiders, who consume them as economic commodities for business or tourism and the resident insiders who experience them as lived-in environments of place identification. There is also a strong cleavage in the localities experienced as transition and transformation between the new indigenous entrepreneurial class and marginalised groups of beggars and the old.

13 The production of regions in transition

This chapter looks at the production of regions in transition and transformation as multi-scalar renegotiations of place, driven by Western European ideas of democratisation and marketisation and enmeshed with the legacies of communism and revived senses of ethnic national consciousness. It explores the competing representations of regional differentiation in transition. The conjunction of democratisation and marketisation as current processes, interacting with antecedent conditions, has recast the production of regions as a dynamic of spatial modernity in the Neo-liberal Project.

Democratisation: regionalisation and regionalism

As I noted in Chapter 12, the creation of subnational, autonomous, representative institutions is a vital element in transforming hegemonic regimes into democratic ones. Democratisation, through its reform of state public administration, raised the question of redrafting the functions and territoriality of a tier of 'regional' governance between the local and the central authorities. The social and spatial practices of devolving political and administrative responsibility to local communal self-governments have left what Bennett (1998) calls 'a democratic deficit' in this tier. Only Hungary has direct elections for county councils. Romania has county representatives elected by local governments, but county government is headed by a prefect appointed by central government. The function of this tier is no more than to implement and enforce central legislation. Elsewhere the 'regional' tier is a territorial division for central administration or does not exist at all. Central government has a dominant position over fragmented local democratic institutions and there is no democracy in between.

As well as creating a 'democratic deficit', devolution practices have created inefficient public administration. Local commune government is not able to provide services because it lacks financial and administrative capacity. Central government might have the resources but lacks a field administration to provide services. The formation of a regional administration with a democratic mandate is, arguably, the most effective way that this weakness may be acceptably overcome. In any event, the democratic deficit and the weakness in regional administration will provide a topic for the political agenda of regions in transition.

However, in the Marchlands the issue goes much deeper than a question of the politics of public administration. The political rights and civil rights, developing with democratisation, have created a tension in ethnic cultural identity and political representation that is at its sharpest in the production of regions as a renegotiation of place between central and local interests. Most states have ethnic minorities and some contain regions in which minorities with a strong sense of their ethnic cultural identity are numerous, sometimes the majority population. Within the democratisation process this situation has raised questions of human rights, ethnic group rights and regional consciousness. It has not always translated into political movements or nationalist political parties, nor does it necessarily lead to demands for autonomy or independence. However, political parties have been formed around ethnic groups in many countries and in general elections their support has been registered in a strong regional showing that could be the basis for the formation of a regional government and demands for a federal constitution founded on regional cultural identity. The issue is at its most critical when ethnic minorities are concentrated in the border regions of states. In the Marchlands the mismatch between state bounded territory and ethnic settled territory is widespread in border regions.

Jordan (1998) represents the democratic deficit in regional governance as a product of the reassertion of the nation-state and centralist concepts within states that have multi-national or multi-ethnic settlement. All the states have constitutions declaring themselves to be unitary nation-states (Chapter 14). The dominant ethnic group concentrates state power centrally; imposes regionalisation that splits up ethnic concentrations administratively; and opposes federalism that might provide an opportunity for ethnically founded regionalism to become a significant political force.

The formation of regional interest groups or élites, other than those with an ethnic base, is another development from democratisation. Regional groups demanding administrative reform, or stressing cultural distinctiveness or other sectional interests, participated in elections in Poland in 1991 and 1993 (Matykowski and Tobolska 1998). Regional growth coalitions including local self-governments finding themselves in relatively weak circumstances (Chapter 12) sought co-operation with other local governments in regional associations or networks with a variety of political, social or economic objectives. Furthermore, the return of the west as democratisation in the guise of the EC/EU came flush with the idea of 'Europe of the Regions'.

The specific regionalist influence of the EU

The regionalist project within the EC/EU, extended to the Marchlands, has brought a specific regional influence in conceptual models of regional development policy and democratic regionalism; examples of best practice in co-operation for regional economic restructuring and promotion; funds to assist implementation; allied with a demand for harmonisation of regional administration as part of the price of the entry ticket to be issued at some future date. The regionalist project within the EC/EU has become a significant scenario in the objective of European integration and the Euroregion has become its icon in the Marchlands, informing the production of new regions.

Regional Development Policy and Nomenclature of territorial units for statistics (NUTS)

The Treaty of Rome included a reference to efforts to equalise socio-economic conditions across the territory of the EEC for social, economic and political reasons. This was put into practice through a weak but evolving European Regional Development Policy (ERDF), established in 1975, that came to frame its own idea of the region and a complex programme for regional assistance. The regional idea signified a perceived need to assist certain 'backward' or 'depressed' regions in coping with spatial marginality and cultural socio-economic restructuring. At the same time, spatial thinking led to the formulation of NUTS as a system for the collection of empirical data upon which to take policy decisions as the EC structural funds were dispersed across member states.

The Euroregion idea

In the late 1980s the idea of the region was transformed by attention being switched to other kinds of regions in the assertion of neo-liberal values.

> Far from being geographical anthologies of cultural particularism and social obsolescence certain regions . . . now seemed the sharp end of European consciousness: areas of sophisticated technology, environmental awareness and a culture and civil society which integrated the intimate and the cosmopolitan.
>
> (Harvie 1994: 2)

Harvie called these regions 'bourgeois'. They were the neo-liberal growth regions of Western Europe, flourishing in the 1980s, and their politicians became confident and aggressive in the EC arena. The pace of co-operation and integration was quickening following the Single European Act and the Single Market Programme agreed in 1986. Regional politicians in Germany, Spain and France, soon followed by others, saw their opportunity to assert a new regional voice in the EC. They grasped the idea of subsidiarity, by which the Commission wanted to ensure that decision-making was devolved to the appropriate level and contested its meaning and application with politicians from member state governments. In this they were aided and abetted by the European Commission and its beaurocracy. The idea of a Europe of the regions sprung on to the political agenda as an alternative to a Europe of nation-states. The region as a project of European federalism thus extended the EC idea of region, allied it with neo-liberalism and appropriated its meaning as the essence of the European experience and consciousness. The idea of the Euroregion synthesised a sense of place. It captured the meaning of the region as imagined, constructed and promoted, in the political arena of current locational, socio-cultural, practices that energised the project of European integration.

The Association of European Border Regions (AEBR)

The third role in this regionalist scenario was played by the AEBR. Conceived by regional planners in 1965, the AEBR was inaugurated in 1971. It was driven by a perception of the border regions as economically backward and peripheral within the territorialisation of Europe imposed by the nation-states. Its main practical task was to formulate expertise to overcome the problems of backwardness and peripherality through institutional co-operation. The organisation evolved, by enlargement, adding new border regions, and also by intensifying its institutional network of relationships with the Council of Europe, the Council of European Municipalities and Regions and the European Commission. At the same time it advanced its main practical task and its project Obervatory for Cross Border Co-operation (LACE) was funded by the Community Interreg I (1989–94) and Interreg II (1995–99) programmes (AEBR 1996).

However, the Interreg Programme, like previous international co-operation projects, requires regional institutions to take initiatives to bid for funding and technical assistance. Some regions of Hungary and the Yugoslav Republics of Slovenia and Croatia participated in the Alps-Adria Working Community before 1989. The Tatra Mountains co-operation project across the Polish–Czechoslovak border was also established before 1989. Since 1989 the Euroregion idea, primarily advanced by bilateral and multi-lateral cross-border schemes, has flourished in the Marchlands.

Marketisation: economic restructuring and the idea of a neo-liberal growth region

Marketisation, through commodification, is shifting the formation of economic regions towards mechanisms of growth and decline that reflect neo-liberal values, priorities and conceptualisations of success. This process is at the basis of the idea of the neo-liberal growth region (Allen, Massey and Cochraine 1998). Regions are becoming defined in contrasted spatialities of neo-liberal growth processes that shift the balance from fixed bounding to multiple functional linkages and contact networks. In general terms this means four interlinked changes. First, it means the pace at which traditional, heavy industries and old labour relations practices could be cleared out. Second, it means the pace at which consumer-orientated, higher technology industries, tertiary activities, financial and business services and new labour practices could be introduced or expanded. Third, it means the setting up of new types, intensity and reach of contacts. Fourth, it means setting up and operating organisations for promoting regions. This process of marketisation was constrained or promoted by western perceptions of existing regional mixes of sectoral branches, enterprise organisation, labour market characteristics, settlement patterns, population dynamics and the cohesion of regional, socio-economic networks. Thus, new economic regions emerged out of old ones.

In this sense, transition might be interpreted as a special case of the wider international or global processes promoting sectoral restructuring whereby 'smokestack' production industries were giving way to 'sunshine' consumer industries. Investment was attracted to or generated where a potential for growth in expanding sectors was perceived by investors as high. In contrast, investment was withheld or withdrawn where predominantly declining sectors were perceived by investors. The critical factor became the combination of new growth mechanisms that reformed regions as new syntheses of socio-economic practices and perceptions, in relation to and in contrast with other regions.

The same perception of transition has adopted the promotion of SMEs as a regional growth strategy. The EU PHARE programme and some government regional policies following them have supported the development of SMEs. This approach was seen as a successful response to economic restructuring of old industrial regions in the west and, therefore, potentially, something to build on in the east. Smith (1998) suggests that this practice equates socialist large-scale industry with western Fordism and in so doing misunderstands the nature of socialist large-scale industry. Smith points out that whilst in the west mass production was based on intensive accumulation, in the east it was based on extensive accumulation. Whilst this is true, as I showed in earlier chapters, socialist economies in some countries did move to more intensive forms of production. In these cases, the multi-plant, multi-enterprise structures have some common features with western Fordism. Smith also points to the contrast between the operation of SMEs in western style industrial districts, and the practices in districts of a supposed similar nature in Slovakia.

These new region forming processes incorporated the mosaic of successful commodified and competitive localities, but regional transition was more than the sum of the formal or functional grouping of such localities. Enterprising local governments and local business communities, even the practitioners of wild entrepreneurialism, did look for co-operative ventures with other localities with economic promotion and profit in mind. These associations formed new networks that included successful localities as well as unsuccessful ones because they responded to internal perceptions of regional issues. Within newly forming regions, therefore, there were localities in which commodification was advanced and others in which it was retarded.

The specific combinations of region forming mechanisms of growth in the process of commodification led to a compartmentalisation of new regional economies. Hamilton (1999) assigned this to a multi-structured separation effect; on the one hand between formal, informal, commercial and barter sectors, to which can be added wild entrepreneurialism, and on the other to separation between commodified and state sectors. It reflected major variations in commodification between and within sectors. Privatisation of the large-scale state enterprises in industry, transport and telecommunications proceeded more slowly than privatisation of small-scale state enterprises engaged in service provision. Contemporaneously, within the large-scale industries, consumer goods and service enterprises were privatised more quickly than mining, heavy industry and defence. Foreign and domestic investors in privatisations were picking prospective winners by applying neo-liberal criteria to enterprises and so shaping regional growth mechanisms.

This new picking and mixing shaped the formation of regions in which new foreign owners of enterprise brought and consolidated new patterns of linkage and contacts with new places, whilst new local owners sought to make, diversify and extend contacts, forming new networks. They resituated newly forming regions in multi-scalar relationships. Most notably they reorientated, intensified and extended the reach of contacts. Regions produced by advanced commodification reflect and benefit from the intensification of contacts with mature market economies. Regions with less advanced commodification retain old linkages with places that are still commodifying slowly. The external contact networks shaping region formation intensify place cleavage between commodified and less commodified regions. They transform the processes of regional identification and regional identity.

Regional development in the Marchlands

The winds from the west, bringing ideas of democratisation and marketisation, have roared through the regions of the Marchlands. Democratisation and marketisation have brought a neo-liberalist veneer of similarity to the new processes of region formation. They have been diversified by a lattice of specifics: in state transformation and regional policies; in inherited societal, cultural, legal and infrastructural conditions; and in the nuances of local collective and individual responses to the new circumstances.

Regional differentiation

Regions are being produced along three new primary axes of differential development.

(a)   Commodification reconceptualises the purpose of enterprise. It shifts practices to new forms of private and public organisation, redefining and resituating regions in a multi-scalar complex connectivity from localities to global economic relationships.

(b)   Cultural identity is redynamised by democratisation. It reworks social and cultural identification, within and across state boundaries.

(c)   EU conceptualisations and practices are promoted as models of regional development. These elicit and support local initiatives, within and across state boundaries.

Representing regional differentiation as socio-economic transition

Regional differentiation related to marketisation has been approached from empirical and theoretical starting points that, most often, have focused on regions within single countries. Empirical approaches have monitored regional differences in development through features such as demographic processes; labour market characteristics and unemployment; foreign capital investment; and privatisation. Theoretical starting points have been models of neo-classical convergence, Myrdallian core and periphery divergence and neo-Marxist uneven development (Pickles and Smith 1998). These approaches have generally accepted state administrative areas as their regional framework, although this is increasingly unrealistic when regions are perceived as relational places, within a complex multi-scalar connectivity. Regulation theorists have also been prominent, sometimes stressing socio-economic and production network conceptions of regional organisation, rather than territorial delimitations (Begg and Pickles 1998). Regional policies have been framed by national governments as responses to perceptions of growing regional disparity caused by transition (Batchler and Downes 1999). Several alternative imaginary future regional socio-economic development scenarios have also been suggested for some countries, for example by Gorzelak (1996) for Poland and Enyedi (1994) for Hungary.

The initial experience of many regions was economic collapse, de-industrialisation, plant closures, unemployment and social disruption. This was especially the case in those regions closely dependent on the economic organisation most typical of communist industrialisation. Such regions were separated from market relations and dominated by large-scale industrial enterprises engaged in mining, heavy engineering, steel, chemicals and armaments production. Similarly, those regions that developed under communism as peripheral, branch-plant economies have also suffered badly. Sometimes new foreign owners and sometimes new indigenous managements have continued production at core plants but closed branch plants. These processes have contributed to marked regional contrasts within most countries, which reflect the broad patterns of inherited conditions as re-evaluated by the priorities of neo-liberal perspectives.

Regional differentiation within states has been represented as marked by: a region's potential to attract FDI; its inherited societal, infrastructural and environmental conditions; the initiative of local and regional élites; location in national regional systems; and the pace of state governments in framing regional and regulatory legislation. These approaches imply that regional differentiation is highly dependent on the state context but that location in the Marchlands as a whole has also been considered as important.

Hamilton (1995) conceptualised transition processes as diffusing commodification eastwards, unevenly, across the whole of Europe. Two other attempts to extend regional analysis beyond the boundaries of a single state were by G. Gorzelak (1996 and 1998) and G. Horváth (1995). Neither Gorzelak nor Horváth conceptualised transition as primarily a process of diffusion, and both used intra-state administrative boundaries as the basis of their analysis in a somewhat questionable, empirical methodology. Gorzelak constructed regional patterns of transformation measured by regional differentiation of demographic processes; labour market features and unemployment; foreign investment; and privatisation. He explained contrasts by reference to 'regional potential' for the adoption of new conditions including: the settlement system; qualitative and quantitative features of the labour market; presence of R and D centres; infrastructural standards; and accessibility to international centres of innovation and capital. This is a kind of Darwinian approach; a sudden change in environmental conditions finds some regions fitter than others in the accommodation and implementation of new current processes. It echoes the winners and losers' discourse at the regional scale. On the basis of this approach, Gorzelak represented transformation in 'Central Europe' as a process of regional differentiation that could be understood in terms of core/periphery contrasts. In particular, he recognised a Central European 'boomerang'-shaped area on the western edge, where transformation processes were concentrated; specific major centres and cores of transformation; and potential and smaller cores of transformation. The pace of transformation was fastest in the west. To the east was a marked periphery bounded by an 'eastern wall' (Figure 13.1)

Horváth constructed 'East–Central Europe' as regions comprising groups of intra-state administrative divisions. He represented them as 'backward', 'depressed' and 'leaders of transformation', but had a considerable number in an 'other' category. His emphasis was on the acceleration of uneven development with regional 'flagships' of restructuring formulating ambitious development programmes contrasted with losers, whose economic restructuring faced major problems and was inevitably slower than that of the leaders. This approach was really based on generalisation from a series of state units whose experience could not be directly compared because of a lack of compatible statistics. His map shows the situation separately in states that are located next to each other, but the regions appear to have no interconnection (Figure 13.1).

Regional GDP per capita (PPP), measured against the average for the ten Marchland states that are candidates for EU membership, is shown in Figure 13.1. This provides a comparative measure of regional development across the whole of the area, complementing and extending Gorzelak's and Horváth's approaches. I have classified regions as leading, intermediate and trailing in development. I have superimposed onto this framework the main features of Gozelak's 'Central Europe' and Horváth's 'East–Central Europe'. Together these ideas construct a regional development space that has a marked northeast, southeast gradient from the southern section of the development axis. Only the Warsaw and Bucharest city regions stand out from this. Prague is the most prosperous region. Only the city regions of Bratislava, Budapest and Warsaw reach more than 80 per cent of Prague's GDP per capita value. Of fifty-two regions, forty-three are below 50 per cent of the Prague figure and eleven of these are below 25 per cent. Old industrial regions and the easterly agricultural regions are the least well off.

I have also constructed a regional differentiation map of Ukraine, using data from the *Ukraine Yearbook 1999* (Figure 13.2), that shows three types of region. In this case the basis is a composite rank score. The results are not comparable with other countries. There is a marked regional cleavage between the east and the west, with the more prosperous areas in the west and the least prosperous in the east.

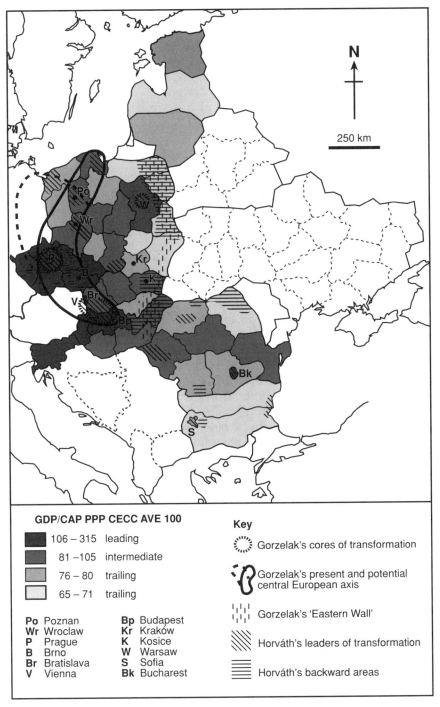

GDP/CAP PPP CECC AVE 100

- 106 – 315 leading
- 81 –105 intermediate
- 76 – 80 trailing
- 65 – 71 trailing

Po	Poznan	**Bp**	Budapest
Wr	Wroclaw	**Kr**	Kraków
P	Prague	**K**	Kosice
B	Brno	**W**	Warsaw
Br	Bratislava	**S**	Sofia
V	Vienna	**Bk**	Bucharest

Key

Gorzelak's cores of transformation

Gorzelak's present and potential central European axis

Gorzelak's 'Eastern Wall'

Horváth's leaders of transformation

Horváth's backward areas

Figure 13.1 Regional economic development: EU applicant states, 1995–97.

Source: Data from 'Statistics in Focus', Theme 1–1/2000 (eurostat 2000).

I undertook a more detailed analysis of regional patterns in Belarus. Belarus is divided into six administrative regions, with the capital, Minsk, making a seventh. The population of these regions was broadly similar in 1991. According to official figures from the *Yearbook of Belarus 1999*, economic conditions across these regions were generally similar at the end of 1997, but Minsk City faired better than elsewhere. The situation reflected the slowness of transition away from state control. In 1997, state-owned enterprises still accounted for 65 per cent of investments in Belarus as a whole. This ranged from 55.5 per cent in Grodno to 76.6 per cent in Vitebsk. For several economic indicators Minsk City is separated from Minsk County that surrounds it, but for the important measure of money incomes the two are combined in official figures. Regional economic differentiation within this framework can be imposed by ranking the regions using various indicators and summing them to make a composite rank score. Whilst this is a simple and well-established method, it probably masks as much as it reveals. Nevertheless, regional contrasts were constructed after this fashion and from it the regions can be classified as leading (Minsk City and Minsk County), intermediate (Vitebsk, Gomel and Mogilev) and trailing (Brest and Grodno). However, these differences and any conclusions drawn from them must be treated with caution (Figure 13.2).

Regional trends in the number of private farms, in Belarus, during the 1990s confirm the pattern of regional economic differentiation (Figure 13.3a). Whilst the Brest region is slightly exceptional, the regional trends in the number of private farms are similar. They show in each region a rapid increase in the early 1990s, followed by a levelling-off, and then a reduction in numbers after 1996. However, the number of private farms separates the regions pretty much as classified by the economic indicators. Even so, the average size of private farms, whilst slightly different between regions, was maintained at about the same area in each region throughout the period. This reflects the government's control of the process, through the allocation of land for private farming.

Another broad sign of place contrasts in Belarus is given by the provision of telephones (Figure 13.3b). The pattern is interesting. The predominance of the city of Minsk and the rural–urban contrasts are strikingly apparent, within similar regional trends. In 1990 fewer than one in five rural households had a telephone. This had increased to one in three households at the end of 1997. So, in 1998 rural provision of telephones was just slightly below what urban provision had been in 1990. However, by the end of 1997, between 60 per cent and 70 per cent of urban households in all the regions had a telephone. This figure meant that by the end of 1997, the urban centres in the regions had almost reached the level of provision in the city of Minsk in 1990, whilst in the city itself, some 90 per cent of households had a telephone by then. Communication by telephone was much more widespread in the late 1990s than it had been in 1990, but the relative differences between rural areas, urban regional centres and the city of Minsk had remained virtually unchanged.

The production of regions represented as economic differentiation in Belarus in the 1990s remained embedded in a vision of state dominated

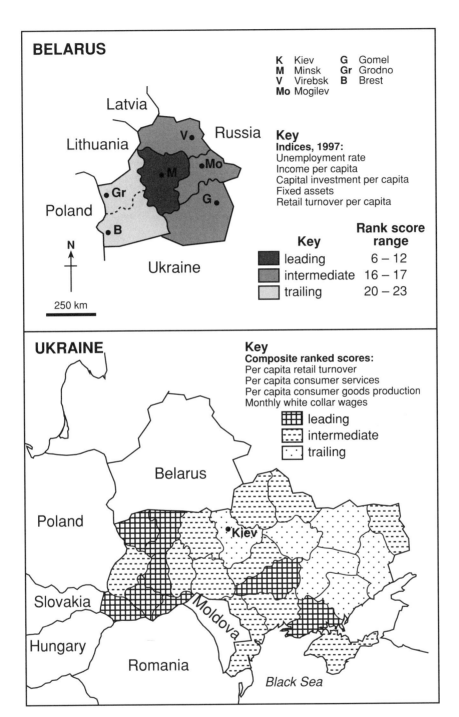

Figure 13.2 Regional economic development in Belarus and Ukraine.

Source: Data from *Yearbook of Ukraine 1999* and *Yearbook of Belarus 1999*.

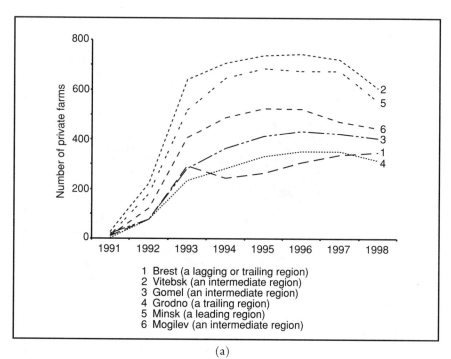

1 Brest (a lagging or trailing region)
2 Vitebsk (an intermediate region)
3 Gomel (an intermediate region)
4 Grodno (a trailing region)
5 Minsk (a leading region)
6 Mogilev (an intermediate region)

(a)

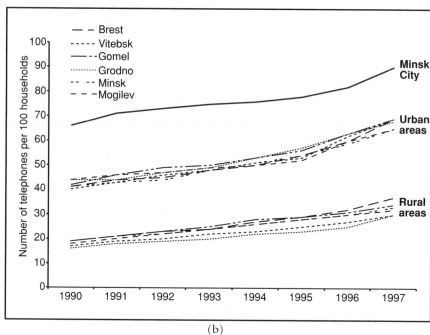

(b)

Figure 13.3 (a) Farm privatisation in Belarus; (b) telephone provision in Belarus.
Source: Data from *Yearbook of Belarus 1999*.

investment, perceived as sharing equally in a limited prosperity and experienced as slow changing.

Regional policy: regions perceived as areas in need of assistance

Transition has been perceived to cause new regional disparities within countries, between cores and peripheries, eastern and western regions, urban and rural areas and in specific problem regions. These inequalities have been met by the formation of regional policies that include reterritorialisation, financial area targeting and the designation of special development regions (Batchler and Downes 1999). Batchler and Downes compare the institutional structure, policy stances and debates surrounding regional policy in the transition countries. Policies have aimed to assist the development of market economies in all regions and alleviate the most socially disruptive aspects of transition. It is clear that all the governments perceive regional policy as a prop for regions that are failing to meet the demands of marketisation. Regional development policy-makers perceive regions in transition within the conceptual framework of winners and losers. This has raised again the old debates setting equity against efficiency in state centred policy-making. However, the new ideological framework inhibits the formulation of a policy that conceptualises regions as inter-related places within state space to which appropriate resources can be directed. This throws policy-makers, encouraged by EU models of regional policy, into the competitive mode, framing assistance in terms of programmes that incorporate bidding procedures. Batchler and Downes (1999) see dealing effectively with the EU demands on regionalisation and regional policy as a problem for Marchlands governments. The programme format and bidding culture may not meet the needs of government, but careful adaptation may be an effective guide to slowing the pace at which regional disparities are increasing. However, progress has been slow. There has been much conceptualising, but little practical implementation in most countries. Budgets for regional assistance are very small. National expenditure on regional policy ranged from around 8 m Euro in Estonia to 82 m Euro in Hungary. This implied per capita expenditure of between 1.3 and 8 Euro (Batchler and Downes 1999). Though caution should be exercised in the comparison, this figure compared with FDI stock in Estonia amounting to about 1,600 Euro per capita and in Hungary to about 1,900 Euro per capita. It is unlikely that, given this discrepancy, regional policy is having much effect on regional development as compared with FDI.

Representing regional differentiation as cultural identity

Regional differentiation on the basis of cultural identity is not directly related to current diffusion processes, but to the mismatch between state bounded territory and ethnic settled territory newly sharpened by transformation processes. The twentieth century has seen a long process of shifting settlement and state boundaries to bring them into alignment. Or, more accurately, shifting people to make ethnic populations fit state boundaries drawn in 1920.

This continued after the collapse of communism (Chapter 14), but the map of ethnic settlement is very complex and there are many enclaves of ethnic minorities in the Marchlands states. The sharpening of ethnic regional consciousness has been related to five principal processes:

(a)   the strength of ethnic nationalism among the 'native' population of the nationalising state;
(b)   the size of the ethnically distinctive minority population in the state;
(c)   the spatial distribution of the minority – the dispersal of ethnic populations within a state, except when concentrated in urban centres, reduces ethnic consciousness, whilst a large regional cluster, especially in a rural border location, increases it;
(d)   the political tactics and strategies of individuals, and regional élites, who manipulate ethnic consciousness for their own purposes;
(e)   the role of the external, homeland state.

Underpinning this is the idea of historic cultural regions that permeate identity despite the constant shifting of settlements, migratory movements and political boundaries.

Understanding current region forming processes associated with ethnicity, especially regionalism, is helped by the ideas of Brubaker (1995), though this was not his specific aim. Brubaker suggested a dynamic relationship between the triad of ethnic minorities, nationalising states and external homelands, in which each of these elements is seen as a relational field. Within each field there are various competing political stances and the relationships between the fields provides a dynamic that affects relationships within each of them. The outcomes of this dynamic can be myriad, ranging from extremes of promoting conflict to promoting co-operation. The particular outcomes are shaped by broad, historical, socio-economic and cultural structurations, but not predicated upon them. The relational dynamic is driven by current configurations of forces and by perceptions and representations of reality just as much as by reality itself. It is, therefore, potentially subject to élite manipulation, from within any or all of the fields, as well as the popular mythology of ancient or more recent times. Brubaker's analysis is cogent and will be referred to again in discussing the break-up of Yugoslavia. Brubaker's ideas can be used to situate cultural region forming processes in a dynamic relational setting, but I think they undervalue the details of territoriality and place identity as factors in shaping regional consciousness. Furthermore, they may be more powerful in explaining conflict than other conditions.

There are many distinctive ethnic groups whose settlement as minorities is an important element in region forming processes. In Poland, Lithuanians, Ukrainians and others are present along the eastern border and there is an important German minority in Silesia. These groups have been active in cultural associations and in political organisations. Poles are a large regional group in Lithuania. In Bulgaria there is a large Turkish minority in the northeast and the southern regions, which keep alive cultural practices and vote for ethnically based parties in elections. However, two nationalities,

Russians and Hungarians, are of special significance because they make up substantial minorities in several countries.

Russians are settled in all the states newly independent from the Soviet Union. They live principally in urban areas and in border regions to the east (Figure 13.4). They came to these territories as settlers under czarist and Soviet Russification policies. They have a long history of settlement in some areas, but their numbers were significantly increased in the Soviet period. Their status has changed from that of a privileged minority in the Soviet period to a more uncertain present with the gaining of independence from the Soviet Union. In all the new states the presence of Russians is a source of ethnic and linguistic tension and cleavage that is sharpening regional identity in transformation. Anti-Russian sentiment in the Baltic Republics has led to dis-enfranchisement and other civil rights discrimination. However, many Russians here seemed more concerned to prosper in business than to worry about ethnic issues. In Ukraine's independence referendum, Russians voted for independence though not so totally as ethnic Ukrainians. Since then regionalism, linguistic issues and ethnic questions have been more important than economic ones in shaping attitudes and have opened up a regional cleav-age between the east and west of the country (Khmelko and Wilson 1998).

Hungarians form significant minorities in the border regions of the states surrounding Hungary. They have been settled in these areas for centuries and have a close identification with them as historic regions of Greater Hungary centred on the Carpathian Basin. Their separation, endured since the Treaty of Trianon, 1920, has reinforced their sense of regional identity and been further embedded by constant disputes with new national majorities. They have also a mature experience of independent political and group cohesion in regions outside present-day Hungary. Hungarian regional consciousness is at its strongest in Transylvania, Romania, where it clashes with an equally strong identification among Romanians. The Hungarian government sup-ports a television station, Duna Television, whose primary function is to pro-vide culturally based programmes to help maintain a strong sense of ethnic identity among Hungarian communities living in surrounding countries. The Hungarian government also supports a standing conference for delegates of Hungarian organisations of civil communities that have parliamentary or regional mandates in neighbouring countries. Furthermore, in Spring 2001, the Hungarian parliament debated a new 'Status Law' that would give wide-ranging support to Hungarians living in nearby countries, whilst restricting their residence rights in Hungary. The aim seems to be to maintain a presence in a wider Hungarian culture area.

Representing regional differentiation as EC/EU models

Regional differentiation on the basis of EC/EU models is complex. Because the governments of most Marchland states were keen to become members of the European Union, the Commission's stance put upon them the need to harmonise their regional systems with the EU NUTS system and to create a regional development and environmental policy that conformed to

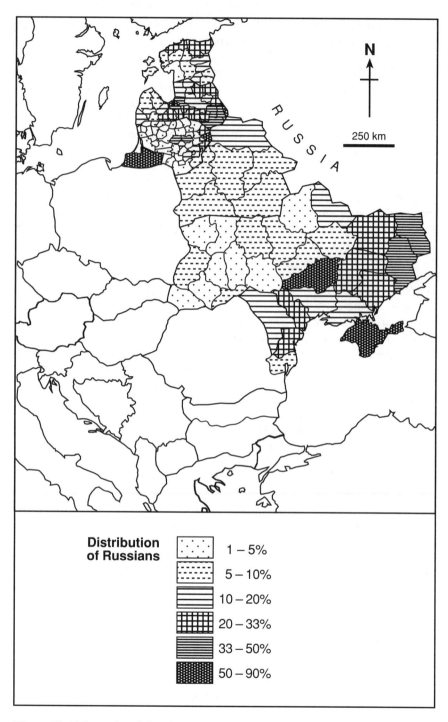

Figure 13.4 The regional distribution of Russians in the Marchlands.

Source: Adapted from Pándi (1997) *Kösztes-Európa*, Osiris, Budapest.

EU standards. NUTS is devised for current statistical purposes and has three levels relating to the area and population of 'regions'. The states most likely to form the first wave of new entrants to the EU have included a strong reference to the need for regional harmonisation in the collection of statistics in their regionalisation policy. However, following the precedents of existing member states, aspirant members have moved relatively slowly towards harmonisation, yet at a different pace in different states, and maintained national traditions in regionalisation. The picture is equally mixed in respect of regional development and environmental policy, but applicant states have made generally very slow progress, especially in the area of environmental legislation. All are dragging their feet in these areas, because they prefer to prioritise economic performance.

Conditions that produced the strong, 'bourgeois' Euroregions within the EU are absent in the Marchlands. Regional economies, even where collapse has been relatively speedily turned around, are weak and the 'democratic deficit' undermines this Euroregion model. Some new states, however, such as Slovenia, may soon conform to this model type as a single region. In contrast, the conditions that produced cross-border co-operation in the west are widespread in the Marchlands. The communist legacy of deprived investment was most extreme in border regions. Closed borders not only condemned border regions to a meagre share of scarce investments, but also prohibited attempts to ameliorate problems arising from communist forms of economic development. Some of the most horrendous environmental conditions were concentrated in border zones, the most devastating located in the 'Triangle of Death' around the junction of the Polish, GDR and Czechoslovak borders. The continuing saga of the conflict between Hungary and Slovakia over the Bös–Nagymaros Dam Project is another such example.

Since 1989 some regional élites perceived their plight of economic backwardness, peripherality and marginality inherited from communism as accentuated by marketisation. They have sought, with the technical and practical assistance of EU programmes, to establish institutional networks to promote socio-economic development in cross-border co-operation. Border regions are peripheral to state organisation, underdeveloped and have more than their share of pollution. Cross-border ethnic settlement adds a further potential for co-operation. Cross-border co-operation has caught the mood of the Marchlands. The term Euroregion has taken on a generic sense in some press and academic circles signifying any kind of cross-border contact and co-operation (Figure 13.4).

Regional development ideas were promoted by the EU in the Marchlands within its PHARE and TACIS assistance schemes. Regional Development Policy and Cross Border Co-operation (CBC) were two elements of PHARE that promoted and gave practical help in transforming Communist Regional Planning into Regional Policy for a democratically based redistribution of resources. At the same time, and perhaps with more long-term significance, these ideas and practices are extending the idea of 'Europe of the Regions'. The LACE–TAP (Technical Assistance and Promotion for Cross Border Co-operation) – PHARE CBC scheme is within the Interreg II Programme extended to 2001. These efforts have been concentrated along the EU's

N

Baltic
Sea

ESTONIA

LATVIA

LITHUANIA

BELARUS

250 km

POLAND

10

9

UKRAINE

1

CZECH
REPUBLIC

8

7

SLOVAKIA

6

MOLDOVA

AUSTRIA

3

4

HUNGARY

SLOVENIA

2

5

ROMANIA

Black
Sea

BOSNIA

Adriatic Sea

YUGOSLAVIA

BULGARIA

ALBANIA

MACEDONIA

12

11

GREECE

1 Swedish–German–Polish Austrian–Czech Euroregion Corridor Euroregions: Pomerania, Viadrina, Spree–Neisse–Bohr, Neisse, Elbe-Labe, Erzgebirge, Egrensis: Bayrischerwad/ Bohmerwald	2 Slovenia 3 West Pannonia 4 Danube–Ipoly (planned) 5 South Pannonia 6 Carpathian (the Hungarian county of Szolnok is in both 5 & 6) 7 Tatry 8 Upper Silesia-Moravia	9 Bug 10 Nieman 11 Association of Rhodopy Municipalities 12 Nestos-Mesta ⬭ other cross-border co-operations

Figure 13.5 Euroregions in the Marchlands.

Source: Adapted from Dingsdale (1999b) 'Redefining Eastern Europe', *Geography*, vol. 84(3), 205–208.

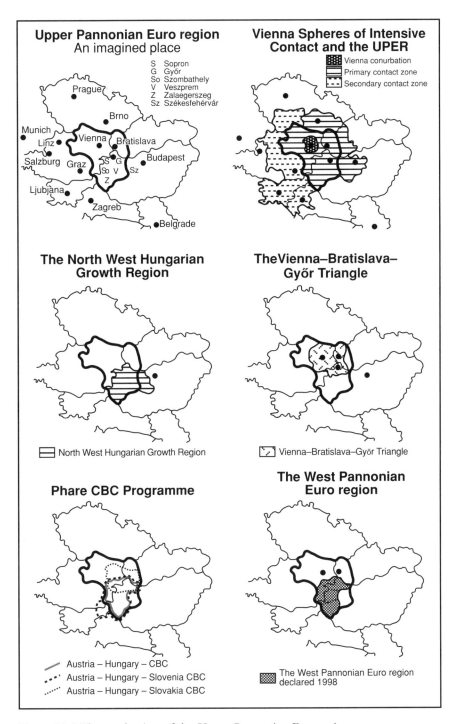

Figure 13.6 The production of the Upper Pannonian Euroregion.

borders with the Marchlands, but the CADSES programme of Interreg II/C enlarges the compass of the EU's imagined regional futures across the Marchlands.

Case example: The production of the Upper Pannonian Euroregion (UPER). A super region in the making?

All regions in the Marchlands have been produced by developments in the material practices as well as representational and conceptual discourses of democratisation and marketisation. There has been, first, an emphasis on issues of socio-economic development and standards of living. For some regions a second layer has been an ethnic cultural regional identity, primarily expressed in interest group or élitist politics, as a struggle within state and nation building. Whilst in others marketisation, sometimes layered by cultural identification, has been permeated by a third layer, an engagement with the European regionalist project, through the Euroregion concept. Region forming processes are at their most complex when these three layers are present together.

Drawing primarily on the work of the research team of the Hungarian Academy of Sciences, Transdanubian Research Institute (Nári and Rechnitzer 1999), I will explore the production of the Upper Pannonian Euroregion as a case example (Figure 13.6).

The Upper Pannonian Euroregion has been produced in the 1990s as an imagined region of development by a process of complex reciprocity between new transition practices and discourses. It has taken its place architecture from new material and spatial practices; new perceptions and representations of those practices as emerging new regional structures; and new conceptual regions of development. Key dynamics in the process have been:

(a) The opening of the Austrian–Hungarian border. This led to new cross-border contacts, the intensification of Vienna's sphere of influence, the economic development of western Hungary and the multiplication of contacts in the Vienna–Bratislava–Győr urban region.

(b) Austria joining the EU in 1995. This led to the introduction of EU concepts and practices of regional development, the strengthening of existing PHARE contacts that had begun in 1989, the introduction of new types of co-operation and the enlargement and defining of territorial organisation, leading to the establishment of the West Pannonian Euroregion.

(c) The next dynamic will be the entry of Hungary, the Czech Republic and Slovenia into the EU, when existing border constraints will be removed. All of the trends that will be discussed are likely to be accelerated by this step.

The centrality of Vienna

Vienna is at the heart of the idea of the Upper Pannonian Euroregion. Vienna has not only a primary sphere of dense contacts extending into western Hungary, western Slovakia and eastern Czech Republic, but also a secondary sphere extending into Slovenia and Croatia. There is a long-term development strategy to strengthen the city's primacy in Central Europe. The new societal and spatial contacts of the 1990s following the opening of eastern borders, particularly the Austrian–Hungarian border, confirmed the predominance of Vienna's competitive position over its potential international rivals, Prague and Budapest (Rechnitzer 1999).

Economic and market development: the opening of the Austrian–Hungarian border, the development of northwest Hungary and the Vienna–Bratislava–Győr Triangle

The opening of the Austrian–Hungarian border and the development of northwest Hungary

The opening of the Austrian–Hungarian border was the first breach of the Iron Curtain. There had been earlier building of cross-border contact, but it was after 1989 that personal, commercial and administrative contact multiplied rapidly, demanding new patterns of life from the borderland inhabitants (Nárai 1999). Northern Transdanubia became Hungary's most dynamic growth area outside Budapest. Foreign investors selected northwest Hungary and multi-national corporations invested heavily in such places as Székesfehérvár (Chapter 12), Győr and Szentgottard, but most districts enjoyed high levels of investment. Hungary's westernmost border counties were transformed from among the least developed to the most developed in the country. Austrian investments are particularly strong in the north west. Whilst Austria is only the third ranking investor country in Hungary as a whole, it is predominant in the most westerly counties. Local coalitions played an important part in creating suitable conditions for investment, setting out industrial and business parks and running promotional campaigns.

There was a rapid growth of indigenous enterprise and entrepreneurship. Women suffered a loss of rights that they had possessed in the socialist period and their position in the labour market was initially weakened. However, many women responded positively to the new situation emerging as entrepreneurs especially active in tourism, service and trade activities related to borderland location (Szörényi-Kukorelli 1999). The labour force drawn from agriculture has needed retraining to undertake new industrial work. However, unemployment has always been below the national average and some structural labour shortages are occurring (Csapó 1999). Cross-border commuting has become important in the labour market. In 1998 an interstate agreement came into operation regulating the amount and conditions of cross-border working and defined the territorial extent of the border region. Cross-border communication and co-operation between enterprises was, however, still quite low. Labour costs were the main advantage of Hungarian enterprises, but the quality and level of service offered by Austrian producers was noted as an advantage. Hungarian enterprises generally thought that new contacts were beneficial, but Austrian enterprises had some doubts. Such perceptions among the border entrepreneurs will need to be overcome by regular meetings of Chambers of Commerce and local government representatives. Few SMEs, Austrian or Hungarian, seemed to have much grasp of the implications of Hungary's joining the EU and were poorly prepared for the event (Dory 1999).

The prosperity of northern Transdanubia is attracting immigrants from elsewhere in Hungary. The immigration of younger individuals and families is balancing the indigenous natural decline and changing the population dynamics of the western border counties. Perhaps the newcomers will bring new attitudes overcoming the constraints to development that persist in the 'them' and 'us' attitudes shown by some long-standing residents on both sides of the border (Hardi 1999).

In this respect the regional press played a rather ambivalent role. The regional daily press showed an inclination to report border matters in the context of national and European events rather than in local terms (Izsák 1999). Only rarely were actual local events along the border reported. So, local residents were less well informed, from the press, with regard to local effects of the cross-border contacts than they were about the broader, especially regional, implications of the changes. Paradoxically, local isolation from wider matters was bridged, but local isolation from local issues was increased. It was a kind of regional territorial shift of local cultural meaning. At the same time cultural affairs relating to borderland transformations formed the largest single category of issues reported. The production of the wider regional consciousness was encouraged in cultural dimensions. In this sense the press helped to produce the wider region as a place of cultural identification.

The bias towards reporting from the national perspective was the source of a pessimistic shift in the press accounts of regional co-operation. From the beginning of the 1990s, meetings between officials from Austria and Hungary advancing progress in regional co-operation were reported. Following Austria's joining the EU there was a period of enthusiasm for co-operation. However, it soon became clear that there were numerous national conflicts of interest between Austria and Hungary with regard to Hungary's entry to the Union. The press view of regional co-operation was coloured by these. It regarded issues such as Hungarians working in the Austrian black economy, the spreading of crime and Austrians buying land in Hungary as needing solutions at national level. Though some of these questions had a particularly regional and local resonance, they were reported in national terms and a period of coldness towards co-operation became prevalent in the press (Izsák 1999).

Austrian–Hungarian contacts have promoted economic development. Contacts with Slovakia and Slovenia have been more limited, putting a brake on the pace of economic development despite all the countries participating in the Central European Free Trade Association (CEFTA). Slovakia has recovered only weakly from economic collapse. Its strongly nationalist government, in power until late 1998, introduced democratic and economic reform rather slowly and was accused of 'cronyism' in its privatisation policy. Economic development in the border regions of Slovakia–Hungary was constrained by political and cultural problems. There was political conflict over the Bös–Nagymaros Dam Project, a scheme intended to generate electricity, regulate river floods and improve navigation on the Danube, but widely criticised in Hungary as an environmental disaster. Slovak discrimination against its Hungarian minority soured relations until 1999. The economic policies of the Slovak government also slowed foreign investment, but Austrian investors were prominent.

The economic policies of the Slovene government constrained foreign investment with privatisation practices designed to encourage domestic acquisition of privatised state assets. Slovenia's economy developed strongly even without foreign investment, but was further strengthened when foreign investment was permitted following EU prompting. Slovenia became more outward looking, and joined the Visegrád Association and CEFTA to the benefit of cross-border contacts.

The Vienna–Bratislava–Győr Triangle Urban System

The centrality of Vienna to the UPER was touched on above. There has emerged a strong axis of development between Vienna and Budapest that has been perceived as an 'innovation zone', and within it the Vienna–Bratislava–Győr area has been called a 'golden triangle' (Rechnitzer and Döményová 1998). The cities are located within 100 kilometres of each other.

The transformation of northwest Hungary, noted above, has been especially intense in Győr-Sopron-Mason County, which is now the most dynamic county in Hungary after Budapest. There have been particularly high levels of foreign investment and a rapid privatisation of state enterprises, accompanied by the establishment in Győr of the regional headquarters of the principal banking, insurance and commercial enterprises. Bratislava as the capital of Slovakia has been the country's main centre for foreign investment, localising $2.5 bn and concentrating the rapidly developing service sector activity. One in five of Bratislava's residents is a graduate and there are very high activity rates. A high proportion of professional employees contribute to making Bratislava's GDP per capita almost four times the Slovak average and one of the highest in Central Europe. Economic development in Vienna and eastern Austria has been stimulated by increased contacts with the Czech Republic, Hungary and Slovakia. This happened particularly after 1994 as Austria emerged from recession, and economic recovery in Hungary, the Czech Republic and Slovakia gathered pace. Austrian joint ventures, industrial and commercial investments increased cross-border contact and trade agreements negotiated by EFTA came into effect (Barisitz 1996).

Transport infrastructure has been improved. The cities are linked by recently completed motorways and EU funding has been used to upgrade the Vienna–Budapest railway line to permit speeds of up to 160 km/h. The completion of the Rhein–Main canal linking the Danube to the Rhine has benefited the river port traffic of the cities. However, the Bös–Nagymaros Dam Project intended to improve Danube navigation was criticised. Vienna's international airport, situated to the east of the city, is more easily accessible from Bratislava and Győr than are either Prague or Budapest airports. Personal experience confirmed that new Hungarian practices on the Austrian border, introduced in March 2000, have speeded up cross-border passage.

Administrative development and institutional co-operation

Co-operation between the national governments of Austria and Hungary, who formed the Austrian Hungarian Development Committee, was supplemented by county and local government co-operation. The provincial government of Burganland on the Austrian side and the county councils of Győr-Sopron-Mason County and Vas County with the city governments of Győr, Sopron and Szombathely on the Hungarian side founded the Frontier Regional Council in 1992. The Province of Lower Austria, Vienna, Slovakia and Győr-Sopron-Mason participate in the working Committee of Danubian Provinces. Official co-operation has been established between the towns of Sopron, Wiener-Neustadt and Eisenstadt and between Győr City and Oberpullendorf and between Szombathely and Graz. Meanwhile the Hungarian government responding to the EU requirement for regional harmonisation formed the counties of Győr-Sopron-Mason, Vas and Zala, each equivalent to NUTS II areas into the NUTS I Western Transdanubian Region. These developments are underpinned by the emergence in northwest Hungary of sound institutional governance. Local government and county government are playing an important part. The northwest has proved to be the most politically stable area of Hungary showing higher turnouts than other regions and consistent voting patterns in the three general elections that have been held since 1990 (Kovács and Dingsdale 1998).

The PHARE CBC: the mirror of Interreg II

Whilst opening up to the international market place and co-operation between multi-level officials were taking shape, a third mechanism was also driving development forward. The PHARE CBC programme for the Austrian–Hungarian border region covered the Austrian Province of Burganland and the Hungarian region of West Transdanubia. The programme and its projects won 35 m ECU funding and were designed to assist five principal categories of development. The first covered studies for regional and spatial planning. The second permitted improvements in technical infrastructure, such as river and air transport, communications and public utilities. The third was concerned with economic development, establishing industrial parks, and promoting trade co-operation and tourism. The fourth was designed to assist human resource development including vocational training and school and university co-operation. Finally there were projects to promote environmental protection and nature conservation (Hörcher, 1998).

Two small additional programmes were also funded. They explored tri-national co-operation possibilities. The PHARE CBC Hungary–Slovenia–Austria extended the Austrian–Hungarian regional programme into the western region of Slovenia as far as Maribor. The Hungarian–Slovakian–Austrian programme covered the northern section of the Austrian–Hungarian programme but extended the area north west to Vienna and north east to the western region of Slovakia.

The foundation of the West Pannonian Euroregion

The foundation in 1998 of the West Pannonian Euroregion was the unifying culmination of the strands of administrative development. With increasing administrative contacts at all levels, the NUTS I regions of Western Transdanubia in Hungary and the Burganland in Austria, which had been combined to create the PHARE CBC programme, formalised co-operation, on the Euroregion model, as the West Pannonian Euroregion. With the entry of Hungary into the EU, the common border between them should be removed and more opportunities created for integrated development. Harking back to 1920, the Burganland dispute will be recalled (Chapter 5), but this time in a spirit of association and co-operation.

My discussion of the production of the Upper Pannonian Euroregion has pointed to specific places of development as transformation. Vienna's centrality to the Central European urban system, hampered before 1989 by international relations, has been stimulated and consolidated by new societal and spatial practices following the opening of borders. However, as the discussion in the following chapter will show, Vienna does not show world city formation trends as strongly as Budapest (Taylor 2001). The rapid economic restructuring of northwest Hungary has been driven by its synthesis of cross-border enterprise, investment and institutional co-operation, political stability and promoted welcome to global investors. A regional consciousness was encouraged by the press. The intensification of contact in the Vienna–Bratislava–Győr urban region has been particularly noted as a perceived consequence of new and variegated contacts. The regional concepts and good development practices of the EU have been grasped by local initiative, and are exemplified in the formation of the West Pannonian Euroregion. Thus the Upper Pannonian Euroregion is a place of the 1990s, it is an imagined place, as yet without a fully fashioned unity of structural, institutional or functional coherence. Whilst it exists as a vision of the future, it is still in the course of construction from diverse perceptions and discourses that engaged a myriad of emergent experiences of new societal and spatial practices.

Conclusion

Transition has been produced as a range of regional transformations. Regions are being produced in newly negotiated multi-scalar, socio-economic and cultural development relationships by contested regionalisation and regionalism, but demand regional initiative. Regions have been experienced as syntheses of diversified socio-economic collapse and dynamism, revived cultural identification and extensions of EU paradigms, anticipating Europe of the regions. They have been perceived through the lenses of theoretical models and empirical studies. They have been imagined places, as projected future alternative development scenarios. They are mirrors of competing neo-liberal and inherited spatialities.

14 The production of states in transition

Modernity, place and the Neo-liberal Project are at their closest expression in the idea of the Public Limited Company (PLC)-state. This chapter looks at the form and process of modernity in the Marchlands through the Neo-liberalist Project's attempt to capture space in the concepts and practices of the PLC-state, the 'hollowed' welfare, nation-state. How could a neo-liberal state emerge from a Party-state? With what alternatives might it compete? It concentrates on the economic imperative in the context of nation building and state building in the formation of place. It first considers the way that the boundaries of the new states were constructed. It then examines the cultural, societal and locational practices, perceptions and conceptions that were used to modernise the new states and the uncertainty of the neo-liberalist state as the outcome. The idea of the state is produced at a different scale on the register of multi-scalar place relationships from localities or regions, but its synthesis and othering are part of the same differentiating processes of spatial modernity.

The idea of the PLC-state

The PLC-state is the state where business activity is individual, entrepreneurial, driven by free capital markets; where a competitive, bidding culture is the dominant mind set; where public services are delivered by private enterprise; where governance is entrepreneurial and governments see their main task as providing the best possible institutional framework for private enterprise; where individualism is universal, there is no such thing as society; and the state, its localities and regions are commodified, packaged for presentation, cartoned for consumption. The boundaries of governance and power are blurred. Does or did such a state exist? Not in lived experience and therefore not perception. Certainly in conception it existed when, as British Chancellor of the Exchequer, Kenneth Clarke discussed UK Ltd; and probably as a Britain of utopian representation in the mind of the former British Prime Minister Margaret Thatcher.

Whether or not the PLC-state would flourish, transforming the Party-state and the nation-state, is an open question. What is not in doubt is that the most obvious fact of the Marchlands in the 1990s is that there are more states. The eight countries of Eastern Europe have been joined by new ones with the collapse of the three communist federal states, the Soviet Union, Yugoslavia and Czechoslovakia. The collapse of the Soviet Union created, for the Marchlands, six new states – Estonia, Latvia, Lithuania, Belarus, Ukraine and Moldova. The death of Yugoslavia gave birth to five new countries – Slovenia, Croatia, Bosnia-Hercegovina, Macedonia and the new Yugoslav Federation. The 'velvet divorce' of Czechoslovakia created two new countries. With the exception of Yugoslavia this process was relatively peaceful. The most remarkable aspect, however, is that territorial boundaries have remained in the same alignment, they have just changed their status. The new countries have been created by adopting boundaries that were earlier republican boundaries within the dismantled federal states.

Each country, old and new, was now independent, with responsibility for its own economic and political system, its own aspirations for international relations and its own sense of cultural identity. Each country was responsible for shaping its own internal geography and finding its position within continental and global space. The visionary ambition of all the reformers was to introduce democracy, the rule of law, civil society and the market economy. Each of these demanded the construction of complex institutional relationships not present in the communist system. Whilst this conception was, perhaps, highly laudable the reality was a maelstrom of constraints imposed by the legacy of communism and the plethora of conflicting examples brought by the return of the west. These conditions meant a strategic institutional context of democratisation and marketisation forming policies in a new spatiality upon, and from the remnants of, various forms of Party-state. The specifics of each country's situation were therefore different, but most élites faced, or perceived that they faced, in different combinations, three pressing issues. First, the need for state building. Second, the need for nation building. Third, the need to recover from economic collapse, by building a market economy.

Communist antecedent conditions drew a broad line of difference between the countries newly independent from the Soviet Union and the others. The Soviet Union had functioned in two ways at the same time. It had been on the one hand an empire and on the other hand a totalitarian state (Motyl 1998). In the empire the Russian centre controlled a non-Russian periphery. Central Russian organisations ruled Russia and the whole Union, whilst non-Russian organisations, dependent on the centre, administered the republics. At the same time the totalitarian state controlled public life. The successor countries therefore were neither states – just bits of a state – nor democratic, civil societies. The other federations, Yugoslavia and Czechoslovakia, were not imperial in character. Elsewhere the old states were Party-states, as discussed in earlier chapters. Within these states old élites remained in different degrees attached to the ideas and practices of the Party-state.

The return of the west brought a conceptualisation of the state as a nation-welfare-state that had been developing within the neo-liberalist agenda. This meant that recent policy-making had been situated in practices and conceptions within neo-liberal ideas that have been perceived as an attack on the state, referred to as 'hollowing'. The hollowing of the

Western European state embraced three primary challenges. First, there was the challenge from global capitalism, promoting the state's withdrawal from the ownership of enterprise and regulation of market activities, together with entrepreneurialism in providing public services. Second, there was the challenge from the EU, participation in which implied a supranational perspective that seems set for enlargement into the Marchlands. It is certainly preceded by the influence of its conceptualisations and practices. Third, there was the challenge of regionalism rooted in a strong sense of regional consciousness, sometimes reinforced by ethnic identity. In Chapters 12 and 13 I discussed the outcomes of these processes in localities and regions as important place conjunctions in multi-scalar place relations.

Democratisation and state building

Constitutions: presidential versus parliamentary

A first step in state building was the drafting of a new constitution. This was of key importance because it set the framework for legal and regulatory mechanisms for democratic societal and cultural relations. The élites had to construct new constitutions either by adjusting the old ones or drafting completely new ones. In some cases interim constitutions had been drafted at the time of the short struggles for independence and more permanent ones introduced when full independence was achieved.

In the democratic aspect of constitution drafting, an important distinction has been made between parliamentary or presidential styles, because this sets out the basic structure of élite power relations. Dostál (1997) cites Linz and Stepen's assertion that a parliamentary system is more advantageous than a presidential system in laying the foundations for the consolidation of democracy because of its efficacy, its requirement for the practice of forming majorities and its capacity for terminating crises of government before a crisis of régime. Presidentialism, it is claimed, reduces the need for multi-party systems that engender disciplined political structures. In short, parliamentary systems, at least under immediately post-communist conditions, are potentially more politically stable than presidential systems. Dostál classifies the constitutional status of the post-communist states into presidential, mixed and parliamentary (Table 14.1). In the Marchlands there are representatives of each type, with parliamentary systems prevailing. Dostál, citing Easter, points to the tendency for presidential systems to be preferred by old élites as a means of holding on to state power in the process of régime breakdown. It is certainly true that the five states adopting presidential systems in the Marchlands had a strong continuity of old élites or strong individual leaders drawn from old élites, whilst those adopting parliamentary systems had more representation from new élites, often drawn from dissident circles. The particular system chosen reflected the struggles of the élites during the process of achieving independence.

The constitutions were strongly centrist in tone. The new élites wanted to establish strong central administrative and political institutions in what were

Table 14.1 Constitutional styles

Parliamentary	Mixed	Presidential
Czech Republic	Poland	Romania
Slovak Republic	Lithuania	Belarus
Hungary	Macedonia	Ukraine
Bulgaria		Croatia
Estonia		Yugoslavia
Latvia		
Albania		
Slovenia		

Source: Dostál (1997).

wastelands of institutional forms and frequently personal integrity too. Whilst the old states had well-established central institutions and their principal task was to ensure a transformation to democratic forms, the new countries were extremely insecure as polities, especially the successor states to the Soviet Union. Ukraine, the most populous new state and territorially the most extensive, faced the emergence of regional élites that merged economic and political control and seemed to have criminal style associations using assassination in their competition for power. These regional 'clans', fuelled by corrupt practices among officials and by gangsterism, proved stronger than the political centre in Kiev preventing the government from establishing a strong judicial system or exerting political control (Varfolomeyer 1997). All the new states faced the danger of 'Zaireisation', but this was especially serious in Ukraine (Motyl 1998). Consequently, Ukraine did not finalise its constitution until 1996, much later than other states and then only after the president called for a referendum on independence.

Democratic transition and the consolidation of democracy

I noted in Chapter 13 the absence of democratic representation at a 'regional tier' of governance consequent, in part, on the perceived need for a strong centralised system. The discourse on democratic transformation in the Marchlands after 1990 draws on the discourse developed for the transformation of the southern European dictatorships to democracy in the 1970s. In particular, the concepts of 'democratic transition' and 'democratic consolidation' inform the discourse on democratisation. The former carries the idea of establishing institutions, whilst the latter means the permeation of democratic practices and ways of thinking throughout society. It is generally accepted that Hungary, Poland, the Czech Republic and the Baltic States have made sound progress to consolidated democracy by the turn of the millennium. They have held regular elections, changing governments in peaceful transfers of power, and political élites have accepted multiparty oppositional politics and judicial and media independence. The same could not be said for Slovakia, Serbia or Croatia where government politics remained largely founded on asserting ethnic national collectivism. The situation improved in Slovakia after the defeat of Prime Minister Mercier in 1998 and in Croatia when a coalition of the previous opposition parties won the presidential and parliamentary elections following the death of President Tudjman in 1999. The ousting of President Milosovic from government in Serbia in late 2000 may offer hope of a more democratic future for Yugoslavia. In Bosnia the role of the UN High Representative has stifled local political development and the same may become true of Kosovo (Bidaleux 1999). However, Agh (1999) sees the Central European states, Hungary, Poland and the Czech Republic, despite different starting points, as having reached the brink of democratic consolidation in the late 1990s. Full democratic consolidation will be the task of the next ten years. Ulran and Plasser (1999: 128) extending the field to Ukraine, Romania and Bulgaria echoed this conclusion, 'democratic attitudes have taken hold, with country specific variations, among a sizeable majority within each country', but they also mark signs of problems in cultures of policy and process.

Bidaleux (1999) points to Linz and Stepan's view that democratic consolidation has been impeded by a 'civil society versus state' discourse in Central Europe. So deep was the suspicion and mistrust of the Party-state embedded in the popular consciousness that forming democratic parties and democratic states would be a far from easy task. Contrasts have been drawn between the idea of 'truth' in ethical civil society and 'interests' in democratic political society. Whilst ethical civil society draws on local traditions that shun authoritarianism, it does not have a connotation of liberal democracy. There is no tradition of any kind of sustained democratic experience across the Marchlands on which the transformation processes could draw.

State building and new external contacts – resituation in international place relations

The constitutional arrangements, democratic transition, the consolidation of democracy, the emergence of civil society and the peaceful transfer of government in Marchlands countries were key 'internal' processes that formed the fabric of state building. The new states, as a measure of statehood, sought to identify and resituate themselves in the broader international arena. The governments reconstructed political, economic, military and cultural international relations. The collapse of the CMEA and the Warsaw Treaty Organisation faced the successor states with complex negotiations over economic and military assets. They have sought membership of continental and global organisations such as the European Union, the Organisation for Economic Co-operation and Development (OECD), the World Bank and the International Monetary Fund, the North Atlantic Treaty Organisation (NATO) and the Council of Europe. The new states had to develop diplomatic relations in their own right, establishing embassies and receiving ambassadors from other states in the international system and joining the United Nations. International recognition was often not gained easily and conferred an essential attribute of statehood. Not all have been accepted into these organisations.

New organisations have been founded by the states themselves in international co-operation agreements. Among these Belarus, Ukraine and

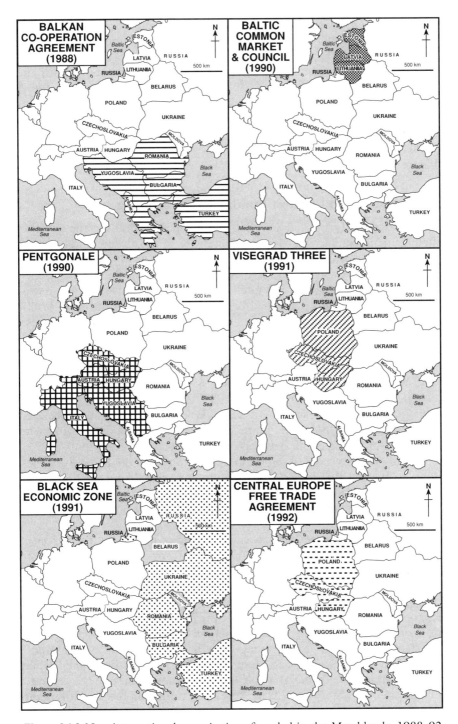

Figure 14.1 New international organisations founded in the Marchlands, 1988–92.

Moldova signed up to the new Russia centred Commonwealth of Independent States. Hungary, Poland and Czechoslovakia founded the Visegrád Association in 1991 and the Central European Free Trade Association (CEFTA) in 1992. The original Visegrád Three became Four with the Czechoslovak 'velvet divorce' of 1993. Later Slovenia joined the CEFTA and there are other applications pending. Romania, Bulgaria and the Baltic States are keen to join. Other newly developing organisations are the Balkan Co-operation Agreement (1988), the Baltic Common Market and Council (1990), the Pentaganale (1990) and the Black Sea Economic Zone (1991) (Figure 14.1).

State building and nation building: nationalising states

Closely connected with state building and democratisation was the question of ethnic nationality to which fleeting reference was made above. The struggles for independence give some clue as to the process of state formation. In all of the new states the primary driving force behind the popular fronts, mobilised by the local cultural intelligentsia, leading independence movements against the Party-state, had been national self-determination. Whilst there had been demands for economic restructuring, political openness, personal freedom and environmental improvement, the primary rallying call had been around the idea of national independence. The symbols of opposition to the Party-state had been national symbols, ranging from the national flags, to such relics as the right hand of St Stephen, founder of the Hungarian state in AD 998. I witnessed, in 1989, among a huge crowd the first public parade of this relic for fifty years. Where war occurred, it occurred over mismatches of politically bounded and ethnically settled territory and had aimed at forming ethnic national exclusive occupancy. Zaprudnik (1993), contrasting the slow pace of reform in Belarus to that of the Baltic Republics, points to the fact that in the Baltics communists like democrats believed in national culture and the need for national independence; in Belarus the Party conceived the Republic as a province of the Soviet Union and opposed nationally inspired change.

In defining 'the other' against which to mirror nationality the successor states of the Soviet Union had little difficulty in putting the role onto Russians. The Baltic States were particularly ready to do this and they all, in different ways, excluded Russians from the new state's institutions. In Lithuania and in Moldova there had been a tendency for the ethnic minorities to side with the old régime against the dominant nationality's pressure for reform and independence, fuelling national exclusivism. In Yugoslavia the new state's élites could cast any number of ethnic groups in the role of 'other'. For the Slovaks, the Czechs and Hungarians were ready-made 'others', whilst the Hungarians already had this role for Romanians because of the centuries old dispute over Transylvania. The Bulgarians turned on the Turkish minority against whom to emphasise otherness. For Hungarian nationalist elements support for the Hungarian minorities outside Hungary, especially those in Slovakia and Romania, but also those in war-torn Yugoslavia, defined the 'other'. Faced with assertions by S. Milosovic about the status of borders in Yugoslavia, the late J. Antál, first post-communist prime minister of Hungary, reminded him that all borders may be subject to review, echoing Hungarian attachment to the territory of Great Hungary lost at Trianon.

Jordan detected nationalist tendencies behind the strongly centrist constitutions adopted by the new states. Romania had the most explicit centrist and nationalist phrasing, declaring the state to be 'a unitary, indissoluble nation state, the sovereignty of which is the right of the Romanian nation' (Jordan 1998: 266). Jordan noted similarly explicit statements in the Latvian and Lithuanian constitutions. Against this, most constitutions were less explicitly nationalist. Some appeared to promote a multi-ethnic citizenship. Ukraine's, for example, declaring it to be 'the state of the whole population of all nationalities'. He also found nationalist leanings among the statements on the rights of ethnic groups. These were further emphasised by political practices. I noted in Chapter 13 Jordan's view that regionalisation had been designed to separate ethnic clusters administratively and prevent ethnic regionalism in politics. In the Baltic Republics, citizenship rules have been drawn up that emphasise nationality. These circumstances led Jordan to state that only Hungary, the Czech Republic, Slovenia and Bosnia-Hercegovina had been able to cast off a strong sense of the nation-state. As noted also in Chapter 13, Hungary was the only country to devolve democratic representation to the second tier authorities, elsewhere devolution to local self-governments had created very weak local institutions. Moldova represented a particular case, where ethnic minorities had been able to force a reluctant dominant, Romanian, ethnic group to recognise the autonomy of the Ragausian Republic (Ragausians) and the Dnester Republic (Russians).

As noted in Chapter 5, the idea of the nation-state was first adopted after the First World War. It was imported from the west, adapted and imposed by relatively small bourgeois élites in post-imperial conditions. Many new élites followed historical precedent because they too perceived themselves to be in a post-imperial situation. For some new countries, such as the successor states of Yugoslavia, 1990 offered an opportunity to succeed in gaining national independence when they had failed in 1920. The new élites of the Baltic States were powerfully conscious of the illegality of the Soviet annexation of their territory under the Stalin agreements of 1944. These were conditions that certainly would put an emphasis on nation building by élites, whether embarked upon with genuine or cynical intent. Historical memory remained strong too, in the sense of defining national borders and homelands, where boundaries had been fluid and changeable.

Powerful current reasons conspired with these memories of history to forge a strong sense of national identity that had been supressed under communism. In the first years of independence as small states with fragile polities and economies, a sense of national cohesion was a source of strength. At the same time the heritage of the Party-state had not prepared the élites or the general public for the practices of democracy or civil society. The general grasp of the meaning of these ideas was far from complete. Under these conditions a tendency to overzealous nationalism and centralism is readily understandable.

The west wind of neo-liberalism

The purveyors of the PLC-state, or at least bits of it, stepped down from the aeroplane in the capitals of the Marchlands before the ink was dry on the death certificate of communism. The Neo-liberal Project, its practical advocates and its theoreticians of modernisation, encompassed in what has been called 'transitology', offered a stock of policies as the optimum means of creating the conditions for capitalist accumulation (Smith 1997). This would provide the Marchland states with macro-economic measures such as fiscal austerity, privatisation, liberalisation and reduction in public expenditure, which it was claimed would set them on the road to becoming modern neo-liberal states. In the debate that surrounded this set of policy practices, issues such as the pace of change, the timing of policy implementation and the type of privatisation were central questions. The approach can be seen as imposing a set of standard 'outsider' practices and ideas that took little or no account of the particular societal relations of the transition. According to some authors, it does not therefore provide either a useful empirical account of the economic processes underway in the Marchlands or a strong theoretical understanding of them (Smith 1997). However, as Smith concedes the myriad of local and regional economies, and the varied emergent forms of capitalism, are increasingly subject to the imposition of formalised neo-liberal discourses conducted by global institutions. The World Bank, the IMF and the EU are successfully imposing their particular form of regulatory order. The regulatory practices and discourses that make up that order have their principal resonance as mediated through the state policy apparatus.

Narratives of economic collapse, recovery and growth

The perception of different countries, as written in the narratives of major economic indicators, recounts contrasting experiences in transition. I will pay attention to just four main themes by way of illustration – first, GDP change and then FDI, privatisation and economic growth. However, inflation, currency rate policies, unemployment, industrial output, social security, economic activity rates and many other policy fora are also integral dimensions of the story.

Economic growth as GDP change

Charting economic collapse, recovery and growth by means of annual change in GDP shows how the experience of each country has been different (Table 14.2). Annual GDP change reflects several short-term features of transition. Among these the most important are: existing levels of economic development; depth of association with the CMEA; the differential dates of economic collapse associated with the collapse of communism and associated events; and reorientation of trading and trading alliances. The westernmost states were most separate from the CMEA trading bloc and experienced a less profound collapse. They collapsed earlier and embarked

Table 14.2 Changes in real GDP 1999–2000

Country	1991	1992	1993	1994	1995	1996	1997	1998	1999	2000
Albania	−27.7	−7.2	9.6	9.4	8.9	9.1	−7.0	8.0	8.0	7.0
Belarus	−1.2	−9.6	−7.6	−12.6	−10.4	2.8	11.4	8.3	1.5	1.5
Bosnia	−20	n/a	n/a	n/a	21	69	30	18	12	n/a
Bulgaria	−11.7	−7.3	−1.5	1.8	2.1	−10.1	7.0	3.5	0.0	12.0
Croatia	−21.1	−11.7	−8.0	5.9	6.8	6.0	6.5	2.3	−0.5	1.0
Czech Republic	−11.5	−3.3	0.6	3.2	6.4	3.8	0.3	−2.3	0.0	1.4
Estonia	−13.6	−14.2	−9.0	−2.0	4.3	3.9	10.6	4.0	0.0	4.5
Macedonia	−7.0	−8.0	−9.1	−1.8	−1.2	0.8	1.5	2.9	0.0	3.0
Hungary	−11.9	−3.1	−0.6	2.9	1.5	1.3	4.6	5.1	3.0	3.5
Latvia	−10.4	−34.9	−14.9	−.6	−0.8	3.3	8.6	3.6	1.5	3.5
Lithuania	−6.2	−4.3	−16.0	−9.5	3.5	4.9	7.4	5.2	−2.5	3.5
Moldova	−17.5	−29.1	−1.2	−31.2	−3.0	−8.0	1.3	−8.0	−5.0	−1.0
Poland	−7.0	2.6	3.8	5.2	7.0	6.1	6.9	4.8	3.5	5.2
Romania	−12.9	−8.8	1.5	3.9	7.1	4.1	−6.9	−7.3	−4.0	1.0
Slovak Republic	−14.6	−6.5	−3.7	4.9	6.9	6.6	6.5	4.4	1.8	2.0
Slovenia	−8.9	−5.5	2.8	5.3	4.1	3.5	4.6	3.9	3.5	3.5
Ukraine	−11.6	−13.7	−14.2	−23.0	−12.2	−10.0	−3.2	−1.7	−2.5	1.0

Source: ERBD Transition Report 1999.
2000 figures forecast Business Central Europe Annual 2000.

on recovery sooner. They achieved stability, followed by generally sustained growth, within which the imprint of the intensification of general and specific protectionist policies by the EU is registered. The economic collapse of the Soviet Union had a significant impact on them as it disrupted supplies of raw materials and energy. The spread eastwards took collapse to states more closely bound to the CMEA; their collapse was later, more profound and their recovery slower and more erratic. The Baltic States who vigorously sought to reorientate economic contacts found it difficult to break away from the geo-economic bindings of Russia and registered the Russian crisis of 1997 most noticeably in the sequence, but none of the Marchlands countries went unscathed. Only Hungary and Poland have been perceived, by the late 1990s, as being on the brink of sustained economic growth. This implies that their economies are encompassed by the world short- and medium-term cyclical patterns of economic change, signifying integration into the global economy. The other countries remain tied to specific transition forms of economic change.

FDI, privatisation and economic growth

The World Bank, most western advisers and local supporters, promote FDI as beneficial in economic restructuring, by providing much needed investment capital and transferring technical know-how. FDI, they argued, if properly utilised in privatisation programmes, can upgrade technology, improve management skills, overcome chronic cash flow problems and open up world markets, all of which promote economic growth. Those who oppose FDI perceived investors as foreigners who repatriate profits and stifle local business initiative. They argued that selling off economic assets to foreigners may result in countries developing low cost, low wage economies that risk having investment quickly withdrawn if events take a turn for the worse.

The conflicting arguments of this discourse left governments, privatisation agencies and foreign investment agencies with difficult policy decisions. How open should they be to foreign investment? How best could they tackle privatisation? How could the country's potential be best represented to investors? At the same time, how did foreign investors perceive the different countries? It is clear that foreign investors did perceive the different countries as having different strengths and weaknesses as targets for investment. Poland, for example, is highly regarded for its market, the Czech Republic for its strategic location and Hungary for its investment climate and labour force (Pye 1997). But this is pretty much how the countries have been marketed by the agencies.

The openness of countries to FDI and the level of interest shown by foreign investors resulted in very different amounts of FDI flows in the different countries, and very different timing of investments. However, FDI investment patterns generally reflected major privatisation sales. Vastly different total FDI figures are also somewhat ameliorated when they are measured against population, giving a per capita figure (Table 14.3). The story of FDI, however, very pointedly directs attention to the countries of Central Europe, who have left far behind the newly independent states within the CIS and the Balkans, with the small Baltic economies some-where in between. At the same time, FDI draws attention to the strongly European dimension of the return of the west. In all the countries of the Marchlands the sources of FDI were predominantly European. The European Union countries dominate FDI stock in 1998 in Marchlands states. Least in Moldova, where it accounts for 23 per cent, it rises to 87 per cent in FYR Macedonia (Table 14.4). Not shown on Table 14.4, because they lie within the EU category, are some macro-regional effects. For example, Greece (39 per cent) is the leading source of FDI in Macedonia; Sweden (32 per cent) is the leading source in Estonia; Russia (29 per cent) in Moldova. In contrast American investment (42 per cent) is greater than EU investment in Croatia; Kuwait (21 per cent) is the single biggest investor in Bosnia-Hercegovina, closely followed by Turkey (12 per cent) reflecting the Moslem contact. Croatia (17 per cent) reflects close political interests in Bosnia-Hercegovina. Germany alone has investments in all the countries.

Measuring FDI is notoriously difficult, but it is widely held by global insti-tutions to be beneficial to transition economies. Yet Hunge (1996) found no strong connection between FDI and economic growth. The amount of FDI received by individual countries did not correlate with the pace of eco-nomic growth. Different privatisation and FDI policies did result in very different ownership structures in different countries. However, these formal ownership structures were less important than change in corporate gover-nance, or adapting production and marketing to new demands. Neither was

Table 14.3 Foreign direct investment stock, 1998

	FDI stock ($bn)	Population (m)	FDI per capita ($)	GDP per capita ($)
Czech Republic	12.8	10.3	1,243	5,479
Hungary	20.0	10.2	1,961	4,730
Poland	35.7	38.5	925	3,887
Slovakia	1.8	5.4	333	3,793
Slovenia	2.6	2.0	1,300	9,779
Albania	0.423	3.2	132	930
Bosnia	0.604		144	1,089
Bulgaria	1.5	8.4	178	1,315
Croatia	2.5	4.8	520	4,820
Macedonia	0.242	2.0	121	1,548
Romania	4.2	22.5	187	1,695
Yugoslavia	2.5	10.6	236	1,205
Estonia	2.5	1.5	1,666	3,593
Latvia	2.1	2.4	875	2,622
Lithuania	2.2	3.7	594	2,890
Belarus	0.456	10.2	44	1,396
Moldova	0.330	4.3	76	432
Ukraine	2.6	50.1	51	846

Source: *UN World Investment Report 1999.*

Table 14.4 International sources of foreign direct investment (per cent), 1998

	CEE including Russia	European Union	Japan	Switzerland	USA	Kuwait	Turkey	Other	Total
Belarus	11	65	—	3	17	—	—	4	100
Bosnia-Hercegovina	23	30	—			21	12	13	100
Bulgaria	2	63	—	4	7	—	2	22	100
Croatia	3	41	1	6	42	—	—	8	100
Czech Republic[a]	2	81	1	2	6	—	—	8	100
Estonia	2	77	—	5	5	—	—	11	100
Hungary[a]	1	59	2	3	15	—	—	21	100
Latvia	13	53	—	4	11	—	—	20	100
Lithuania	3	57	—	—	16	—	—	24	100
Macedonia[a]	3	87	—	—	—	—	—	10	100
Moldova	32	23	—	6	19	—	—	21	100
Poland[a]	1	77	—	3	12	—	—	7	100
Romania	4	60	—	2	7	—	5	21	100
Slovakia	10	71	—	—	11	—	—	9	100
Slovenia[a]	15	75	—	4	5	—	—	—	100
Ukraine	7	25	6	6	18	—	—	43	100

Source: *World Investment Report 1999*.
a = 1997.

Satisfaction with transition and transformation

The makers of the negotiated revolutions were naively idealistic as regards the pace of change. They lacked experience of democratic politics and often mistook rhetoric for reality, institutional form for substance. The example they were offered from the west was hardly helpful. The general public were shocked by the experience of transformation. They had no experience of liberty and many individuals in private and public life interpreted it as licence. The re-election of reformed communist parties to government in the second round of general elections in several countries reflected the fear of disorder, the problems of individual responsibility and a reassessment of the late communist period against the new reality. Reformed communists accepted democracy, ruled accordingly and accepted defeat when it came. Many opinion polls have been taken in an attempt to assess the degree of public satisfaction with transition and transformation. Table 14.5 shows a snapshot of the situation in 1998. It suggests that whilst in most countries the new regime was preferred to the old, respondents are even more optimistic about the future. They also reject non-democratic forms of government. However, corruption is perceived to be much more prevalent than earlier.

Table 14.5 Satisfaction ratings

Country	Satisfaction ranking % approval		Corruption in 1998 relative to 1999 % of replies		
	Current regime	Communist regime	Higher	Same	Lower
Poland	66	30	51	37	12
Bulgaria	58	43	71	26	3
Czech Republic	56	31	70	24	5
Romania	56	33	84	13	3
Hungary	53	58	77	20	3
Slovenia	51	42	59	28	14
Slovakia	50	46	81	15	4
Belarus	48	60	70	11	5
Ukraine	22	82	87	25	2

Source: Bidaleux (1999).

there strong evidence that FDI promoted innovation. The picture depended far more on specific sectors, companies and countries (BCE May 1999). Different countries have adopted different policies towards research and development funding and this is important in innovation patterns. However, foreign investors do not tend to repatriate their profits, but reinvest them, suggesting that medium- to long-term benefits will emerge.

Reports in the business press suggest that restructuring takes time and an early start, enhanced by FDI in the early phase, may be a winning formula. It can build up a head of steam that influences foreign investor perceptions. The Hungarian government, the front runner in early FDI, committed itself to selling state assets to foreigners, but hedged this with an emphasis on new green field investments. Old company names were retained, but there were new companies behind them. In 1998 foreign owned companies produced 35 per cent of GDP and Hungary was closely integrated into European production networks. Hungary's exports, like those of the Czech Republic, were in high value sectors and many went to the EU (BCE 1999). FDI also flooded into Poland, but was less on a per capita basis than went into Czechoslovakia, Hungary and Slovenia. Poland's recovery was more particularly associated with domestic SME development and it still has potential for foreign investment in the privatisation of some major state assets. Some SMEs were linked into FDI as suppliers to major foreign corporations. However, Poland's exports of high value goods are relatively low and less EU orientated, so the country is more open to transition shocks and currency problems than either the Czech Republic or Hungary (BCE 1999).

Sequential development

In seeking to achieve democracy, rule of law and a market economy in conditions of societal and cultural turmoil, reformers and new government élites had a complex task. Not only would they encounter entrenched opposition from vested interests, but also the achievement of some objectives had to precede others upon which to some extent they depended. Change would be sequential because some desired ends preconditioned others (Motyl 1998). The relationship between objectives had an ill-formed logic that needed to be fashioned into ordered experience. I shall call this process 'sequential development'. Democracy and civil society could only exist where the rule of law was established. The rule of law implied some kind of state, though the form of the state was more open; states can exist without either nations or markets. Thus as Motyl (1998) points out in Ukraine, for example, state building and nation building were possibilities. State building was a necessary first step that nation building could help to consolidate by promoting popular cohesion and consensus. The same was equally true of other countries, but in all of them the meaning of nation and citizenship needed particular definition or democracy and civil society could be endangered. The implementation of markets might follow.

In these circumstances the clamour from western advisers for radical economic transition showed that they had little understanding of the real situation and little thought for non-economic conceptualisations of development and modernisation. They were blinded by their neo-liberalist world view. They were seized by enthusiasm bordering on evangelism, which was tinged by euphoria at the 'victory' of the west in the Cold War. From this triumphalist stance they promoted the neo-liberalist agenda. They could not see that their prioritisation of economic change amounted to a demand for, in Motyl's phrase, 'revolutionary voluntarism'. Such a demand was not a normal aspect of western practice and was, in post-communist conditions, generally impractical. Many promoters of the neo-liberal agenda were also blinded to the delicate relationship between the reform objectives. Bidaleux put it very gently, referring to:

> the imprudence of western advisors who try to give economic restructuring priority over the consolidation of democratic procedures which can help to control, regulate and confer legitimacy upon privatisation and marketisation and the new public and private enterprises to which such transformations give rise.
>
> (Bidaleux 1999: 32)

The logic of 'sequential development' was that western ideas would only become embedded in societal and cultural practices, perceptions and conceptions of the state as a place, if rule of law, democracy and civil society were established to give marketisation legitimacy.

I would like to make very clear that this concept of 'sequential development' is not in any way akin to the idea of stages of economic development that implies a set sequence of events. Indeed, the point is that each country had different conditions and so the experience of transformational sequential development was different too. It does mean that the development of democratic institutions and civil society proceeded quickly and effectively. It also means that the strides through transformation should not be judged primarily on economic criteria. In this respect the demands of neo-liberals to do this had to be rejected. Most moderate leaders were able to achieve what was possible in progressing to democratic and market institutions and withstand criticism that market reform was moving too slowly.

The institutional equation

The states of the Marchlands were, in the 1990s, emerging in practice and discourse as a product of institutional shift through democratising, nationalising and liberalising processes. Dostál (1997) conceptualised the early phase of transformation as an 'institutional equation'. The institutional equation examined the relationship between modernisation, which was a measure of antecedent levels of development; democratisation, which was a current process measured as constitutional selection and the consolidation of democracy; and economic liberalisation, which was a current process measured as internal and external practices of economic relations (see Box below). The institutional equation concerned two sets of relationships: (a) those between antecedent conditions and current processes; and (b) those between different current processes. It examined modernisation and democratisation as determinants of economic liberalisation to assess the claim that democratisation was crucial in mediating the effects of existing modernisation on economic liberalisation. Dostál's approach situated economic liberalisation within the maelstrom of inherited conditions and current processes. This allowed a differentiation to be made between states emerging from the Party-state on

The institutional equation

Concept	*Measured as:*
Democratisation	(a) A choice between presidentialism and parliamentarianism, with parliamentarianism posited as associated with new élites and presidentialism preferred by old élites seeking to retain state power.
	(b) Consolidation of democracy measured as respect for political rights and civil liberties as monitored by Freedom House Institute.
Economic liberalisation	Privatisation of large- and small-scale enterprises; enterprise reconstruction; price liberalisation and competition; trade and foreign exchange dealing; bank reform and securities markets; non-bank financial institutions; effectiveness of legal rules on investment.
Modernisation and development conditions	(a) Economic indicators: per capital GDP at purchasing power parity; share of industry in GDP; share of agriculture in GDP; level of urbanisation; earner dependency ratios of total population.
	(b) Extent of trade with CMEA.
	(c) Dominance of titular ethnic group.
	(d) Population size.

Specific values of these indicators were applied to the LISREL statistical model.

the basis of a synthesis of institutions. At the same time, it tested empirically one of the principal theoretical claims of the neo-liberalist agenda.

Dostál concluded, first, that democratisation had a positive effect on economic liberalisation; second, that the advantage of higher levels of inherited modernisation was mediated by the particular character of democratisation. Democratisation was good for economic liberalisation, but parliamentary democracy was especially good for the most developed countries. Dostál also concluded that the Czech Republic, Hungary, Poland, Slovakia, Slovenia and Estonia were close to the institutional standards of the European OECD states.

Path dependency

The concept of path dependency (North 1990) forms an intermediate model of change between the general model of antecedent conditions/current processes and the ideas of the institutional equation and sequential development. North was concerned with the relationship between institutions (the rules of the game), organisations (the teams in the game) and individuals (who make up the teams). North, thus, specifies three categories within the process of development, who as human agents I have earlier theorised as possessing different spatialities. Path dependence postulates that the development path of a specific institution, economy or society is shaped by (a) its historical, cultural and societal experiences and legacies and (b) contingent factors, working randomly, that favour the adoption of particular solutions to problems constructed by organisations and individuals. Change is usually incremental, but radical changes in formal institutions do not alter informal networks and historically driven attitudes, behaviours, values or perceptions held by individual and organisational actors and decision-makers that retain an influence in the short and long term.

The idea of path dependence is relevant to understanding the transition and transformation processes that are producing states within the Marchlands as part of the process of spatial modernity. The trajectory of each country, as I have illustrated, has been a particular synthesis of state building, nation building and economy building, which has been conditioned by the particularities of the interaction of current processes and contingent responses shaped by long-term historical values and experiences, and the more recent legacies of communism. The trajectory of the state as a place produced by societal and cultural particularities in the scalar register of place relations has been a primary factor in dispersing the effect of neo-liberal attitudes and values. The trajectory of the state has shaped the experience, perception and conceptions of transition and transformation as spatial modernity. There has been a divergence between the trajectory of states through transition and transformation that can be related to their path dependency. However, this divergence is also strongly related to their resituation in multi-scalar place relationships. This factor is understated in the idea of path dependency. In other words, path dependency offers interesting insights into the particularities of states, but undervalues locational dimensions of place synthesis and the effects of the multi-scalar properties of place.

Case example: The Yugoslav conflict: the problem of the map

The death throes of the Federal Republic of Yugoslavia were the birth pangs of five new states. The Yugoslav conflict was a very complex event. It was a civil war at several scales; an ethnic conflict between groups with a history of enmity; a struggle for political authority in a post-communist vacuum; a war of secession; a conflict along an economic divide; and a conflict along a cultural divide. Depending on how the relationship between these identifiable elements is structured, it is possible to perceive contrasting realities of how the event has unfolded and the relative importance of the different factors.

As I showed in Chapter 5, the idea of Yugoslavia was the idea of a South Slav state, but this failed and the first Yugoslav state was a multi-national kingdom that immediately faced ethnically based political problems. It was terminated by the Second World War. The second Yugoslav state was created after the Second World War and headed by Marshal Tito. It had two objectives: to change the self-identity of the ethnic nationalities into one of being Yugoslav and at the same time to get them to think of themselves as socialist. The two objectives might work in the same direction or they might double the difficulty of the task. Under the charismatic leadership of President Tito there was a kind of unity. After Tito's death the old centrist, devolutionist regional conflict resurfaced, this time within the ranks of the Yugoslav League of Communists. Throughout the 1980s the Republican Parties became increasingly powerful at the expense of the Federal Party. S. Milosovic, a Serb, first of all tried to gain control of the Yugoslav League of Communists in order to re-establish centralised power. Having failed to do this, he played the nationalist card and became the strongest challenger to the federal structure, ultimately leading to the downfall of the communist state. The conflict began in Kosovo where the Albanian majority wanted to regain the autonomy taken from the province by Milosovic when he first came to power.

A British newspaper as early as 1981 had carried a cartoon showing the break-up of Yugoslavia into its constituent republics, on the death of Tito. This was indeed prophetic because this is just what happened. The territorial integrity of the republics has been maintained as the Federation of Yugoslavia has disintegrated. The most likely reason for this is the federal constitution of 1974. This constitution, among other features, was aimed at forming a policy for the 'nationalities' of Yugoslavia. It recognised six nationalities – Serbs, Croats, Slovenes, Montenegrins, Moslems and Macedonians. The territory of the federal state was divided into six republics. Each nationality received its own republic. Serbia was the Serb Republic, Croatia the Croat Republic, Bosnia was the Moslem Republic and so on. Two 'minorities' were also recognised, the Albanians and the Hungarians. Each of these was given a territorial homeland, an autonomous region within Serbia. It was also recognised that the republics were not settled solely by one nationality. On the question of leaving the Federation, the right to secede was vested in the republic. However, if a national group forming a minority in the republic wished to stay in the Federation, it could declare its intention to do so (Englefield 1992). How this would be achieved was not made clear. It seems likely that those who drafted the Constitution did not envisage this happening. In such an event the only way to resolve the issue was by force of arms. This Constitution thus set up the structures for Brubaker's (1995) triadic relationship – nationalising state, ethnic minority, external homeland – within the Federation of Yugoslavia. The political practice in which it operated, however, was different. The territorialisation and place identity, not explored by Brubaker, were the vital elements of the dynamic.

The sequence of events in the disintegration of Yugoslavia is closely related to the problem of the map of bounded territory and settled territory. Figure 14.2 illustrates the complex mismatch of bounded territory and settled territory in Yugoslavia. It shows the key to understanding the process of

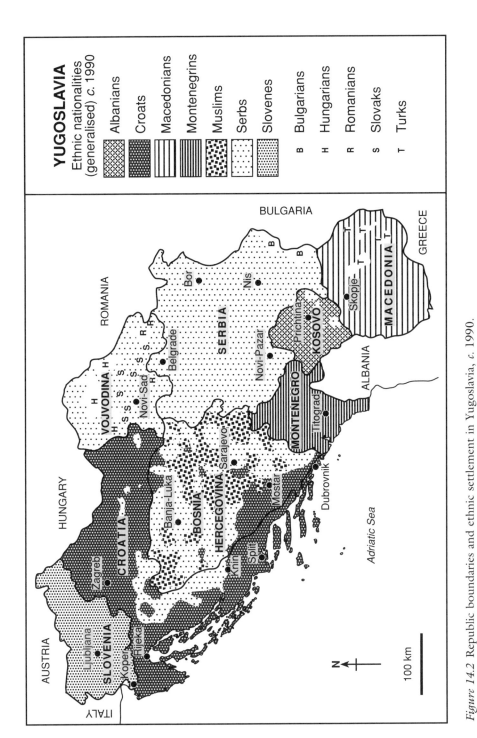

Figure 14.2 Republic boundaries and ethnic settlement in Yugoslavia, *c.* 1990.

Source: Dingsdale (1999d) 'The Problem is the Map', *Lock Haven University International Review*, 19.

disintegration. The first to leave the Federation was Slovenia. The secession resulted in just nine days of sporadic military conflict. The reason for this is demonstrated by the map. About 90 per cent of the population of Slovenia were Slovenes. They amounted to about 94 per cent of all the Slovenes living in Yugoslavia. In other words, the bounded territory of the Republic of Slovenia and the settled territory of the ethnic Slovenes were, more or less, coterminous. Add to this that very few Serbs lived in the republic and you can see that in the given political circumstances there was no problem with the map. The dynamic triad was incomplete; there was no significant ethnic minority.

Croatia was the second republic to secede. Here the situation was different. About 75 per cent of the inhabitants were Croat, some 11 per cent were Serb, the remainder belonged to a number of other ethnic groups. So, there was a mismatch between bounded territory and settled territory, but it is relatively simple. The Serb population was largely concentrated in the Krajina in the west with smaller groups in Eastern Slavonia in the east. In this situation when Croatia declared independence from Yugoslavia, the Serbs living in Croatia declared their independence from Croatia. In so doing, both groups could reasonably claim to be acting within the terms of the Constitution of 1974, the legal framework for events. Serbs could also point to the fact that the new Croatian Republic was denying them citizenship. The armed conflict in Croatia lasted for one year. The UN intervened and UN forces were deployed to create UN protected areas around the territories occupied by the Serbs. UN mandated agreements lasted until 1993. Soon after they ended, the Croatians regained Krajina by force. The triad operated powerfully here, and was effectively underpinned by a constitutional dimension.

The third republic to secede was Bosnia-Hercegovina. Here the mismatch between the bounded territory of the republic and the settled territory of Serbs, Croats and Moslems was at its most complex. The republic was designed as the Moslem homeland. They were the largest single ethnic groups, but made up only about 39 per cent of the inhabitants. Croats made up about 18 per cent and Serbs 32 per cent. Furthermore, Moslems were settled in a scatter of villages; they had no consolidated territory. Serbs and Croats, on the other hand, were concentrated in particular areas, as well as being intermixed with Moslems across the republic. When Bosnia-Hercegovina declared its independence, the Serbs refused to join in. They wanted to stay within the Federation. By this time Milosovic had embarked on a Greater Serbia policy. He was determined to extend his control to all Serb settled territories. President Tudjman of Croatia had also embarked on a Greater Croatia policy, to extend Croatia to all Croat settled territories. These two groups would fight hard to gain control of Bosnia and would commence 'cleansing' the territory of other undesirable ethnic residents. In this situation the war in Bosnia lasted for four years. The problem was the map. Lord Owen, the EU negotiator, in many interviews stated that 'the problem is the map'. Many maps were drawn, and the 'international community' was involved in many ways before a lasting, fragile, peace was achieved with what really might be interpreted as a partition of Bosnia, despite the facade of an elaborate structure of all-Bosnia institutions.

Brubaker (1995) did not apply his triadic relationship to Bosnia, yet it is valuable in understanding events in the very complicated set of the relationships. The Moslems, the largest single group, were not excessively aggressive in trying to nationalise 'their' state and were further weakened by their lack of a consolidated territory. The Croats and the Serbs were strong minorities within Bosnia. Furtermore, both the Croats and the Serbs had strong external homelands, on which they could call for support. The Moslems on the other hand were weak within the state, and had no strong outside group upon which they could call.

The complexity of the relationship between bounded territory and settled territory was an important factor in the sequence of events surrounding the collapse of Yugoslavia. It affected the duration of the conflict and the range of the actors involved. It mismatched places of identity and places with identity. The foreign press reported on the conflict daily, but they did so from a western perspective that contained no understanding of the place identity that underpinned the mismatch of the map, except in the most superficial way.

In the spring of 1999 Milosovic began an attempt to 'cleanse' Kosovo of undesirable ethnic Albanian residents. The demand for the restoration of their autonomy by the Albanian majority in Kosovo was met with a brutal attempt, by the Serbs, to drive them out or kill them. The intervention of NATO has prevented this. The international community has ruled out independence for the time being, though strongly supports autonomy within Yugoslavia.

Conclusion

Is the state dead by hollowing? In the Marchlands it is not. The formation of new states has been a major theme since 1990. They have been formed within existing boundaries. In navigating a routeway from the Party-state through transition and transformation, the countries of the Marchlands have been stumbling in a maelstrom of visions of the future, perceptions of the past and experiences of the present that have reflected the idea of hollowing, but in ambiguous and distinctive ways. No trajectory is a simple pathway from the Party-state to the neo-liberal state. Neo-liberalism has pretty well seen off the Communist Project, but it has found a more durable opponent in the Nationalist Project, and the idea of the nation-state has retained its strength.

The Party-state was obsessed by official control, regulation and authority. The neo-liberal state, the hollowed state, is the antithesis of this. The boundaries between public and private, official and unofficial, formal and informal, internal and external relations, individual liberty and social consensus in the legitimisation of governance and power are blurred and eroded, more open to the assertion of individual self-development. Yet the subtleties of 'such distinctions were crucial to the rise of limited constitutional government, civil society and liberal democracy in the West' (Bidaleux 1999: 48). Most élites have only a hazy vision of the future, a confused perception of the present and, in many cases, a reluctance to leave the experience of the past.

In their uncertainty, for most élites, the idea of the state as a place, bounded and meaningful, has offered an anchor of stability.

Concepts such as path dependency, sequential development and the institutional equation help in understanding how a particular state's trajectory was structured in relation to the adaptation of western institutional forms of governance. These included aspirations of EU membership, and World Bank and IMF demands that promoted the spatial modernity of neo-liberalism. The particular pathway negotiated between state building, nation building and the recovery from economic collapse conditioned, as I showed in previous chapters, the experience of localities and regions and introduced new contacts with states in continental and global spatiality. States were produced as particular places, differentiated from each other and each resituated differently in the multi-scalar place relations of spatial modernity driven by the neo-liberalist project.

15 The Marchlands in the production of the New Europe

This chapter looks at the Marchlands in the New Europe. The return of the west is producing a spatial modernity of diverse experience in localities, regions and states. Looked at through the lens of Europe, new imaginings, perceptions and experiences, producing the Marchlands in the 1990s, come into perspective. The first part of the chapter discusses some competing and contrasting visions of Europe's future spatial development drawn up in the early 1990s, followed by a review of the main processes shaping change in practice. The second part provides an account of the perceived spatial reconstruction of the Marchlands into four, new, subcontinental places resulting from the operation of these processes on insider and outsider spatialities.

Imagined New Europes – competing and contrasting visions as development scenarios

Integration in Western Europe and disintegration in Eastern Europe between 1985 and 1992 created in the early 1990s a strong sense of a New Europe. The third New Europe of the twentieth century. This New Europe struck up an immediate ambiguity. The meaning of the New Europe, as it brushed into western popular debate, was the new form to be taken by the EC as it prepared to become the European Union by the signing of the Treaty of Maastricht. In this popular, western, spatiality the Marchlands remained a largely unknown place. However, the east, blank in their spatiality, was to become elaborated by new opportunities to extend westerners' experience by travelling there. For many in the Western European business community the blank space was newly perceived as a business opportunity. For western political élites, the east became a place to recover for western liberal democracy.

Among western geographers, the discourse on the New Europe could construct a broader vision of imagined European development scenarios. The production of New, imagined, Europes could precede those constructed on perceptions of any new spatial and social practices that might emerge. So in the early 1990s a new imagined Europe was produced in a discourse of speculative development scenarios for the new millennium.

Smith (1993) envisaged three possible scenarios of geo-political development. Each of them perceived the meaning of Europe as embedded in the norms, values and experiences of Western European countries and sketched in development scenarios that would, in different ways, bring the countries of the east into new place relations with this western core. The first scenario was of the emergence of 'a common European home', stretching from the Atlantic to the Urals. It recalled the earlier vision of General de Gaulle and its revival by M. Gorbachev (Chapter 9). The second scenario Smith called 'back to the future'. This echoed the flawed nationalist solutions of the past. It implied a fragmented and conflict-ridden east of Europe that would sit uncomfortably next to a peaceful and integrated west. The third scenario he termed 'not quite all Europe'. This scenario centred on the EU project. It envisaged an enlargement of the EU eastwards, but not all the eastern countries of the continent would qualify for membership. Not all would be deemed to be 'European'.

Other western academics who imagined a Europe beyond the EC/EU anchored their imagined development scenarios even more firmly upon the EC/EU and the west than did Smith. Among these were Masser, Sviden and Wegener (1992) and Dunford and Kafkalas (1992).

Masser, Sviden and Wegener (1992) envisaged Europe as it would have developed by 2020. The Europe they conceptualised was constructed on three assumptions and three scenarios. The three assumptions were, first, that 'Europe' would be larger than the current EC. It would include EFTA countries, and some Eastern European countries, within a European federation of 400 or 500 million people. Second, that there would be a European government with a president, a cabinet and a parliament, but the federation would have more or less autonomous countries. Third, that there would be peace in Europe. These three assumptions underpinned three inter-related scenarios. The first scenario had economic growth as its main objective. The second scenario had a reduction of inequalities as its main aim. The third scenario had environmental improvement and the quality of life as its primary motive. Masser and colleagues thus imagined a utopian meaning for 'Europe', that is territorially larger from the EC, but is the outcome of the values, attitudes and practices of the EC. It includes most, but not all, of continental Europe. Not all of the eastern continental countries, it seemed, were perceived as within the meaning of Europe.

Dunford and Kafkalas (1992) elaborated one scenario they referred to as a 'Greater Europe'. They conceptualised this Greater Europe as an enlarged zone of commodity exchange without supranational social regulation. Germany and the EC – possibly differentiated into an internal core and periphery – would be at the centre. Surrounding this centre would be a zone made up by EFTA members and perhaps including former Austro-Hungarian countries. Countries historically associated with the Ottoman Empire would be excluded. The EC may close its borders to insulate itself from ethnic and national conflicts to the south east; implying fragmentation and disruption would be experienced in southeast Europe. This vision of the development of the New Europe represented the salience of current commercial and economic meaning. It was a Europe of free-market capitalism driven by the global Neo-liberal Project. Yet it also recalls, in the minor key, the historical cultural configurations of European and Moslem civilisations that, as I have shown in Chapter 2, shaped the spatial order of continental Europe for many centuries. It also reflects the traditional western perception of southeast, Balkan, Europe as unruly.

The predominance of the EC/EU in the meaning of Europe in the future was not shared by everybody. A 'fantasy' of 2020 (Figure 15.1) imagined a Europe divided geo-politically along historically and ethnically rooted lineaments into a series of federations and unions. The EU and its enlargement are absent from this scenario. Ten of the twelve member states plus Malta, some subdivided, have formed the United States of Western Europe, whilst Denmark has joined the Scandanavian–Baltic Union and Greece is a member of the Balkan Union. The Baltic Republics are in the Baltic Union, which includes Kaliningrad, returned to its German name of Königsberg, whilst Belarus and Ukraine are with Russia, members of the Slavic Union. An enlargement of the old Austro-Hungarian territories into Switzerland, Poland, Romania and Moldova makes up the Central European Union.

The EU Commission was also active in the realm of imagined future Europes. However, the Europe imagined here was different from those so far discussed. This imagined Europe was within the geo-political agenda that was backed by the exercise of real political power. This Europe had greater practical potential. Agenda 2000 (1993) and Agenda 2000+ (1997) set out imagined scenarios of 'Europe's' future enlargement and development. They envisaged a phased enlargement, preceded by a programme of assistance for economic reconstruction and democratisation to overcome perceived problems; projections of increased connectivity in migratory movements, trade and transport flows and the elaboration of transport and environmental scenarios. The European Spatial Development Perspective (ESDP) was accepted at Potsdam in May 1999 as an important step towards European integration. The centrepiece of the ESDP is the perception of a 'pentagon' of wealth and economic dynamism delimited by Hamburg, London, Paris, Milan and Munich. At present the ESPD only covers the existing EU territory, but accession countries will be informed of its progress, as will the Council of Europe, itself engaged in preparing a pan-European spatial development perspective (Faludi 2000). ESDP will be associated with the Interreg programmes for 2000–2006 constructing a strong conceptual lattice for development that has an extension into the Marchlands in the action programme for Central–Adriatic–Danubian–Southeast Europe (CADSES) being developed within Interreg II/C.

A particular, and more sinister, vision of the Marchlands in Europe's future development, which came from the east, was rooted in a nationalist and imperial scenario. It was sketched by Vladimir Zhirenowski, the Russian nationalist leader. Zhirenowski's vision divided the Marchlands between German and Russian spheres. Between them Poland is reterritorialised, losing western lands but gaining the Lvov region of Ukraine. Hungary was enlarged into Transylvania and Bulgaria was enlarged to gain territory from all its neighbours. This mirrors the historical situations of great power agreements and small nation nationalism concertinaed as a reconstitution of political

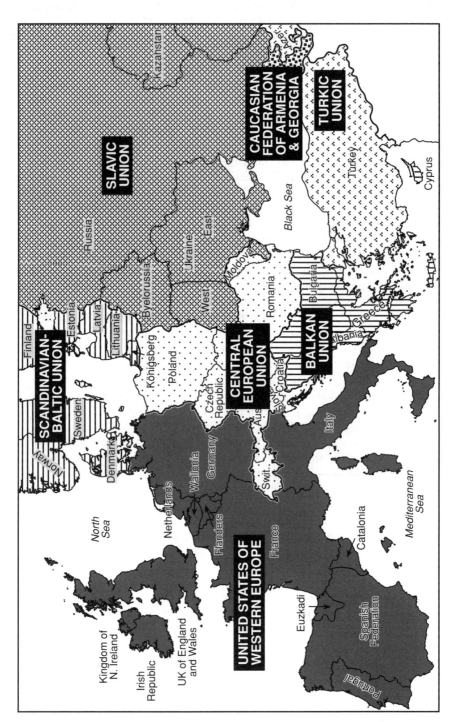

Figure 15.1 Europe 2020: a fantasy.

relations. It carries elements of the Nazi–Soviet Pact, the Nazi war-time grants of territory and the traditional association of Russia and Bulgaria. It gives salience to the outsider perception of the Marchlands as a place of outside domination.

The visions of Europes just discussed produced a discursive Europe of competing imagined spatialities. Yet all of them, originating in traditional European cultural and historical experience and discourse, have envisaged a reconstructed pan-European spatial architecture and resituated the Marchlands in a new imagined European conjunction of spatial modernity.

Spatial and social practices and patterns of European development

The imagined Europes discussed in the last section were constructed from four primary themes: first, the internationalisation of economic activity; second, the revival of nationalism; third, the idea of the Common European Home; fourth, the idea of the primacy of Western Europe and in particular of the European Union. The EU is the current icon of Europe as modernity. These themes will form the framework within which, in the next section, I will discuss the spatial and socio-economic and cultural practices and patterns that are producing the Marchlands within the New Europe.

The internationalisation of economic activity, regional wealth and urban development

Transition, as I have shown in the preceding chapters, has produced the Marchlands as a place of strongly contrasting, local, regional and national experiences. The internationalisation of economic activity through the spread of capitalist commodification, the disbursement of FDI, the kindling of business enterprise and the redirection and intensification of trade have produced distinctive multi-scalar place relations. Taken together they have had particular implications for wealth creation and distribution.

Regional patterns of prosperity

Grimm (1994, cited in Burdack, Grimm and Paul 1998) using average monthly income pointed to thresholds of wealth in Central and Eastern Europe. One threshold ran along the eastern border of Germany, Austria and Italy. West of this line mean monthly income was over 2,000 DM. A second threshold ran along the eastern border of Poland, Slovakia, Hungary and Slovenia. Within the middle zone formed by these two thresholds, mean monthly income was about 500 DM. East and south of the second threshold line mean monthly income was below 100 DM.

Another way of measuring wealth distribution is provided by EU figures for regional GDP per capita purchasing power parity (PPP). These figures, referred to in Chapter 13, permit further comparisons across Europe as a whole. Examined by comparison with Western Europe, using regional GDP per capita, the contrasts within the Marchlands seem much less stark than they have so far appeared. The primary contrast, taking a European perspective, seems to be in a contrast, west and east of a prosperous north–south axis, layered with a core–periphery pattern around the pentagon (Figure 15.2)

Regional GDP per capita (PPS 1995–97 average) figures drawn up by the EU, set against the EU standard, show the ten Marchlands countries that are applicants for EU membership to be remarkably uniform in prosperity – and poor. Of fifty regions, at NUTS 2 equivalent, forty-three lay between 24 per cent and 60 per cent of the EU average. The 1999 European Spatial Development Perspective was not concerned directly with the Marchlands, but identified a pentagonal core of wealth and dynamism in Europe (Faludi 2000) edged by London, Paris, Milan, Munich and Hamburg. It is underscored by intense transport and communications connectivity and, though occupying only 20 per cent of EU territory and with only 40 per cent of EU population, generates more than 50 per cent of EU wealth (GNP). The eastward gradient from Europe's pentagon and axis of wealth is very steep. Only the capital city regions of the westernmost states of the Marchlands came near to the EU average GDP per capita, and only Prague exceeds it (Figure 15.2). The Czech Republic, Slovenia and the westernmost regions of Hungary just match the poorest, most peripheral areas of the existing member states of the EU that are much more distant from the pentagon.

Urban hierarchies, hinterlands and hinterworlds

The construction of Europe as a network of city hierarchies and hinterlands provides a functional alternative spatiality contesting the formal patterns of administrative regions in representing and understanding practical and material trajectories of European spatial development. In Chapter 8, I discussed contrasting views as to the similarities and differences of urbanisation under capitalist and socialist development. Whichever side is taken in that debate, it is certain that between 1945 and 1990, two separate urban systems emerged in the two halves of Cold War Europe.

During the 1980s the cities of Western Europe underwent economic restructuring under pressures of changes in industrial technology and organisation, communications and information systems and governance. They entered into a competitive mode, as neo-liberal economic ideas turned them into commodities, whose representatives re-imaged and marketed them. They competed as centres of entrepreneurial initiative to acquire command and control functions in national, European and global urban hierarchies. This trend accelerated into the 1990s, but now the cities of the Marchlands, particularly the capitals of the westernmost states, were drawn into the competitive hierarchy.

Enyedi (1998b), who argued that socialist urbanisation in Eastern Europe was similar to capitalist urbanisation, thought that, because of this underlying similarity, Budapest, Prague and Warsaw would be quickly reintegrated into the European urban system. Each of these cities has particular advantages for participation in the intercity competition, but is weakened by poor

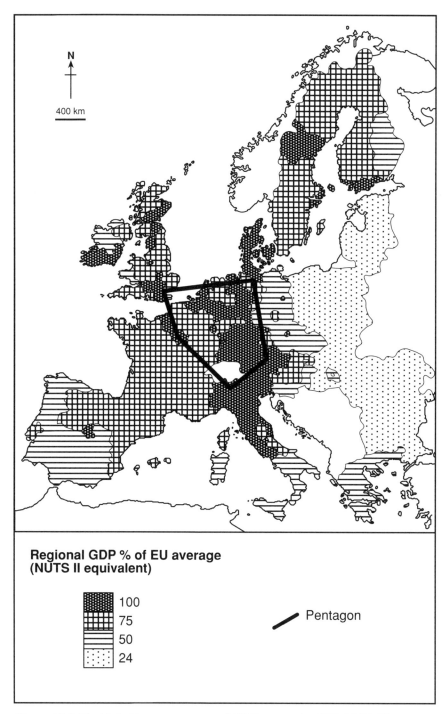

Figure 15.2 Wealth in Europe: patterns of GDP per capita PPP, 1995–97.

Source: Data from 'Statistics in Focus' Theme 1–1/2000 (eurostat 2000).

Table 15.1 Relative strengths of Marchland cities as business centres

Capital	No. top 100 firms (by sales 1998)	No. top 30 banks (by assets 1998)	Stock market capitalisation ($US bn 1999)	Daily business volume (average $US m 1999)
Warsaw	38	13	22	62
Budapest	14	4	13	52.5
Prague	20	5	12.4	14.2
Ljubljana	4	1	2.5	1.5
Bucharest	4	1	0.4	0.4

Source: *Business Central Europe Annual 2000.*

Top 100 firms and 30 banks refers to organisations registered in specific countries but compared across the Marchlands as a whole.

infrastructure and transitional societal tensions in comparison with Western European cities. Berlin as capital of a reunited Germany has great competitive potential for growth. Bratislava, Zagreb and Ljubljana are smaller and weaker. Further east, north and south, capitals such as Riga, Vilnius, Belgrade, Bucharest, Sofia and Kiev are much more weakly connected to the European urban hierarchy. They are far less internationalised and have very poorly developed infrastructures as well as being relatively slower to adopt new practices. The European urban hierarchy and a Europe of city regions is being formed in and around the pentagon referred to above. Marchlands capitals are very small players in this European urban hierarchy and the experiences of regional and smaller cities following communist development contrast dramatically (Chapter 12).

Recently published research on European cities based on a statistical analysis of forty-six firms in corporate producer services has produced a spatial order of European cities under globalisation that is constructed from three main ideas. These are, world city formation, a typology of European cities as service complexes and urban hinterworlds. Beaverstock, Smith and Taylor (1999) studied 263 cities across the globe, in search of 'world city formation'. They detected world city formation in 142, but classified just fifty-five as world cities. These they divided into three levels of relative importance, alpha, beta and gamma. Of the fifty-five world cities, twenty-two were in Europe. Three Marchlands cities, Warsaw, Budapest and Prague, appeared in the lowest, gamma, class. A fourth, even weaker category, of cities that were showing only 'some evidence of world city formation', included Bratislava, Bucharest and Kiev.

In a further analysis of the data, Taylor and Hoyler (2000) extended the work to produce a typology of European cities as corporate service complexes. They recognised five types: major spine cities; minor spine cities; outer triangle cities; Eastern European cities; and British Isles cities. These different types formed a spatial structure of European cities around the historic north–south spine of European urban development, from London, through the Netherlands, Germany and Switzerland, to northern Italy.

Budapest and Vienna were situated in the 'outer triangle', and Warsaw, Kiev, Bucharest, Prague and Bratislava were in the Eastern European category. Taylor and Hoyler point out that whilst their typology turned out to be geographically specific it was not necessarily helpful in the discussion of urban systems and city hierarchies.

However, Taylor (2001) extended the analysis of the data still further. This time he took up the idea of city spheres of influence or hinterlands and introduced the new idea of urban hinterworlds. Because of electronic communications, all world city hinterlands overlap with all others; there are no boundaries between them. The difficulty of boundlessness is overcome with the concept of hinterworlds that indicate the scale and nature of service provision from specific world cities. Taylor argued that world city formation is 'an intercity process' and that behind every successful world city there is a broad and intensive hinterworld. Although Taylor did not put it this way, the analysis permits the conceptualisation of Europe as the urban hinterworlds around the twenty-two world cities of the continent that must form a hierarchy because they have properties of multi-scalar intensity. There were, after all, in Europe cities on different levels, four alpha cities, four beta cities and fourteen gamma cities whose hinterworlds will differ in scale and intensity. Taylor notes that there is a 'European bias towards London, especially conspicuous in Eastern Europe. This regionality relates to the role of cities as pan-regional gateways . . . London operates as a gateway to Eastern Europe' (Taylor 2001: 57). The European urban system forms a regional bloc, a hierarchy of cities and hinterworlds. Taylor argues that this is not a Europe of cities to rival a Europe of nation-states. It is a spatial order of cities in a space of flows and networks of globalised region building. Nevertheless, from the point of view of spatial modernity it is a competing spatiality of Europe.

The revival of nationalism

Many contentious domestic and bi-lateral issues in the Marchlands were related to the revival of nationalist consciousness. Smith imagined among other European futures one that he called 'A Europe back to the future'. It was a conception that carried:

> a salutary reminder that the most striking feature of the rebirth of civil society in the east is the salience of ethno-regional division . . . a political ideology rooted in a tradition of European thought . . . given concrete meaning in the twentieth century in the Wilsonian ideal of self determination.
>
> (Smith 1993: 94)

Nationalism sparked off anti-democratic movements and has been interpreted as part of Eastern Europe's perpetual attempts at catching up with the west (Dogan 1997), though this interpretation should be treated with caution.

Nationalism is a dangerous force, but I think it has been overstated in the process of resituating the Marchlands in the new European modernity. National sentiment was a powerful source of symbolism in the demand for independence from the Soviet Union. It was fiercely present in the collapse of Yugoslavia and it coloured the split-up of Czechoslovakia. Nation building has accompanied state building (Chapter 14). A new nationalism was a response to Soviet imperialism and the suppression of national identity and national ambitions for statehood. There was some feeling of completing unfinished business as in Yugoslavia. Some governments adopted national discriminatory practices, as in Slovakia; enforced national and linguistic tests for citizenship, as in Latvia; and introduced nationally orientated school curricula. There are rightist political parties in all countries, but they have little popular support. As development has occurred, the strength of nationalist tendencies has abated somewhat. There is still, however, a problem in the age-old prejudice against the Roma (gypsies).

The common European home and the European Security Agenda

This conceptualisation of Europe is associated with the Gaullist vision of Europe 'from the Atlantic to the Urals'. De Gaulle's vision incorporated a Europe of nation-states, following French cultural leadership and was invented to counter American influence. Life was breathed into its reincarnation by M. Gorbachev who reflected on a cultural and historical meaning of Europe whose peaceful development embodied a spirit of diverse wholeness. There is some ambiguity in the interpretation of the purpose of this revival. Gorbachev's vision may have been designed to reconnect the Soviet Union/Russia with Europeanness – however ambivalent that had been (Chapter 2). He may have sought to open up links in order to gain technical assistance for Soviet *perestroika*. He may have been attempting to reconstitute the superpower dwarfing of Europe. He may have been seeking to separate the Western European and American dimensions of Atlanticism. He might have been redefining Soviet relations with Eastern Europe, in the light of German reunification. He may have been resituating the Soviet Union in global space – he also referred to the Pacific and the sub-Arctic as 'common homes'. All of these elements may be interwoven into his complex sense of Europe.

Though it began in an appeal to a sense of cultural and historical identity, Gorbachev's invocation of the common European home became in practice an issue of the European security agenda, most especially for the successor states of the Soviet Union. What was to replace the NATO–Warsaw Pact confrontation that had been the perverse, yet effective, framework for security in Europe? Whilst the Conference on Security and Co-operation in Europe was already in existence, the new turn of events put the heart of the issue in the enlargement of NATO or some alternative structure of a collective European security community. This was a very delicate process that had to balance the aspirations of Marchlands states, many of

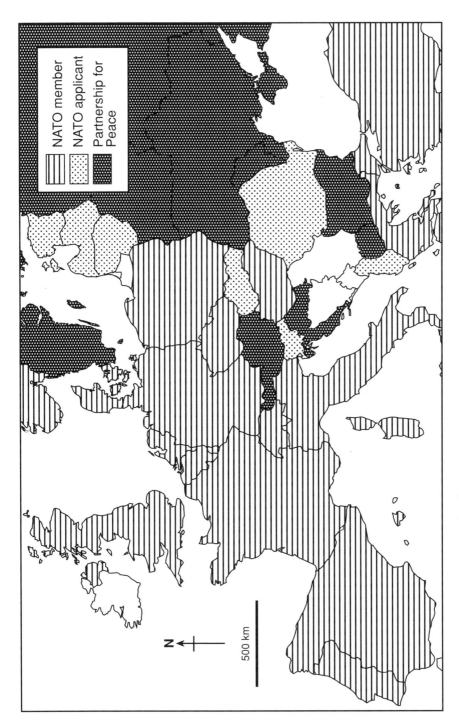

Figure 15.3 NATO and the Partnership for Peace.

Source: NATO.

whom clamoured for NATO membership against the security worries of Russia.

The outcome has been a process of negotiated enlargement of NATO, within a wider system, the Partnership for Peace (PP) and the Euro-Atlantic Partnership Council (Figure 15.3). Enlargement brought in Poland, Hungary and the Czech Republic who became members in March 1999. A special and distinctive partnership was arrived at with Ukraine. Croatia was admitted to PP in May 2000. The process continues under the Membership Action Plan (MAP). The MAP incorporates an annual review of aspirant members' preparations for entry covering economic, political, defence, resource, security and legal dimensions, and offers feedback advice on progress. A substantial report assessing development will be submitted to the NATO Summit in 2002. NATO enlargement is aimed at creating a more stable Europe within a single security architecture. NATO Secretary General Lord Robertson links the process to the enlargement of the EU: 'The prospect of NATO and EU membership has already prompted many Central and Eastern European countries to solve contentious bilateral as well as domestic political issues' (Robertson 1999: 14).

The European Union Project and the meaning of Europeanisation

Graham perceives the EU as 'an essentially centripetal force of integration, imposed on past political and economic heterogeneity' suggesting an homogenisation, despite which 'Europe remains characterised by manifest cultural diversity . . . a fragmentation of identity and allegiance, largely shaped by contested readings of the past. How – or even – can these apparent contradictions be reconciled?' (Graham 1998: 1). I think that Graham and others are wrong to see the EU and cultural diversity as 'contradictory'. The European Union is the most successful of all projects that have attempted to unite Europe. In popular discourse, even to some extent in official and academic discourse, the EU has become the meaning of 'Europe'. There is some justification for this, because it currently most nearly approaches the synopsis of the idea of Europe as modernity – recalling Heller's formulation 'Modernity, the creation of Europe, itself created Europe' (Chapter 1). It is the current high symbol of the tradition of Europeanness as modernness. As such it is the heir not only to European economic and political identity, but also to cultural identity. Restated in the phrase of the former German Chancellor Helmut Kohl, it is Europe as 'Unity in Diversity' and proclaims an acceptance of Graham's cardinal conclusion:

> It is clear that diversity or heterogeneity is the key to Europe's human geographies and that any efforts to evoke a European level of consciousness must be vested in notions of plurality (and also) recognise that a policy for European union involves . . . the inclusive principles of convergence and cohesion.
>
> (Graham 1998: 311)

I think that in practical experience and in discourse the EU Project has these notions and principles as its centrepiece, but achieving them is far from easy in the context of neo-liberal modernity. The practicalities of economic and political integration in a culturally diversified Europe do result in tensions and conflicts as Graham and Hart (1999) have pointed out. These may be brought under particular strain by the enlargement into the east.

The EU is a key development dynamic producing European spatiality. It is through the EU that the development relationship of imagined, perceived and experienced place is most clearly seen to operate. The EU conceives the Marchlands as part of Europe, but perceives them as flawed Europe and so engages in practices that will remove the shortcomings to create a new experience. The EU is developing its vision of Europe by reterritorialising the Marchlands within its policies on assistance for economic restructuring, immigration and enlargement, but there is some fragmentation of the processes. A coherent spatial development conception is lagging behind other sectoral activities.

Economic restructuring and the enlargement of the European Union into the Marchlands

The collapse of communism took the EC by surprise. At first there was no co-ordinated policy as each member state responded in its own way. Member governments' responses reflected their stances in debates that had been present in the EC since its inception. These debates negotiated 'widening' the extension of the powers of the European Commission, 'deepening' the shift from co-operation to integration and 'enlarging' the number of members and the area covered. The Treaty of Rome says that any European democratic country can join the EC/EU. But who decides whether or not a country is democratic or European? This goes to the root of the meaning of 'European' and who decides what it is? In practice, these days, it is the EU and its member states.

The EU Project promoted the image and idea of Europe as modernity. In the EU's institutional spatiality the Marchlands were perceived as barbaric, lacking the civil society of the west; inferior, lacking the democratic, political institutions of the west; and backward, lacking market economies. In short, the EU, as an institution, regarded itself as the current banner carrier of western modernity, and perceived the east in traditional mythological terms. This conceptualisation has only slowly changed and with it attitudes and policy positions have changed too.

A major feature of the EU spatiality was a distinction between Russia and the CIS on the one hand and the other post-communist states on the other hand. This has led to contrasting policy approaches. The EU approach to Russia and the CIS has been to offer advice but relatively little material support. Support has tended to follow the sequence of presidential elections. The west, the EU and the USA, was keen to support President Yeltsin, perceived as the westernising reformer, at election times, but its interest has waned in between. A separate assistance package, Technical Assistance for the Commonwealth of Independent States (TACIS), was arranged to help economic restructuring. In 1994 the EU concluded Partnership and Co-operation Agreements with Russia and Ukraine, but these were different from the agreements negotiated with other countries, reflecting the contrasting course of transition.

For the non-Soviet Marchlands states the EU has fashioned a different but broadly common policy sequence, moving from assistance to co-operation. The Poland et Hongrie Assistance à la Reconstruction Economique (PHARE) was developed out of a scheme agreed at the G-7 summit held in Paris in July 1989, at which the Commission had been asked to co-ordinate international assistance to Poland and Hungary. This clearly had the underlying agenda of separating Poland and Hungary from the other socialist states, because they were, at that time, showing liberalising tendencies in response to popular unrest. The programme was soon extended to other countries as their communist governments collapsed. PHARE is a wide ranging scheme providing technical assistance for public sectors such as health and education, for infrastructure developments, economic sectoral reconstruction and the development of small and medium sized enterprises. More specialised financial investments were made the task of the newly founded European Bank for Reconstruction and Development. Assistance is given as loans from the European Investment Bank. In addition bilateral aid is given separately by member states.

Following the 1988 Common Declaration between the EC and the CMEA, the Commission signed bilateral trade and co-operation agreements with individual East European countries. Between 1988 and 1992, thirteen agreements were signed. The economic effects of these agreements were limited by technical conditions, but had political importance for East European countries because they implied the EC no longer viewed them as state trading countries. At the same time the agreements increasingly included 'political conditions' (Kramer 1993), by which the Commission encouraged political plurality, the rule of law and minority rights.

So rapid was the pace of change that the trade and co-operation agreements were soon inadequate for the conduct of EC relations with the Marchlands' emerging states. The Commission introduced 'European Agreements'. European Agreements were a special form of association agreement including provisions for political and cultural co-operation as well as economic, trade and financial co-operation. The Agreements required a commitment on the part of the Marchlands country to take concrete steps towards a market economy and plural democracy. The Agreements envisaged the formation of a free trade area with the associate country over a period of ten years. Whilst these agreements did not formerly imply 'enlargement' into the countries involved, they certainly encouraged those in the Marchlands who had EU membership as a goal.

EU policy has been proactive and reactive with regard to the Marchlands countries. It has stimulated change and it has rewarded what it perceives as the right kind of change. Poland and Hungary were selected for assistance through PHARE when they showed signs of liberalisation, encouraged

Agenda 2000 annual reports and the Copenhagen criteria

EU spatiality perceives the Marchlands as divided into two prime areas. The first consists of ten countries whose applications for membership are likely to be accepted, in the medium or long term; the second consists of those who are not likely to be accepted. The first group is further divided into two: the leading contenders, that is Hungary, Poland, Slovenia, the Czech Republic and Estonia, are already advanced in negotiating membership; the second group, Bulgaria, Romania, Latvia, Lithuania and Slovakia, are not likely to meet the criteria in the medium term, but have just been upgraded to start negotiating for membership soon.

The EU issues annual progress reports on the ten applicants that assess their transformation along three dimensions of development – political, economic and the Aquis-EU law, these are known as the Copenhagen criteria. Whilst this is narrow and perceives development from a particular stance, it is a neo-liberal influenced stance. The EU is a political and economic project that is challenging the idea of the state by promoting a single European market and a state borderless Europe. The assessment of the 'progress' of the applicant states to membership is therefore an assessment of their progress towards a particular variant of the neo-liberalist agenda. Progress towards membership also acts as an incentive to political and economic reforms in applicant states. It may seem hypocritical that the EU measures progress against 'a Utopian model of liberal democracy' (Bidaleux 1999: 25) and statehood that the members of the Union are themselves leaving behind. However, it is not so surprising when it is remembered that the Union's enlargement is conceived in terms of states and the member states remain the strong players in the evolution of the Union.

by Gorbachev's policies in the late 1980s. Poland, Hungary and the Czech Republic have also received much greater funding under PHARE than other countries. Those who were most willing to change and achieved stable change were rewarded with encouragement, more funds and then more responsibility. So they moved along a path to partnership that might ultimately lead to the greatest reward of all – membership of the EU.

The next enlargement will be the fifth and it will be the most difficult so far. Even compared with Spain and Portugal, who also made the transition from dictatorship, the economic and democratic institutions of many Marchlands countries are fragile. At the same time the EU has become much more integrated. The Maastricht and Amsterdam treaties have given form and substance to new frameworks of integration that have put more demands on member states. Yet it is important that progress to enlargement runs at a pace that does not discourage would-be new members, or lead to cooling for the idea among member states. The EU issues Agenda 2000 annual progress reports on the ten candidate countries, measuring progress against the Copenhagen criteria and there are clear differences between the preparedness of these countries. So much so that from January 2000 the EU abandoned the simultaneous negotiation with Poland, Hungary, Estonia and Slovenia, and now anticipates accepting new members individually when they are ready. At the same time, and given the greater flexibility, the EU members agreed to open formal accession negotiations with Bulgaria, Latvia, Lithuania, Romania and Slovakia.

International east–west migration in Europe

Patterns of international migration in Europe are resituating the Marchlands in European place relations. The collapse of communist governments sent shockwaves of fear through Western European governments who anticipated a huge influx of migrants from Eastern Europe. The feared invasion from the Marchlands after 1990 did not materialise, though the numbers migrating increased sharply (Doomernik 1997). The Western European governments' record in dealing with refugees from the Yugoslav conflict was not impressive and put pressure on some countries within the Marchlands, especially Hungary whose government had to deal with that stream in the form of 'crisis management' with little western help (Dingsdale 1996). A steady stream of Albanians seek entry to Italy. Romania is a source of many illegal immigrants passing through Hungary and the Czech Republic in an attempt to enter the EU.

Whilst the EU feared an influx of migrants, the governments of the Marchlands states also saw a danger in the new situation. They feared a migration to the west that would separate out their best educated, youngest and most energetic citizens, who would seek better pay and conditions working in the west. Such a 'brain drain' might denude the new democracies of their most needed talent. Many individuals did migrate, but not in the numbers feared. Western policy was aimed at an orderly, managed migratory movement (Rédei 1995). Migration patterns also resituated the Marchlands in place relations in another way too. Incoming migrants made the more westerly countries of the Marchlands places of net immigration for the first time in their history (European Parliament 1998).

The 'new' subcontinental regions

The individual, organisational and institutional spatialities described above are reterritorialising the Marchlands, transforming their place particularities, differentiating and resituating them in the New Europe. Taken together these processes are producing four distinctive subcontinental areas in material practices and in discourses of transformation and transition. In this section I will describe the syntheses that are reconstructing and redefining the territoriality and the meaning of what was before 1990 Eastern Europe and the western Soviet Union. I will identify four newly forming subcontinental regions as Central Europe, Southeast or Balkan Europe, Baltic Europe, and Eastern Borderlands (Figure 15.4). The socio-cultural construction of these areas, the bounding, scaling and naming, is, as ever, contested.

Central Europe

Business Central Europe, a monthly business magazine widely read across the whole of Europe, imposes a vague spatiality under the Central Europe of its title. It carries reports from all the former socialist countries and consistently refers to 'the Baltics', the Balkans', 'southeast Europe' and other places without showing a strong concern for conceptual distinction. Its editor, an outsider, seems to take it for granted that these different places exist as part of Central Europe and are embedded in the spatiality of its readers. It is difficult to say the extent to which this influences or mirrors the spatiality of the outsider and indeed insider business community.

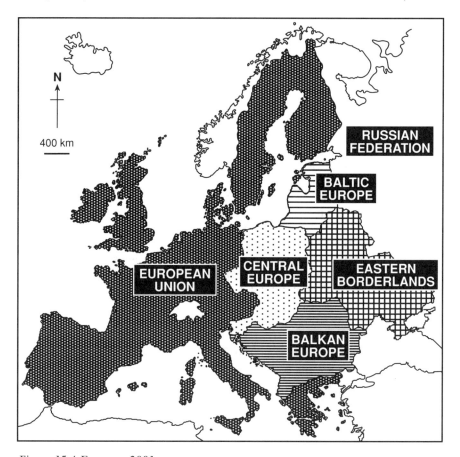

Figure 15.4 Europes, 2001.

Source: Adapted from Dingsdale (1999b) 'Redefining Eastern Europe', *Geography,* Vol. 84(3), 208.

To Rey (1996: 49), another outsider, identifying a 'new Central Europe' after the fall of the Iron Curtain, it also seemed 'in a way' that the former Eastern Europe had become Central Europe because it was situated between Western Europe and Russia. But she soon refines this, 'for the time being', to the Czech Republic, Slovakia, Hungary and Austria. Poland she excludes on the grounds that its state profile, influenced by its Baltic position and the Berlin–Moscow axis, differs from that typical of Central Europe.

In the perception of the Hungarian geographer Gy. Enyedi (1998a: 9), an insider, 'post communist Central Europe is deemed to consist of the Czech Republic, Poland, Hungary, Slovakia and Slovenia'. However, he thought this to be a limited definition, which he adopted because research findings were comparable, and he declared that he intended no political implications. Another insider, the Polish geographer Gorzelak (1996), also perceived a new Central Europe in terms of the states of Poland, the Czech Republic, Slovakia and Hungary.

My Central Europe is made up of the countries of Poland, the Czech Republic, Slovakia, Hungary and Slovenia. It is constructed from historical and cultural identities, the imaginings of dissident intellectuals, and societal and spatial practices that are transforming its internal particularities and resituating it within the New Europe. This follows the Central Europe of the insiders as touched on above. Poles, Hungarians, Czechs, Slovaks and Slovenes identify themselves as Central Europeans and conceptualise Central Europe in this way. Croats, too, identify themselves as Central European and their country as within Central Europe. Outsiders, however, observing the transition trajectory of Croatia have been cautious about including Croatia in their idea of the new Central Europe.

The idea of Central Europe was resurrected initially in the mid-1980s by Polish, Hungarian and Czech writers, some of whom were emigrés. It conveyed a strong sense of cultural identity, and was stridently anti-Russian in sentiment (Garton-Ash 1986). Though it was never given territorial definition, and rejected by some insiders (Enyedi 1990b) it was closely associated with the Habsburg monarchy and promoted primarily as a way of culturally distinguishing these countries from southeast Europe (Delanty 1996). Since 1990, Central Europe has been taking shape in a more practical way, too. The first post-communist governments of Poland, Czechoslovakia and Hungary formed the Visegrád political association and the Central European Free Trade Area (CEFTA) (Figure 14.1, page 231). They were soon joined by Slovenia. The countries have been eagerly adopting current western forms and see themselves as 'returning to Europe', and their historical antecedents from which they were cut off for almost fifty years after the Second World War.

The countries that make up Central Europe have been the biggest beneficiaries of foreign investment among the Marchlands countries. They were perceived by the global business community as the most progressive in shifting from state socialism, and the most likely to successfully rejoin the capitalist world economy. Poland, Hungary and the Czech Republic were the most attractive countries for direct foreign investment receiving a total of US$53 bn between 1989 and 1999. They have also led the league table of financial assistance from the EU and were the first to sign 'European Agreements' with the EU. Following the EU's publication of 'Agenda 2000', including its strategy for the next enlargement, they became the leading contenders for accession. Spurred on by the EU they have been reforming their administrative and legal systems to bring them in line with EU protocols. The steps they have taken to co-operate with each other in political and economic matters – the Visegrád Association and the CEFTA – were also prompted by considerations of EU entry. They have made the most rapid progress towards the introduction of stable democratic institutions and human rights protection.

These countries are increasingly bound together by intensified trading links through the CEFTA, but they also have close trading relations with Western Europe. Germany is the dominant trading partner of Poland and the Czech Republic and is also the leading partner of Hungary and Slovenia.

Slovakia has lagged behind in this and its increasing contact with Russia between 1993 and 1998 weakened its position in the group. Slovakia and Slovenia together received $4.4 bn of FDI during the 1989–99 period. Slovenia, after a slow start has been accepted into the economic and political associations of Central Europe and been invited to make an application for membership of the EU. Slovenian political parties are united in looking north and west, seeking to promote the country as a bridge between east and west. In contrast the authoritarian government of Croatia did not find much effective support among foreign governments for its political or economic policies. Since the death of the nationalist President Tudjman in 1999 the new government has been pursuing reform policies that have been making international friends and Croatia has been accepted into NATO's Partnership for Peace.

Thus, Central Europe is being produced in discourse and in material practices by a synthesis of historical–cultural traditions, practical economic and political associations and intensified links with Western Europe that differentiate it from other areas. For insiders and for outsiders alike, the idea of Central Europe carries a positive meaning in adopting and adapting to the return of the west. Poland, the Czech Republic, Hungary and Slovenia in particular are perceived as shifting from a status as the western edge of the east to being, once again, the eastern edge of the west.

Southeast/Balkan Europe

When the *National Geographic* magazine published a map (*National Geographic* 2000) that represented Hungary as in the Balkans, several Hungarian friends of mine were shocked and angered. The Balkans, within their personal spatiality, was a place of conflict, even barbarism, far distant from their image of Hungary. Hall and Danta (1996) described the reconstruction of the Balkans as 'a new geography of Southeast Europe'. They envisaged a reconstruction around emergence from 'the Yugoslav vortex', and perceived Albania, Bulgaria, Romania and Hungary as Yugoslavia's 'neighbours in transition'. The place of Croatia is at present uncertain, territorially and meta-phorically straggling the border of Central Europe and Balkan Europe.

Balkan Europe consists of the successor states to the communist Federa-tion of Yugoslavia, excepting Slovenia, together with Bulgaria, Romania and Albania. These countries have become associated with Greece and Italy (Hall and Danta 1996). They have also forged links with Turkey.

Historically this was the most complex territory in Europe. It lies at the junction of Western European, Eastern European and Islamic cultures, but is predominantly Eastern European and Islamic (Figure 2.1, page 18). Individuals and institutions in the Balkans were least touched by the spread of Europe as modernity. For the western outsider the Balkans became iden-tified as a place of unrest and fragmentation. The term 'Balkans' came to signify instability and conflict. The construction of this meaning was a complex nexus of experience and invention. Insiders held tightly to ethnic identities that separated rather than unified them. The idea of Yugoslavia, of a South Slav unity, never took hold. In the census of 1981 only 22 per cent of the Yugoslav population identified themselves as Yugoslav, the rest clung to their traditional ethnic identity. The Orthodox Christian tradition of national churches that contrasted with the Catholic tradi-tion of an international church reinforced feelings of difference. Territorial conflict was enflamed by claims and counterclaims. When, after the break-up of the Federation of Yugoslavia, the Republic of Macedonia was declared, it enraged the Greeks for whom the idea of Macedonia signified their homeland.

In this complex and volatile cultural mixing, the chief response to the collapse of communism was the resurgence of ethnic patriotism and nation-alism. It was, in the Balkans, where the imagined future tied to the flawed solutions of the past (Smith 1993) was most sharply realised. The Yugoslav conflict was complex (Chapter 14). Whilst political aggrandisement underlay the Greater Croatia and Greater Serbia projects and 'ethnic cleansing', it is clear that the idea of 'place identity' was also deeply embedded in the psyche of most groups. This sense of place was a factor not completely understood by those western politicians who sought to bring the warring parties together (White 1996). Their understanding of the Balkans was in terms of its cultural invention in western thought reincorporated into the new geo-political framework (Todorova 1994). In this it was in marked contrast to the spatiality of the insiders.

The post-communist governments of Yugoslavia, Albania, Romania and Bulgaria retained the old repressive methods inherited from communist and pre-communist days. The legacy of authoritarianism went deep. They found it difficult to grapple with economic instability. Economics were peripheral to their concept of self-identity, which was more consciously political and ethnic in its priorities. They promoted an aggressive nation-alist sense of identity in unstable political conditions. Albania remained in a state of lawlessness, armed conflict erupted in summer 1997 and again in summer 1998. The Serbian President Slobodan Milosovic, having first taken away autonomy from Kosovo in 1989, again in autumn 1998 followed an aggressive military policy of driving ethnic Albanians from their homes. This prompted NATO intervention. With a change of government in Romania in November 1996 some signs of a new beginning were recognised, but the situation reverted to instability when the governing coalition disintegrated in summer 1998. Bulgaria, having suffered unstable political regimes and some economic gangsterism, enjoyed after April 1997, for the first time, a stable reformist government.

Because of the war in Yugoslavia and the slowness of ruling élites to embrace change these countries were unable to attract FDI. Italy and Greece, their nearest neighbours and, at least in the case of the Greeks, cognisant of the Balkan way of thinking, were the leading sources of foreign investment. Balkan States did much better in attracting western government sponsored bilateral and multi-lateral assistance. Bulgaria and Romania signed 'European Agreements' with the EU and negotiations for accession have started. Romania has also joined CEFTA and Bulgaria began negotiating membership in late 1997.

Whilst making overtures to the EU and Central European institutions, Romania and Bulgaria have looked south and east for new international institutional links. In June 1997 the Balkan countries including Greece, an EU member, signed a co-operation agreement in the Greek city of Thessaloniki. All these countries have also joined with Turkey, Ukraine and Russia to form the Black Sea Economic Zone (Figure 14.1, page 231). On the southeastern EU border there is co-operation between Greek and Bulgarian border districts in the Nestos–Mesta Euroregion. Albania has cross-border associations with Bulgaria and Greece and separately with Italy. Whilst at present these associations are not strong they may be a new European region in the making.

Greece and Italy as well as being the major foreign investors are also important trading partners. Membership of the Black Sea Economic Zone does not as yet seem to have created much extra trade. The significance of the role of the CIS in Bulgarian trade can be explained by past associations, but is still nevertheless extremely important. Petrakos (1997a and 1997b) sees a potential new European 'Macro-Region' in the making as economic contacts increase between the Balkan countries.

Thus, despite the war in Yugoslavia and continued unrest in some other Balkan States there are signs that stability is returning. Even so, produced in negative outsider perceptions, as the commonly held idea of 'Balkanisation' seems to convey, southeast Europe may always be differentiated from its Western European neighbours by its propensity for fragmentation and conflict.

Baltic Europe

At the core of the new Baltic Europe is the mechanism of transition and transformation of the new Republics of Latvia, Estonia and Lithuania. This has been perceived as:

> Nimbly and ruthlessly the three states have spent the decade since inde-
> pendence juggling the demands of reasserting national identities with
> the relentless pursuit of a capitalist turnaround.
>
> (Orton-Jones 2001: 70)

This dynamism has been energised by the idea and re-imaging of the Baltic States as leaders in the 'New Hansa', heirs to the commercial and cultural traditions of the historical Hanseatic League. It is an image that fits the times and consolidates a Baltic identity for both insiders and outsiders.

The Baltic coastal districts of Russia are also participating in contacts united by the Baltic Sea. Northern provinces of Poland and Germany, too, are reviving Baltic links. Strong new contacts are being forged with Finland, Sweden and Denmark, all EU members. These new dynamics are resituating the Baltic Republics in Baltic and European place relations.

In the later Middle Ages and the Early Modern period the Baltic States forged dynamic commercial links with Western Europe through the Hanseatic League. Although it was principally a commercial and trading organisation the League also had important cultural and political dimensions (Örn 1994). During the eighteenth century the Baltic States were overrun by Russia and incorporated into the czarist empire. In the twentieth century, the inter-war period brought independence, during which strong links with Western Europe were re-established and progress to capitalist modernity and prosperity was rapid (Chapter 5). After the Second World War eastern influences returned and the Baltic States were incorporated into the Soviet Union. This resulted in the immigration of ethnic Russians whose numbers in 1990 accounted for more than 50 per cent of the population in Riga (Latvia) and was as high as 20 per cent in many districts of Latvia and Estonia (Figure 13.4, page 216).

These historical antecedents created a strong sense of regional identity and association among the Baltic nations. Latvians, Estonians and Lithuanians have a common sense of internal community, fashioned from their place proximity and their Russian and Soviet experience, which overcomes national differences. The Baltic Soviet Republics formed a 'unique cultural unit' in the Soviet Union (Misiunas and Taagepera 1993). It was an 'otherness' that differentiated them from Russia and Russians. This sense of community is now further strengthened by a wider sense of community with the Scandanavian states. The Baltic nations have always seen themselves as a distinctive part of Western European civilisation. A new spirit of Baltic association links the states around the Baltic and reinforces the wish of the Scandanavian states to assist economic reconstruction and strengthen democracy. The Baltic Sea was part of the 'Iron Curtain', preventing normal relations and imposing a 'mental barrier' that has now been removed permitting the reconstruction of ties around the Baltic. The new societal and spatial practices that have emerged since 1990 have broken down that 'mental barrier'. In thought and action Baltic Europe has been vigorously revitalised in the spatiality of insiders and outsiders.

Since 1990 a huge variety of local, regional and interstate ties have been forged around the Baltic (Tivenius 1996). In 1990 Estonia, Latvia and Lithuania concluded the Baltic Economic Co-operation Agreement and a free trade agreement came into force in 1994 (Figure 14.1, page 231). These steps have been important in reworking the sense of Baltic identity, but the most powerful unifying concept, in these days of neo-liberal economic and commercial revival, is that of the legacy of the Hanseatic League. The Hanseatic Confederation was re-established as long ago as 1980 and 208 towns from seventeen European countries joined. The Latvian capital, Riga, proudly hosted the '21st International Hanseatic Days' in June 2001 as part of its 800th anniversary celebrations. However, it is with the idea and image of 'the New Hansa', coined to describe the intensifying network of commercial and cultural links that have been created around the Baltic Sea since 1990 (Örn 1994), that the identity of the Baltic States has been formed. 'The New Hansa' conveys a sense of a Baltic European identity associated with the Scandanavian countries, northern Poland and northern Germany. In 1992 Tallin, the capital of Estonia, hosted a modern style Hansatag, or regional parliament. Mayor Velliste declared 'Our Hanseatic heritage is part of our genetic information like our fingerprint' (Von der Porten 1994: 78).

Economic performance has been mixed and political stability has been achieved only at the expense of disenfranchising some of the Russian ethnic minority population. Lithuania has exploited its warm water ports, for example Klapeida, to handle a proportion of the world trade of the Commonwealth of Independent States (CIS). Latvia has introduced financial reforms that allowed it to become an offshore banking centre and currency market for the CIS (Turnock 1996b). This move is connected to the eastward-looking Russian dominated business community, and the policy has produced short-term financial gains, but the economic and political instability of Russia has made it an uncertain long-term proposition. However, Russia's 1998 crisis caused only fleeting turbulence for Latvia. Estonia has attracted external investment from Finland and joint ventures have been established particularly in the region of the capitals. The future of the Baltic States looks bright to many businessmen:

> A decade of prosperity and imminent EU accession means that every (outside) corporate mind should know what's going on in Estonia, Latvia and Lithuania.
>
> (Orton-Jones 2001: 69)

The Swedish and Danish governments direct their aid programmes for former socialist countries to Latvia, Estonia, Lithuania, Poland and the St Petersburg and Kaliningrad districts of Russia (Swedish Ministry of Foreign Affairs 1996 and Danish Ministry of Foreign Affairs 1996). They also lobbied strongly in the EU for the admission of the Baltic States. Estonia was invited to apply for membership of the EU in 1997, whilst Latvia and Lithuania have had to wait until the beginning of the new millennium to open accession negotiations. The pace of reform seems to lie behind the decision, though it has been argued that the three economies are so small that absorption of all together would not be difficult. The governments look firmly west, even if the business community looks eastwards. The Baltic States have also expressed interest in joining the CEFTA.

The Baltic States, Estonia, Latvia and Lithuania, have received assistance from the EU through PHARE and TACIS programmes. The small size of the countries is reflected in the total sums received, but Estonia compares with other countries across the whole of Eastern Europe in terms of per capita assistance.

The Baltic States are, as a unity, becoming a distinctive place in the spatiality of the European and global business community, and they have a positive image. It is no surprise that it is the legacy of the Hansa that has risen to prominence in the identity of the Baltic States, because it captures and re-presents the image of a successful commercial age. Estonians, Latvians and Lithuanians have a long association, culturally and commercially, with Western Europe. Their experience differentiates them and their homelands from the Central Europeans. They looked north to Scandinavia as well as west. They had closer connections with imperial Russia and the Soviet Union than they cared for. Incorporation into the Soviet Union imposed direct political control from Moscow, integration into Soviet economic planning, and settlement by ethnic Russians, all of which were avoided by the satellite states that now make up Central Europe. On the other hand the Baltic States are strongly differentiated from the Eastern Borderlands, Belarus, Ukraine and Moldova, because they have embraced new ideas from the west, partly from a sense of westernness and partly because they wish to distance themselves from Russia. They have not joined the CIS.

Eastern Borderlands

'Here on the river Bug three countries converge; Ukraine, Belarus (White Russia) and Poland. These are the debatable lands, the flat, vulnerable, steppe country between Germany and Russia ... The Soviet empire has receded ... These lands are debatable once again. The borderlands across which Europe's great empires have ebbed and flowed' (Little 1994). The Republics of Ukraine, Belarus and Moldova make up a new Eastern Borderlands. It is a place whose identity is a residual of the larger communist Eastern Europe, which is changing only slowly. Historically, these states have long been part of Russia and the Soviet Union. Unlike the Central Europeans and the Baltic Europeans there is no sense of a common identity here. Ukraine and Moldova have significant ethnic tensions. The long association with Russia has brought large ethnic Russian minorities into these countries. Moldova is peopled by inhabitants of Romanian stock as well as Russians.

Since 1990, unlike the Baltic States, these Eastern Borderland states have not differentiated themselves clearly from Russia. They are all signatories to the CIS Treaty, the Russian dominated organisation that includes most of the former Soviet republics. At the beginning of 1997 Belarus signed a unifying treaty with Russia. Despite difficulties in dealing with the Russian government, Ukraine remains eastern facing in its international stance.

One unifying common attribute is that political élites were slow in introducing democratic principles into government. The government of Belarus was reluctant to remove the political and economic institutions of state control. In 1996 the president, an ex-communist, forced through parliament a repressive constitution to maintain his power. Ukraine has been characterised by the formation of regional élites that merged economic and political control and even seem to have criminal style associations using assassination in their competition for power. These regional 'clans' proved stronger than the political centre in Kiev and prevented the government from establishing a strong judicial system, exerting political control or creating a political consensus or sense of Ukrainian nationhood. Under these circumstances governments made what progress they could with the introduction of democratic institutions and privatising the economy. A potential ethnic conflict between the Ukrainian majority and Russian minority, the latter mainly concentrated in east Ukraine and the Crimea, has not, however, materialised. Trade within the CIS remains important and new trade with the west is limited (Marples 1994; Varfolomeyer 1997).

Tardiness in introducing political change has discouraged would-be commercial investors. Belarus, Ukraine and Moldova did little to attract outside investment. At the same time western governments were cautious in the extent of their aid assistance. Belarusian élites still look eastward and seek greater association with Russia. Ukraine on the other hand with a sharp ethnic divide is split between a Ukrainian west and Russian east in its aspirations. Moldova has a sharp political and social cleavage based on ethnic composition. Russians seeking independence in the Republic of Transdniestria east of the Dniestre river and the autonomous Gagauz region look east to Russia, whilst Romanians look west to Romania. In 1996 Russia was its principal trading partner followed by Ukraine and Romania. The principal western trader was Germany. Yet despite the enormous size of the German economy, trade with Moldova was only at the same level as Moldova's trade with Belarus and with Bulgaria.

Conclusion

This chapter has examined the production of the Marchlands in the New Europe in the context of the meaning of Europe as imagined, perceived and practised in neo-liberal modernity. The primary question confronting the idea of spatial modernity is: 'Is there a coherent place conceptualisation redefining the Marchlands within the New Europe?' To this question the answer is a tentative yes. The New Europe as the object of a spatial order of differentiated places is being constructed from a subjective mixture of locational, societal and cultural values and attributes as they have been newly imagined, perceived and experienced by insiders and outsiders. There is a broad coherence between the patterns and process forming the spatiality of the discourse on imaginary Europes and the Europes of perception, material practices and experiences. The imagined Europes considered here identified important dimensions and processes of development that have been producing the spatial architecture of the New Europe. However, it is clear that there is not one New Europe, but many New Europes, formed from competing spatialities.

The production of the Marchlands in the discourses and material practices of European neo-liberal spatial modernity is embedded primarily in the institutional spatiality of European multi-national corporations, the EU and NATO. All these spatialities fashion their Europe in thought and action around a wealthy, powerful and dominant west and a peripheral central and east. They perceive themselves, if only subconsciously, as custodians of modern Europeanness and arbiters of those who are, and those who are not, European. In concept and practice these spatialities have been adopted and adapted by many of the élites of the Marchlands, but not uniformly. They have, in some places, encountered strong conservative forces, drawn from attachments to historical, cultural antecedents and more recent political ideological legacies.

Western corporations, governments, EU and NATO leaders have been motivated by the desire to promote western ideas of democracy, security and the market, but they could not be sure that these would take root. The adoption of these ideas owes much to historical and cultural self-perceptions of those individuals and societal élites receiving them. For Central and Baltic Europeans, the new modernity from the west is in keeping with their self-image and their perception of history and culture, from which they construct their spatiality. They readily adopt the new capitalist market and political pluralist forms because it returns them to their perceived historical and geographical trajectory, for which EU membership is the great symbol. The belief that they are Western European permeates the thought and action of their spatiality.

For the heirs of Eastern European and Islamic civilisations, the New Europe is ambiguous. It is perceived, precisely, as a transition to Western European 'modernity', now clothed in the garb of international capital, NATO and the EU. Such 'modernity' is attractive to some individuals and groups, but repulsive to others. Whilst the west looms large in their spatiality, and is perceived by some as liberating, it is perceived by others as threatening and alien. The pertinent question for them is how much of a centuries-old culture, from which they derive meaning and identity, will they have to abandon in order to embrace a modernity which may bring nothing more than superficial consumerism? Is the societal and cultural fabric of the civilisations to be thrown away unthinkingly? It is very unlikely that the traditions of a millennium can be transformed without societal and cultural turmoil. Is transformation really worth the upheavals? Once again, the west is spreading its modernist project eastwards, and rekindling the historic conflicts of identity, place and spatiality. This turmoil is most keenly felt in the Balkans and the Eastern Borderlands. For those on the borders of the European Culture Area, it is not only because of the legacies of communism that the west and its ways of thinking and acting are ambiguous, but also because of the challenge it represents to much older cultural traditions.

Thus, two of four New Europes being produced in the Marchlands exist as multi-structured coherences of development in the spatialities of insiders and western outsiders alike. Central Europe and Baltic Europe have distinct and positive identities in the New Europe as produced from within and from outside. In contrast the Balkans and the Eastern Borderlands have been produced in conflicting western outsider and insider spatialities. The Balkans have a perverse, paradoxical, coherence of fragmentation and conflict. This is perceived from outside as negative, but fits the spatial logic of the strong ethnic place identity in the perception of insiders. Eastern Borderlands is differentiated because it has no such coherence. Contrasts with other subcontinental places stem from the depth of opposition to the western Neoliberal Project among powerful élites, the traditionalist sentiments of much of the general public and the persistence of ties to Russia.

'We Western Europeans used to know where our eastern frontier lay. It lay along the Iron Curtain and it was fixed as firmly in our minds as it was drawn through the heart of our continent – the eastern edge of our western, liberal, free market civilisation' (Little 1994). There is now a new boundary dividing Europe, west from east, and replacing the Iron Curtain. It is a Golden Curtain of wealth and prosperity. The line of this curtain separates Baltic Europe and Central Europe, which lie to the west, from the Eastern Borderlands and Balkan Europe, which lie to the east. It reflects the next enlargement of the EU and the degree of integration into the European economy.

Within the Marchlands, Central Europe is emerging as a 'core' of political and economic strength and stability whose governments are firmly set on framing their economic, social and political policies in terms of gaining early membership of the EU. Poland, the Czech Republic and Hungary are members of NATO. Centripetal forces are drawing Slovenia, Romania and the Baltic States towards Central Europe's international economic organisation. No matter how interim it may prove to be it is perceived as a stepping stone to economic co-operation within the EU. Counterbalancing centrifugal forces are drawing each of these new 'Europes' outwards to its nearest, powerful, external neighbour. Central Europe looks firmly west, Baltic Europe looks north to the Nordic lands as well as to the west, Balkan Europe is looking to the southeast and Eastern Borderlands looks to Russia for new contacts and development.

The newly emerging spatialities producing the Marchlands reconfigure the key defining features referred to previously (Chapter 2) and their elaboration in the Europe of neo-liberalist spatial modernity. New externally generated ideas, perceptions and practices are differentially modified by internal re-evaluation. Responses to the new situation are informed by existing attitudes and values, and inherited institutional and spatial patterns. Current economic and political factors, contingent on Western European integration and neo-liberal ideas, are the main agents of change, but they are conditioned by a resurgence of historical and cultural identities that were suppressed during the communist period.

16 Central and Eastern Europe as Marchlands in the global spatial modernity of the 1990s

This chapter looks at Central and Eastern Europe in the context of the conceptions, perceptions and practices of the 1990s that have been permeated by the idea of the 'global'. There is a new consciousness of the global; the globe as a single place. Place in its material and discursive modes has become understood as globalised. This chapter explores Central and Eastern Europe in the practices and perceptions of five discourses of global relations. These are interpreted as competing conceptualisations encompassed in a new global spatial modernity. The first of these is globalisation and the second globality. These two will be treated as a one-world, whole earth, oppositional discourse; the third is the discourse of emerging markets; the fourth is the demise of the communist global world system and the idea of post-socialist global regions; and the fifth is World Systems Theory. None of these discourses uses the concept of Marchlands, but as I will show in each Central and Eastern Europe are produced as a borderland. Each provides a partial insight into the neo-liberal order of spatial modernity, as it affects the Marchlands. Thus the chapter explores the new particularities of the Marchlands as a place of transition, and resituates them in the maelstrom of place relations in global spatialities of the 1990s.

Globalisation and globality

The exploration of space and the one-world, whole earth discourse

Space flight and photographic imagery have transformed the conceptualisations and symbolism of modernity that incorporated a notion of the global. Photographic imagery brought together global spatiality, in thoughts and actions. Cosgrove (1994) has shown how the image of earth, number 22727 shot from Apollo 17, was a culmination of this imagery and the iconography of modernity. He used this image and an earlier one, Earthrise, shot from Apollo 8, to provide a stimulating discussion of one-world, whole earth ideas. In spatial modernity it provides a framework for the resolution of the space/place transposition at its global limit. It conceptualises and gives meaning to planetary terraquous space as a formally bounded place (whole earth) and as a functionally linked place (one-world) at the upper scalar limit.

The Apollo images derived from the 1960s and their origins can be traced to the challenges and confrontations of the particular circumstances of the Cold War between the competing modernities. As used by Cosgrove, they sharpen the more fundamental ambiguity in the relationship between modernity and the earth as a resource that has come to the fore in the meaning of the Marchlands since the collapse of communism and the shattering of the Cold War's dual global order. The Apollo photographic images represent a great technological triumph. However, they show little of the earth's diversity and so convey a veneered sense of oneness. Thus, they bring together a sense of technology and the global as a single place in spatial modernity.

The ideas of globalisation and globality mirror at the global scale the tensions of individual versus institutional, environmental, modernity. The root of the globalisation–globality discourse is a dichotomy of spatial modernity writ large. It is individual development as expressed through the neo-liberal emphasis on individual, unregulated action versus institutional, environmental, development that takes account of communal interests in the present and in the future. The relationship between globalisation and globality is mediated by the enabling/disabling force of technology.

One reading of globalisation expresses the neo-liberal rhetoric of freedom for personal individual development of thought and action. In neo-liberalist thinking, personal ideas and interest operating through a free market are the basis of economic efficiency. Personal motivation is the only valid, naturally legitimated, platform for action. Communal institutions and the physical environment are objectivised and separated from individual development in thought and action. The crucial new dimension is the dramatic increase in the potential for interpersonal communication and with it the sense of personal freedom from an asserted, supposed, collective restraint imposed by organisational or institutional rules, usually exercised by the state. Generally, however, this ideology is tempered by the need to work through organisational team effort within institutional rule of law. Personal identity is still formed within a dense maelstrom of competing collectivities, as I have shown in earlier chapters. Even the Internet is not the open totality it is often claimed to be, because the supporting infrastructure is not universally available, and accessibility is not cost free.

Globality, promoted by environmentalist ideas and by movements to overcome Third World poverty, expresses a challenge to neo-liberal assertions by championing the ideas of whole earth. It invokes notions of togetherness, a single communal humanity sharing spaceship earth, as a global neighbourhood; of smallness as a separate planet in the vastness of cosmic space; and of stewardship of the earth's environmental resources. It subjectivises individual relationships, stressing empathy with the physical environment. It advocates global consensus in efforts to achieve sustainability in meeting environmental challenges. Technology is needed in the service of community and environment, not in their wilful, selfish, exploitation.

Globalisation – the one-world global spatiality

Globalisation theorists invite us to conceptualise the globe as one-world, perceived as flows of information, ideas, services, commodities and people, and experienced as complex connectivity. Globalisation sets functional spatialities against formal ones and emphasises the former. It can only be emphasis, because such concepts as the global division of labour, contrasts in production and consumption patterns, territorial specialisation, and global city control centres specify particular places. Even though these concepts represent the spatiality of globally organised corporations, marketing global brands and providing global services they imply attributes of formal place differentiation. Globalisation stresses geo-economic functional flexibility against geo-political formal rigidity. It proclaims the hollowing of the state in a predominant capitalist world economy encompassing the globe. It takes its primary cue from internet-mediated financial transactions and telecommunications-mediated experiences that are constantly pressed upon us as the norm of the immediate future if not the present.

Globalisation emphasises a so-called de-territorialisation in global conceptualisations, perceptions and practices as crystallised in the discursive prominence of modern communications technology. The ultimate practice of de-territorialisation has been the Internet, on which somewhere-place, has become everywhere-space, has become nowhere-cyberspace: the Internet is cyberspace, not geographical space. To be sure cyberspace is an interesting space and is now being studied as cybergeographies, even though in April 1999 the on-line population was only 158 million, just 2.6 per cent of the global population (Donert 2000). At the turn of the millennium the Internet and especially its use in commerce, E-commerce, is being given the hype that is a hallmark of the Neo-liberal Project of modernity.

One-world globalisation is the broad brush of spatial modernity. I will touch on three types of flow to illustrate it and the situation of the Marchlands within it: personal communication; FDI and transnational corporations; and global migration streams.

The Marchlands and the explosion of personal communications

The 'explosion' of personal communications technologies is a vital dynamic of neo-liberal spatial modernity. It liberates individual development. More than anything else the possession of a mobile phone, a personal computer and an internet connection creates the excitement of participating in the modernity of individual development. It opens access to friends, information, education, commerce and myriad other opportunities that at once encourage a self-awareness and promote individualness and independence. Or is it really isolation from physical place in a surrounding blank space? The self becomes the only point in a luxury of self-indulgence. At the global scale the practice of this individual spatiality is principally a matter of money within relation fields of collective wealth. It is an opportunity most noticeably afforded in the developed, rich world, less so in the transition world and hardly at all in the developing world.

No matter how you measure the practicality of individual development by reference to personal communications, this global distribution is restated. In the ownership of mobile phones and personal computers, as well as the availability of internet connectivity, globalisation's intensity is embedded in collective wealth. The global distribution of individual communications connectivity mirrors the broad global distribution of wealth. In 1997 in those countries defined as high income by the World Bank there were, on average, 188 mobile phones and 269 personal computers per 1,000 inhabitants. Middle income countries had 24 mobile phones and 32 personal computers per 1,000 population. Low income countries had just five mobile phones and five personal computers for every 1,000 population (World Bank Development Report 1999/2000). In 1999 North America had 94 million internet users, Western Europe 36 million and Asia–Pacific 26 million, of which 20 million were concentrated in Japan and Australia. In contrast, Africa, the Middle East and South America combined had only 6.4 million (Donert 2000).

In the 1990s the mobile telephone became the fashion symbol and a practical mediator of personal communications in the Marchlands. The relatively rapid spread of mobile phones was easily explained by the notoriously slow provision of terrestrial phones and the legendary poor quality of the state-owned telecommunications industry. The ownership of PCs soon took on a similar status and internet connectivity is now spreading. Even so in 1997 there were on average only 28 mobile phones and 48 personal computers per 1,000 population but a notable west–east gradient existed in this distribution. It is also certain that the regional and local distribution is extremely uneven. The number of computers in use is rising by about 18 per cent a year. Hungary and Slovenia have almost reached the levels of Western European states. At the end of 1999 internet users numbered about 5 per cent of the population in the western countries and rose to 20 per cent in Estonia. This compares with 34 per cent in the US, 18 per cent in the UK and 15 per cent in Germany. Mobile phone ownership in 1999 ranged from 0.6 per cent of the population in Ukraine to 18.9 per cent in the Czech Republic (BCE October 2000).

FDI and transnational corporations (TNCs)

In 1998 the Marchlands received an inflow of $16 bn of FDI. This compared with inflows of $85 bn into the Asia–Pacific developing area, $71 bn into Latin America and $8.3 bn into Africa. The Marchlands figure was a 25 per cent increase on 1997 and brought total FDI stock accumulated in transition to $92.4 bn. In 1998 the Marchlands were the source of only $1 bn FDI outflow (UNCTAD 1999). Internal FDI between the countries of the Marchlands was negligible. The figures for FDI flows (Table 16.1) show the dominance of the developed countries of the globe. FDI flows are notoriously difficult to measure yet an assessment of the situation of the Marchlands depends on how involvement is measured. Taken as a simple percentage of total global investment flows, the participation profile is very weak. Between 1995 and 1998 annual inflow was highest in 1995, but only reached 4.5 per

Table 16.1 Foreign direct investment flows

	Inflows (%)				Outflows (%)			
	1995	*1996*	*1997*	*1998*	*1995*	*1996*	*1997*	*1998*
Developed countries	63.4	58.8	58.9	71.5	85.3	84.2	85.6	91.6
Developing countries	32.3	37.7	37.2	25.8	14.5	15.5	13.7	8.1
Central and Eastern Europe	4.3	3.5	4.0	2.7	0.1	0.3	0.7	0.3

Flows per $1,000

	Inflows				Outflows			
	1995	*1996*	*1997*	*1998*	*1995*	*1996*	*1997*	*1998*
Developed countries	9.4	9.5	12.4	—	13.8	14.4	18.4	—
Developing countries	19.3	22.3	26.9	—	9.4	9.8	10.0	—
Central and Eastern Europe	20.6	15.3	22.3	—	0.7	1.4	4.1	—

FDI flows per capita (dollars)

	Inflows				Outflows			
	1995	*1996*	*1997*	*1998*	*1995*	*1996*	*1997*	*1998*
Developed countries	238.6	240.3	309.3	518.3	350.4	264.0	460.2	669.5
Developing countries	23.8	29.8	37.4	35.4	11.7	13.0	14.1	11.1
Central and Eastern Europe	42.3	36.8	55.1	52.2	1.4	3.3	10.2	5.7

Source: *UN World Investment Report 1999*: 20/21.

cent of the global total; outflows were highest in 1997, but amounted to a mere 0.7 per cent of the global total. If inflows and outflows are taken together and measured against per $1,000 GDP or per capita, the profile is different. With regard to GDP in the three kinds of global region the value of FDI flows is roughly very similar. Assessed per capita, FDI Marchlands flows generally exceed those of the developing world, but of course fall far short of the developed countries' figures.

A closer spatial analysis of FDI reveals, however, that the global connectivity of the Marchlands is rather weak. There are three core areas of developed economies – the USA, the EU and Japan. At the centre of these core areas are three global city localities that command and control flows and services – New York, London and Tokyo. These global city localities maintain a twenty-four-hour watch, divided into three eight-hour segments. They are central to three world blocs and are underpinned by what Taylor (2001), using the term 'world city', calls their hinterworlds. The relationship between the Marchlands and these cores is very different. In 1998 the EU accounted for 61 per cent of the FDI stock of the Marchlands, the USA 15 per cent and Japan a mere 0.5 per cent. These core economies concentrate on investments between themselves and after that they tend to focus on their own backyard (Figure 16.1). The Marchlands are closely tied to the European bloc. The Russian Federation accounts for just 1 per cent of FDI stock in the Marchlands.

From the point of view of development the general level of FDI, participation profile is perhaps not the most important factor. The type and purpose of investment is more critical. For example, flows financing R and D initiatives are more desirable than those attracted to finance low wage value or low value added production. My purpose has not been to evaluate FDI in the development of the Marchlands, this is a vastly complicated topic and has been touched on at other scales earlier. My purpose has been simply, if crudely, to situate the Marchlands in global flows. It is clear that the Marchlands participate only marginally in this aspect of globalisation and their situation is highly dependent on the EU.

The Marchlands and TNCs

FDI is undertaken for a variety of reasons and in many different forms. However, it directs attention to transnational corporations and so it is to these organisations that I turn briefly for another measure of participation in globalisation. In the late 1990s there were 850 parent corporations based in the Marchlands and 166,760 foreign affiliates according to the much hedged round figures of the *World Investment Report 1999*. This compared with 59,902 parent corporations and 508,239 foreign affiliates globe wide. Of the top 100 companies in Central Europe, all the Marchlands except Ukraine and Belarus, compiled by Deloite and Touche on the basis of 1998 sales value, 64 were privately owned. Of these 30 were foreign owned by the end of 1999. Twenty-six were owned by foreign TNCs, seven of these were in the top 25 and three of them were part of the Volkswagen group (BCE 2000). Of the top 25 TNCs based in the Marchlands only seven

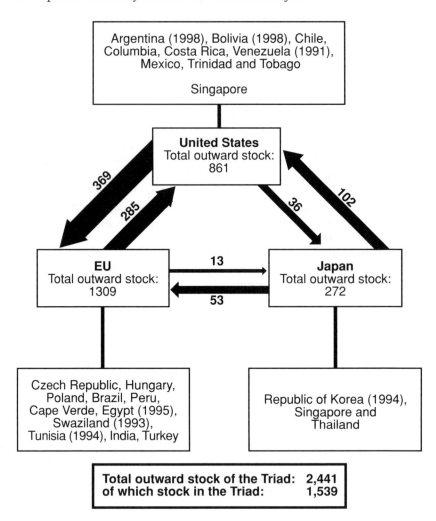

Figure 16.1 Foreign direct investment flows, 1995–98.

Source: *UN World Development Report 1999.*

were big enough to appear in the Marchlands' top 100 companies, but three of these were in the top five, one being foreign owned. It is likely that foreign TNCs do have portfolio holdings in some of these companies, only one of which was not privately owned.

Marchlands TNCs are very small by global standards. In 1998 together they had only $2.3 bn foreign assets, averaging $93 m. The median value for foreign assets was $52 m as compared with $1.3 bn for the top 50 TNCs based in developing countries. Some TNCs were, however, growing rapidly. The Hungarian oil and gas corporation (MOL) increased its foreign assets by 222 per cent in 1998. Total foreign sales of the top 25 TNCs increased to 3.7 bn in 1998. Foreign employment was low. Only five corporations had a ratio of foreign employment greater than 12 per cent, as compared with 35 per cent for the top 50 in the developing countries.

The UNCTAD Transnationality Index (TNI) measures something of the depth of foreign involvement of a TNC. It is composed of three sets of ratios: foreign assets to total assets; foreign sales to total sales; and foreign employment to total employment. At around 31 per cent for the top 25 Marchlands TNCs in 1998, it was lower than the top 50 in developing countries (34.2 per cent) and well below the global top 100 (55.4 per cent). All the global top 10 of 1997 had a TNI above 80 per cent. The TNI median value for the Marchlands top 25 is only 14 per cent, suggesting that the majority are very little transnationalised. In 1997 only the top ranking corporation, the Latvian Shipping Company (LSC), would have made it into the top 50 of the developing countries. Yet the LSC was not big enough to be on the top 100 Central European list.

The purpose of these comparisons is to broadly situate the Marchlands within the FDI flows and TNC profile of globalisation. It is not intended to explain it. However, it seems likely that the situation is consistent with the Marchlands as emerging market economies. Foreign TNC penetration links the Marchlands to globalisation, much more than the other way round. There is no surprise in this. These figures relate to a situation barely a decade after the transition began. This is hardly time for involvement to have become very deep. A much more exhaustive analysis would be needed to offer any detailed explanation.

The Marchlands and global migration streams

Global inequalities in wealth, opportunities for personal development, regional conflicts and political régimes are linked by global migration flows of economic migrants, refugees and asylum seekers. The Neo-liberal Project has resituated the Marchlands in global migration streams that have increasingly become an illegal people trade in the 1990s. Location on the edge of the European Union, the emergence of the Fortress Europe mentality in the EU, the vision of the EU as a golden place in the imagination of economic migrants, and the perception among economic migrants of the Marchlands as an ante-room for EU entry, have been specific factors of neo-liberal global spatiality that have shaped this resituation. The Marchlands thus became a transit route and a transit camp for a very heterogeneous group of migrants from Africa, Asia and South America en-route to the EU. At the same time, for some of these economic migrants and asylum seekers the Marchlands themselves are an intended destination.

EU immigration policy designed to deal with in-migration has adopted policies to defend 'Fortress Europe'. However, the Marchlands have become a conduit for economic migrants from farther afield trying to enter the EU both legally and illegally (Figure 16.2). The increase in illegal and exploitative people trafficking has become of great concern. The Baltic States have a serious problem of illegal migration traffic to Scandanavia organised by traffickers from Moscow (European Parliament 1998). Marchlands states have become a buffer zone protecting Western Europe. Those countries not perceived in the EU as sources for international migrants (Chapter 15) have been allocated the roles of filtering and absorbing global migrants seeking to gain entry to the EU. Economic migrants from farther east are now using established routeways through Poland and are collecting in the Czech Republic and Hungary as final staging posts for a dash into the EU. These countries have been given the task of absorbing those migrants who have not been filtered out by Marchlands countries further east. EU governments have tightened border controls and imposed repatriation agreements that require Poland, Hungary and the Czech Republic to accept back illegal immigrants apprehended after gaining access to the EU. Ukraine is acting as 'a comfortable waiting room to the west' for African and Asian migrants (IOM 1994). Thus increased contact with Western Europe has imposed on the Marchlands the burden of policing global migration through the territory and incurring significant penalties for failure.

The Marchlands countries have themselves been the intended destination of some migrants from further east. Kazaks of Polish ancestry, Russians and Vietnamese, as well as Ukrainians and Belarusians have migrated to Poland (Iglicka 1998). Increasing numbers of migrants are seeking permanent residence in the western Marchlands countries. Many of them are applying for residence after they have initially entered the country on transit visas. The economic strength of these countries and their rapid preparation of legal procedures for dealing with economic migrants, refugees and asylum seekers in line with the EU Aquis, makes them attractive destinations. As EU entry for these countries approaches reality, the further resituation of the Marchlands in the globalisation of people movement can be anticipated.

The spatiality of globalisation

Globalisation is a spatiality of flows and complex connectivity, in which the salient dichotomy is between the global and the local. Dunford and Kafkalas (1992), for example, discuss the 'global–local interplay', 'processes of globalisation' and 'tendencies to localisation of development'. Amin and Thrift (1994) discuss 'the local in the global', where the local signifies cities and regions and the global, worldwide processes. Global externalities are perceived as the prime causal category of local development and experience. Local developments are explained as responses to the effects of global challenges. The term 'glocal' has been coined to mean clever marketing locally in the context of a global village economy. Three 'global city' localities – New York, London, Tokyo – command and control the range and intensity of flows and connectivity. Tomlinson (1999) discussing globalisation and culture writes of the de-territorialisation of culture, by which he means the intrusion of global processes into the shaping of local culture. The attachment of culture to localities has been severed by the advance of global communications technologies.

Figure 16.2 Transit migration through the Marchlands.

Source: Data from *International Migration Organisation Country Transit Migration Reports.*

However, as the above analysis has shown the globalising/localising world have not yet driven out other strands from the spatiality of globalisation. The geography of flows and connectivity may be reshaping the global structures of the bipolar ideological world of the Cold War, the three worlds of development and the triad of cores and peripheries of the capitalist world economy, but it has not yet completely wiped them out. They remain as shadowy spaces, present, for example, in Dunford and Kafkalas' analysis of Europe as a 'global bloc', 'Fortress Europe', in a 'multi-polar order'. There are three blocs of globalisation – 'globalisation is very regional in nature' (Taylor and Hoyler 2000: 176).

The Marchlands in globality – the whole earth

Whole earth globality is a different conception of spatial modernity. It is 'an environmentalist's concept that appeals to the organic and spiritual unity of terrestrial life [that is] drawn toward a transcendent vitalism as a basis for universal order and harmony' (Cosgrove 1994: 290). There has been a strong strain in academic, business and popular discourse to thinking of the globe as supporting a single global environment over which humanity has a stewardship responsibility. International institutions such as the World Bank have incorporated environmental thinking into their spatiality. However, the most environmentally focused and active organisations are non-governmental, such as Friends of the Earth and Greenpeace.

The spatiality of the World Bank includes globalisation, globality, internationalism and localisation. In what it calls 'Protecting the Global Commons' the World Bank declares that certain environmental issues will put all countries at risk if no collective action is taken (World Bank 2000). It identifies these as: desertification, persistent organic pollutants, the fate of Antarctica, the high seas and seabed; ozone depletion; global climatic warming; and threats to biodiversity. The World Bank declares globalisation and localisation to be important development themes, but approaches them within a spatiality structured by international relations of states and markets.

In contrast to the World Bank's perspective, non-governmental organisations (NGOs) have found a specific role between the market and the state in the mobilisation of public opinion in issues that transcend national politics (Taylor 1999). Environmentalism is widely interpreted as a movement against modernity and in particular the consumerism and globalisation of American hegemonic modernity (Taylor 1999; Thomas 1999). It emerged in North America and Western Europe within a framework of established liberal democratic order and two principal strands evolved. 'Deep ecology' led to conservationist and protectionist attitudes. A second strand resulted from a heightened awareness of human influenced environmental forces such as global warming, depletion of the ozone layer and loss of bio-diversity. These issues it was perceived demanded an internationally co-ordinated response. International co-operation between governments has been difficult to achieve. Friends of the Earth and Greenpeace are the best known internationally active environmentalist NGOs and they have achieved some notable successes against TNCs and in influencing governments. These organisations have been faced with a task of mobilising opinion among a public that had no means of directly experiencing or conceptualising the global issues at stake. Consequently they sought to demonstrate that local action would have global impact. 'Think global, act local' is a slogan intended to give the public a sense of effective involvement in activities to overcome threatening global processes and has been persistently part of the vocabulary of the environmentalist movement.

Environmentalist movements in the First World have little resonance in the global south. Because they are formulated around issues of global forces they seem distant from local issues of poverty and livelihood and may be perceived as maintaining past practices that paid no attention to local knowledge, needs or aspirations (Thomas 1999). An enforced sense of dependence and political instability have formed the context of the evolution of environmentalist movements in the south. Consequently they criticise modernity as an 'outsider' imposed development order. One approach rejects development altogether because it is perceived to benefit only powerful élites and penalise the majority whose poverty remains endemic. A second advocates a resource management approach seeking ways of sustainable use of resources. NGOs try to ensure that the environmental costs of large-scale projects are met within the project. Such an approach is supported by First World NGOs (Thomas 1999).

The idea and practice of environmentalism in the Marchlands

Fleischer, Pomazi and Tombacz (1991) condemned communist environmental thought and action as 'environmental nihilism'. It is, indeed, difficult to exaggerate the extent of the environmental devastation that afflicted the Marchlands in the communist period. Every kind of environmental pollution and degradation was present in critical or near-critical form. Air pollution, water pollution and land degradation were widespread as a consequence of communist governments' policies. In addition the poor standard of design and engineering of nuclear power stations, brought into the international arena by the Chernobyl (Ukraine) disaster, threatened many districts. Despite this situation, if the ideas and practices that embedded the environment in communism were nihilistic, then it is the passive rather than the active variety of nihilism that predominated. Even so, there is little doubt that cleaning up the environmental legacy of communism makes the Marchlands one of the global hotspots of multiple environmental problems.

To avoid the mistrust of the population, pollution statistics were kept secret by communist governments. Despite the experience of this appalling environmental degradation, because of the authoritarian control and secrecy the environment was not an issue of prominence in the communist period. However, in the 1980s across Eastern Europe the appalling communist record of environmental devastation in general, and specific projects in particular, became a rallying point for popular opposition movements.

Unlike the environmentalist movements in the First World and the Third World, the environmentalist movements in the Marchlands began as a popular political agenda against communist modernity, not American consumerism. The Polish Ecological Club was founded in Kraków in 1982. The Danube Circle mobilised opposition in Hungary to the construction of the Bös–Nagymaros Dam Project. These Polish and Hungarian environmentalist movements were organised by intellectuals. In Bulgaria the environmentalist opposition to the communist government had its origins in the formation of local groups involved in social disobedience (Pávlinik and Pickles 2000).

The environmentalist movements of the 1980s, in the Marchlands, began what Pávlinik and Pickles (2000) termed the democratisation of nature. The environment has remained an issue of some public concern, but the popular movements have given way to professionalised NGOs. Since 1990 Western European international environmentalist organisations such as Greenpeace and Friends of the Earth have gathered support. The Polish Ecological Club, for example, is now an affiliate of Friends of the Earth. In Latvia, Lithuania, Hungary and Macedonia environmental groups are affiliated to Friends of the Earth. The Regional Environmental Centre for Central and Eastern Europe (REC) was set up in 1992 in Budapest with international funding. In 1997 REC identified and surveyed 3,020 environmentalist groups in fifteen Marchlands countries receiving 1,873 responses. The survey showed that the vast majority of groups were very small and financially insecure, with fewer than twenty-five active members and annual budgets below US$500. There was very little co-operation between the groups except at local level. They lacked expertise and were away from the centres of power. Only 17 per cent of groups saw lobbying as important and only 14 per cent considered protest action as important. They saw themselves as surrounded by a public that lacks awareness of environmental issues and regards the environment as a low priority when compared to material consumption. Yet, the environmentalist groups assessed themselves as quite successful in educating the public, fighting apathy, raising awareness and monitoring changes. However, they have little influence on government. The 1990s saw attempts to consolidate a ground level movement, but as yet this can only sustain local action and so environmental representation is weak (Dingsdale and Lóczy 2001).

Non-governmental organisations have been working for environmental improvement. Other processes that are embedded in the nature of development, as socio-economic and cultural transition, have also reshaped the ideas and practices of environmental relationships in the Marchlands. Among these the most important are commercial investment, external financial assistance for environmental policy and the policy priorities of governments.

Judged from the perspective of commercial investors, environmental improvement must enhance profitability, production efficiency or corporate image. Foreign investors sometimes insisted on government funded measures to improve environmental standards before they would invest significant sums. For example, the purchase of Skoda (Czech) by Volkswagen (German) and Tungsram (Hungarian) by General Electric (US). New parent companies sometimes imposed their standards on companies they acquired, as did the Swiss corporation Holerbank when it purchased the Hungarian Héjoscaba Cement and Lime Company. State-owned companies have also introduced new technologies to improve efficiency. Czech power plants have introduced modern desulphurising equipment for this purpose. New markets into which products are sold have imposed standards of product quality that have demanded investment in more environmentally friendly technology. The Sovnaft Oil Refinery company took such steps when it sought to sell in EU markets. The KGHM Polska Miedz sought to transform its image by installing fluid filters at its plants in Legnica and Glogow (Dingsdale and Lóczy 2001).

One international channel of western government assistance for environmental improvement 'was the Environmental Action Programme'. This plan aimed at funding clean-up operations and introducing new technologies (EEA 1993). Some 'debt for nature swaps' have been arranged. For example, in 1996 a debt reduction agreement was reached between Bulgaria and Switzerland whereby 20 per cent of Bulgaria's debt to Switzerland would be diverted to environmental improvements in Bulgaria (Kolk and Van Der Weij 1998). Meanwhile the EU is contributing about ecu 120 m annually to funding environmental improvements. However, this is less than 0.1 per cent of the estimated ecu 120–130 bn thought to be needed annually if the Marchlands states are to comply with the EU Aquis. It is clear that the environmental bill will have to be met by the governments of the Marchlands.

Governments in the Marchlands face a big task. They need to enact a framework of legislation and then introduce effective enforcement. The enactment might be relatively easy, the enforcement more difficult. Most countries have adopted the 'polluter pays' principle, so that a proportion of environmental expenditure is funded by levies, taxes and fines. At the same time governments must try to avoid the mistakes made by the EU countries. Governments also face the challenge of meeting the EU environmental Aquis. This task will be expensive, and suffers from the persistent practice of the EU in improving the standards that must be met. Whilst adopting the rhetoric of 'sustainable development' Marchlands governments do regard it as being forced upon them from outside and have been slow to enact and enforce environmental legislation. EU Transition Reports constantly highlight the weakness of measures to progress towards the Aquis in this category. Unsatisfactory performance, from the EU perspective, threatened EU accession of front running Hungary.

It is clear that the public insists that economic growth and the extension of material consumption takes precedence over environmental objectives. For this reason, economic development must be a priority for governments in the Marchlands and environmental influences are unlikely to carry much weight. There is, however, a new engagement with the economy/environment relationship. This is difficult to resolve because of the cost of meeting the environmental clean-up and formulating an environmentally responsive development policy. These objectives can only be achieved if the economy grows strongly. It is not likely that outside assistance will be sufficiently large or readily forthcoming. It is an enormous task to dismantle a command economy, institute a market economy and a civil society, and improve environmental conditions all at the same time.

Two themes have dominated the discourse on environmental transition in the Marchlands. One is the relationship between economic development and the environment. The other is the pace of change since 1990. As Pávlinik and Pickles (2000) have argued environmental transition is embedded the broader socio-economic dynamics of transition. The environment and its relation to development is now a part of the spatiality of individuals, organisations and institutions in a way that it was not under communism. However, greater awareness does not necessarily lead to a perception that prioritises environmental stewardship over economic development. In fact the practices adopted by insiders and outsiders that have produced the Marchlands environment in transition show economic growth is a higher goal than environmental improvement. The trends have, however, changed the environmental particularities of the Marchlands and resituated them in global environmental discourses.

A discourse on global poverty and anti-capitalism

A parallel whole earth discourse of opposition to neo-liberal globalisation has gathered around the issues of developing countries' debt and global poverty. The policies of global institutions such as the IMF, the World Bank and the World Trade Organisation (WTO) have been perceived and criticised as bringing debt and poverty to Third World countries. Since 1989, the 'shock therapy' advocated by western minded and funded consultants has thrown 150 million people living in Central and Eastern Europe and the former Soviet Union into poverty. That is more people than the combined population of France, the UK, the Netherlands and Scandanavia (UNDP 1998). Contrasts in levels of unemployment and polarisation of incomes are marked features of transition that have left many in poverty. Even in 1998 substantial numbers of people were living on less than $4.30 PPP per day, ranging from 1 per cent in the Czech Republic and Slovenia to 35 per cent in Latvia and reaching 44 per cent in Romania (Table 16.2) (BCE November 2000).

Table 16.2 Population in poverty, 1998

Country	% living on $2.15 per day (PPP)	% living on $4.30 per day (PPP)
Romania	18	44
Latvia	6.3	35
Ukraine	2.5	29.5
Lithuania	2.5	22
Estonia	2.0	19
Poland	1.5	18
Bulgaria	3.0	18
Hungary	1.5	15.5
Slovakia	2.5	6.5
Croatia	—	3.5
Czech Republic	—	1
Slovenia	—	1

Source: UNDP (1998).

The global confrontation between grass roots groups and global institutions produced Prague as a place of conflict in 2000. The joint meeting of the IMF and World Bank in the city was besieged by protest groups. They protested that the policies of global institutions were stifling local initiative for development in the developing world. However, the groups conducting the protest were outsiders, not many were local activists. The televised images of the street protests in Prague (as in Seattle before them) brought the attention of the wider western public to this discourse.

Chossudovsky (1997) presented a scholarly critique of the global Neo-liberal Project arguing that it was globalising poverty. Within the globalisation of poverty, Eastern Europe and the Soviet Union were perceived as undergoing a process of 'thirdworldisation'.

Chossudovsky argued that neo-liberalism had become embedded in academic and scientific thinking through a distorting ideology and a manipulation of statistics that was globalising poverty. The role of academic and scientific institutions was to produce economists trained in a scholarly discourse of pure theory (theory without facts) and applied economics (facts without theory), whose purpose was to conceal the mechanisms of the global economic system. Furthermore Third World intellectuals, and presumably this included those of the former eastern bloc, were enlisted to support the neo-liberal paradigm. The dominant dogma admitted no dissent, but set up a false debate between the official position and a counterparadigm that engaged a moral and ethical discourse. This discourse distorted and stylised policy issues of poverty, environment and the social rights of women. In practice the counterparadigm did not pose a real challenge to neo-liberalism, but rather evolved in harmony with neo-liberal dogma. This false discourse depends on the manipulation of income statistics to conceal widening disparities and the true extent of global poverty. Crucial in this manipulation is the World Bank's arbitrary definition of poverty as income below the equivalent of one US$ per capita per day. This enables poor populations to be represented as minorities in developing countries.

Within Chossudovsky's account of the globalisation of poverty, the former eastern bloc is perceived as undergoing a process of 'thirdworldisation'. As noted in Chapter 10, Eastern Europe, as part of the communist world system, was within the second world of a global three world spatiality. At the same time it was perceived as in the developed north, in the north/south global spatiality of development. Although generally below OECD members the level of material consumption, standards of health and education, and average income in Eastern Europe were much higher than in developing countries. The end of the Cold War is perceived by Chossudovsky to have caused a fundamental shift in the global distribution of income. The effect of IMF sponsored reform programmes has been to impoverish the states of the former eastern bloc that are now classified by the World Bank as 'developing economies'. This 'shift in categories reflects the outcome of the Cold War and the underlying process of "thirdworldisation" of Eastern Europe and the former Soviet Union' (Chossudovsky 1997: 42). They have now become the objects of the restoration of colonial patterns imposed by the new global Neo-liberal Project.

The spatiality of the globality discourse

Within the globality discourse two forms of a dual spatiality are strongly evident. There is a spatiality of the local and the global and a spatiality of the north and the south, with a shadow presence of the three worlds of development. The duality has occurred in an academic discourse in the 1990s around the ideas of 'ecological modernisation' and 'sustainable development' that also raise the issue of a unified global environmentalist movement. As mentioned earlier environmentalist campaigns have long encouraged the public to 'think global, act local'. Rootes (1999) switched this around when, in discussing the prospects for a global environmental movement, he asked if 'think local, act global' may not be a more useful strategy. He argues that despite appearances effective environmental action comes from educated, élite groups, mobilised for global causes, rather than bottom-up public opinion mobilised for local action. Local action sometimes brings environmentalists into contact with national and global actors. The media in reporting local environmental incidents often place them in a global setting. My point, however, is not concerned with which way round this duality should be operationalised, but rather its salience in discourse.

The contrasting conditions for NGOs in the north and the south have been referred to above. This reflects to some extent the radicalism and experience of activists in the contrasting environmental circumstances of a two-world globality – North and South. There is little reference to the environmental circumstances of the 'East' in transition. It, too, has implications for a unified global environmentalist movement. However, the debate is dominated by First World perspectives and as Van der Heijden (1999) has warned discourses reflecting the pragmatic compromises of the affluent First World may not be readily accepted in the South or in the Transition World. Governments outside the First World often see attempts to impose environmental standards as the same as attempts to undermine their path to living standards that match those of developed countries. Thomas (1999) thinks that only a new conception of progress that can compete with the consumerist and materialist achievements of modernity would suffice. Such a concept would need to satisfy the fundamental duality of modernity, individual development and collective development under diversified cultural conditions. That will be a difficult trick to turn.

The Marchlands and the discourse of 'emerging markets'

In 1998 the value of portfolio investments flowing into the Marchlands fell by 40 per cent on the 1997 figure, to $12.5 bn. Compared with the relatively smooth upward trend of FDI, portfolio investment was more erratic between 1993 and 1998. This pattern is related to the production of the Marchlands in the discursive global spatiality of 'emerging markets' as a specific feature of the Neo-liberalist Project of spatial modernity.

Recently attention has been drawn to a new revisualisation of global space that refers to the categories of 'emerging markets' and 'frontier markets' (Sidaway and Pryke 2000). These authors trace the origins of this revisualisation that 'rescripts' the three worlds division of global development spatiality to the history of the International Finance Corporation (IFC). When established in the 1950s as an affiliate of the World Bank, the IFC was intended to invest in the developing world rather in the manner of a merchant bank. By successfully investing it was meant to encourage private capital to follow. It was, in the Cold War, a weapon of private enterprise and capitalism in the conflict of modernities that aimed at assisting the growth of capital and money markets in Third World countries. It was intended to assert the superiority of capitalist democracies over the alternative state socialist model of development.

The IFC from 1986 has elaborated an index of emerging and frontier markets that was instrumental in defining the meaning of the concepts. The index was applied to certain economies of countries classified by the World Bank as low or middle income, whose financial regulations met certain criteria of openness to foreign portfolio investments. In the 1990s some economies of the Marchlands met these criteria. It should also be noted that Portugal and Greece, both EU members, are also included. Sidaway and Pryke represent the global spatiality of emerging markets as a 'simulation' constructed by the IFC, several private banks, brokerage and investment houses and the specialist financial press. Within this they identify a western construction of non-western places. The idea of emerging markets caught the imagination of investors who became engaged by discourses of exploration, exoticism, unpredictability and wildness that continued and reformulated the colonial and imperil discourses of an earlier age.

Whilst this is their principal thesis Sidaway and Pryke (2000: 195) also recognise that 'a reworked version of development as modernisation is once again virtually hegemonic in the South and has found new application in former communist states'. It is this comment that specifically resituates the Marchlands in the global spatiality of emerging markets. It implies the construction of the Marchlands in a new form of colonial or imperial spatiality, that of the neo-liberal spatial modernity. Colonial and imperial metaphors have previously had their strongest resonance in the Marchlands with reference to Nazi German and Soviet Russian spatiality. These were specific episodes in the wider discourse of modernity as development that informs my understanding of the Marchlands in spatial modernity.

Sidaway and Pryke point to the power of perception in shaping the practices of allocating portfolio investment in the emerging markets. Because there was a scarcity of information on which to make investment decisions, fashion, contagion and promotion drove the practice of investment. In this way the 'operationalisation of flows through emerging markets was generated by geographical imagination and socially constructed processes' (Sidaway and Pryke 2000: 197).

The division between 'emerging' and 'frontier' market status accorded to Marchland economies (Figure 16.3) restates the contrasts already noted as part of the multi-scalar renegotiation of place relationships. Hungary, Poland and the Czech Republic are perceived as emerging markets whilst the Baltic States and Ukraine are perceived as frontier markets. Other states

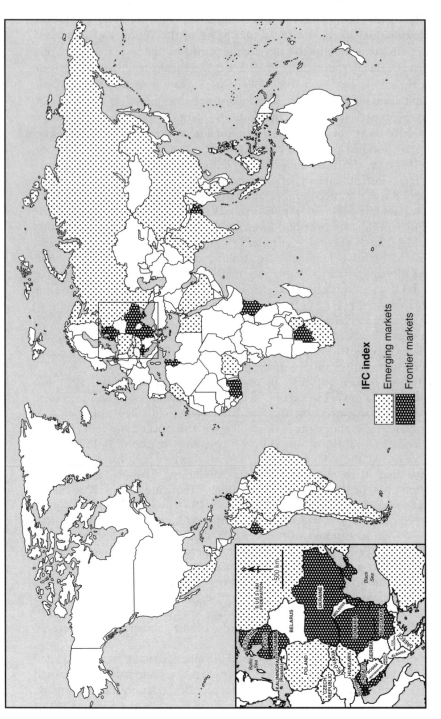

Figure 16.3 Emerging and frontier markets.

Source: After Sidaway and Pryke (2000) 'The Strange Geographies of "Emerging Markets"', *Transactions of the Institute of British Geographers*, New Series, Vol. 25(2), 192–3, © Royal Geographical Society (with the Institute of British Geographers).

are not regarded as in either category. Though they point out that their account is only speculative, the emphasis given by Sidaway and Pryke to geographical imaginings and social construction provides an excellent illustration of the significance of imagined, perceived and practised spatiality that is a key mechanism in my idea of development in spatial modernity.

The demise of the global communist world system and the idea of post-socialist global regions

In Chapter 1, I noted that post-communist, post-soviet and post-socialist labels had been contenders to describe the situation after the collapse of communist governments in Eastern Europe and the Soviet Union. Of these post-socialist has become the most widely used and applied to other global regions as well. This begs the question of what is the meaning of post-socialist? Does it mean when communist governments have been turned out or does it mean the present period after the collapse of the communist world system? The emphasis here is on the comparative pathways out of the global communist world system.

In Chapter 1, I described communism as a world system. I described how the communist world system was enlarged so that by the 1980s it was represented on every continent. In Chapter 11 drawing on Smith (1989), Desai (1992), Levigne (1992) and Michalek (1995) I sketched in the place order of the global reach of the communist world system. I noted the particularities and place relations of Eastern Europe within this global order. The collapse of communism in Eastern Europe and the Soviet Union, combined with the new enlargement of the capitalist world economy, has caused a disintegration of the communist global world system. The work of Desai and Levigne on which I drew in Chapter 11 was actually concerned with the effect of the collapse of communism on the developing Third World. Michelak's work dealt with the resituation of Eastern Europe in the global distribution of trade and aid after the collapse of communism, but also assessed the effect on the countries of the communist Third World.

This has led recently to the idea of post-socialist global regions. Herrschel and Light (2001) propose a global spatial division of five post-socialist global regions: Central and Eastern Europe; the post-Soviet states; China and Asia; Africa; and Latin America. Although these regions are called post-socialist, it will be immediately noted that some of them have states that retain communist governments. Herrschel and Light point to the different ways that socialism was introduced to these regions and the development contrast that I explored in Chapter 11, and wonder if they are leading to differentiation in the post-socialist pathways of the regions as a whole or whether a new state framework of experience is emerging. In broad terms the experiences of these regions does seem to be on a divergent path, only in Central and Eastern Europe has communism been completely displaced. In Cuba and China, for example, the story seemed to be one of a greater economic openness, but the maintenance of strong political control. In practice, in regions outside the Marchlands and the post-Soviet states, communist governments remain in power and are finding new roles in managing economic development. This may perhaps be interpreted as a form of communist evolution, rather than a post-socialist period in these countries. Herrschel and Light suggest this framework as a research agenda to explore the experience of the post-socialist world and, as yet, relatively few ideas have emerged in this discourse.

World systems theory

Most economic geographers did not regard communism as a separate world system during the period from 1945 to 1990, but as a particular form of state capitalism. As noted in Chapter 11, as part of a process of 'European shift' and 'global shift', Eastern Europe became drawn into the capitalist world economy from the 1960s onwards. Thus during that period, in world systems theory, the states of Eastern Europe and the Soviet Union were allocated to the semi-periphery. Bearing this in mind, it is not difficult to understand that when political geography was castigated by Michalak and Gibb (1992) for failing to theorise the collapse of communism, two celebrated geographers, Taylor and Johnston, could reply:

> The current transition going on in the former USSR and its erstwhile eastern allies is . . . not a transition from one mode of production to another, but rather a process of selection for the next patterning of uneven development within the capitalist world-economy. Some parts of this semi-peripheral zone will be peripheralised; other parts will retain their current status; and a small section may become part of the future core.
>
> (Taylor and Johnston 1993: 304)

The above quotation is the starting point for Bradshaw's (2001) recent charting of the transition trajectories of the twenty-seven successor states of Eastern Europe and the Soviet Union. Bradshaw's purpose was to assess 'the efficiency of the world systems approach as a framework for understanding post-socialist transition'. He pointed out that as Eastern Europe and the Soviet Union had been assigned to the semi-periphery before 1990, the question of post-socialist transformation, affecting 400 million people in these states, was 'an important issue in the semi-periphery', and asked if it was 'a major source of "peripheralisation" and social deprivation'. Bradshaw drew on a wide range of reports from European and global institutions for his analysis. Most of these sources have already been mentioned in my earlier analyses, because they were also the basis of my discussion. Bradshaw concentrated his attention on the economic dimension of transition, because world systems analysis stresses economic processes in allocating states to core, semi-periphery and periphery status. However, he recognised that economic processes were only a part of much broader systemic transformations.

Bradshaw examined the twenty-seven transition trajectories under three headings. First, climbing out of 'transition recession'. Second, the progress reports issued by the EU in the framework of Agenda 2000. Third in the recorded patterns of FDI. (See Chapter 14 for a detailed discussion of each of these elements for the individual Marchlands' states.) Under the first of these headings, he concluded that their was no single variable to explain the diversity of state trajectories, but that the experience of the 1990s indicated that many transition economies were bound for the global periphery. Under the second heading, he concluded that Hungary, Poland, the Czech Republic, Slovenia and Estonia were in a good position to join the core. Under the third heading, he concluded that in Central and Eastern Europe privatisation was the main incentive to foreign investors, and that these states had become integrated into the capitalist world economy. In contrast in the CIS states, it was the rich natural resources that had attracted foreign investment, but to a smaller extent, and so the majority of CIS states are still not engaged in the global economy.

He concluded, more broadly, that across these transition economies there was a west–east divide, between Central and Eastern European countries, including the Baltic States, and the CIS. There is in both the east and west a north–south division. Geography, by which Bradshaw seems to mean location, is important, but was not the determining factor. He believed that history, current economics, politics and culture were the critical factors. Furthermore, the pattern was not the result of interaction between the transition economies, but due to the different roles of the states in the 'global economy'.

Bradshaw accepts that the analysis could have been undertaken without mention of world systems theory, but claims that a world systems analysis is a worthwhile framework for three reasons. First, it puts the transition trajectories of post-socialist states in a 'global setting'. Second, it puts 'transition' in the broad perspective of capitalist economic development. Third, it offers a dynamic model of the global economy of which the transition economies are an important part.

I have dwelt at length on Bradshaw's analysis for three reasons. First, it is recent and sets out clearly the world systems approach to the successor states of Eastern Europe and the Soviet Union. However, it is only these socialist states to which the analysis is extended. Other post-socialist states, in the sense of the demise of the socialist world system, are not investigated. Second, it shows how the capitalist world economy may be redistributing these post-socialist states to all the global spaces of world systems theory. Third, and most importantly, its conclusions confirm the more detailed analyses of Chapters 14 and 15 at the global scale. The analysis in those chapters focused only on the states within the Marchlands, whereas Bradshaw included states beyond the boundaries of the Marchlands.

The spatiality of world systems theory

As noted in Chapter 1, and demonstrated by Bradshaw's analysis, world systems theory proposes that there is a single global market, within which there were multiple numbers of states. The states are allocated to three spatial categories: core, semi-periphery and periphery on the basis of the processes going on in them. These three zonal categories are differentiated on the basis of production systems, but came to be interpreted as systematically related by dependence of the semi-periphery and periphery on the core. World systems theory, therefore, advocates a three scale spatiality – global, zonal and state. This spatiality is the outcome of the capitalist world economy whose formation began in the eighteenth century and whose enlargement attained a global reach around 1900 (Taylor 1999). World systems theory assumes that this spatiality will form the framework of dynamic equilibrium for resituating countries in the global market.

Conclusion

What and where are the Marchlands in the competing global spatialities of neo-liberal spatial modernity? Globalisation, globality, emerging markets, post-socialist global regions and world systems theory each describe processes from a global perspective. These processes are changing the place particularities and the place relations of Central and Eastern Europe that I call the Marchlands. Each spatiality emphasises a different aspect of individual, organisational, institutional and environmental development. Globalisation and globality stress a global/local dichotomy within a fading structure of the three worlds development model, and a background of capitalist multi-polar order. Globalisation stresses flows of people, ideas, money and commodities in a complex connectivity of personal, telecommunications mediated, experience of globe and locality in a one-world development arena. Three city localities command and control the range and intensity of these global flows. They form three blocs of globalisation. Globality stresses global environmental challenges and the deepening and widening gulf between rich and poor in a whole earth development scenario. Emerging markets emphasise a new colonialism in a dual spatiality of emerging and frontier markets in the global semi-periphery and periphery. Post-socialist global regions focus on a fivefold global division of comparative pathways out of socialism and towards the new neo-liberal global market. World systems theory describes the resorting of states presently in the semi-periphery, into the three zonal spaces of a single global market.

Each of these spatialities spreads across global space. Each, in its own way, constructs Central and Eastern Europe as a place in its global perspective. Each draws a global boundary through Central and Eastern Europe, but in each the boundary is different. For each of these spatialities Central and Eastern Europe are borderlands. They straggle a major fissure in the global space of neo-liberalism. The spatialities are themselves competing global perceptions. They are best understood as partial discourses of neo-liberal spatial modernity at the global scale. Taken together they produce the idea of the global as a place ordering of global spatial modernity. Within this they produce Central and Eastern Europe as a global Marchlands of neo-liberal modernity.

Finally

Central and Eastern Europe in practice and in discourses are Marchlands. They are disputed borderlands. They lie between Berlin, Vienna, Istanbul and Moscow; between Germany and Germans, Russia and Russians; between Europe and Asia; between West and East and South. Modernity in these Marchlands has been produced as three geo-historical periods, spatial modernities, embedded in multi-scalar place/space transpositions, as local, regional, state, European and global transitions and transformations.

The aim of this book has been to interpret the historical geography of Central and Eastern Europe since 1920 as a reciprocal interplay of geographical space and modernity. The aim has been to chart the nature of modernity as mediated through and by geographical space. To do this, I have introduced the theory of spatial modernity. Spatial modernity theorises an experience of time and space, produced as multi-scalar, meaningful, relational place, by individual organisational and institutional development in thought and action, and expressed as competing spatialities – place orderings of space. Spatial modernity asserts that modernity is an experience of time and space as place development, embedded in socio-cultural practices and discourses. Spatial modernity draws attention to the thoughts and actions, the ideas and practices, that produce modernity as multi-scalar place relationships of development in the here and now: to the spatiality of individuals, organisations and institutions as locational, societal and cultural actors in imagined, perceived and experienced place/space transpositions; to the constructed, contested production of place that imposes order onto a maelstrom of people, ideas, events and artefacts – the nature of the modern experience is an experience of spatiality. Modernity is a unique experience for every individual, but is structured within a nexus of contested spatiality, at the core of which is the space/place transpositional relationship. The experience of modernity is different depending on spatiality. Place makes modernity.

Spatial modernity was used to identify a distinctive, modern geo-historical period in Central and Eastern Europe. The First World War was recognised as having imposed a break on the spatial development of Central and Eastern Europe that inaugurated a 'modern' phase in their place development. This roughly coincided with the third phase of historical modernity as identified by Berman (1983) and Taylor (1999). Berman recognised a first phase of modernity extending from the sixteenth century to the eighteenth century and the second phase in the nineteenth century. The third phase was the twentieth century. Taylor had an approximately similar sequence, identifying what he called three Prime Modernities. These he described as the Dutch-led mercantile modernity, the British-led industrial modernity and the American-led consumer modernity. In the first phase culture and ideology were 'man' centred. In the second phase it was the idea of progress that dominated thinking. In the third phase the idea of development and modernisation tamed the tyranny of novelty. It was in this third phase that modernity as Taylor understood it became global.

Berman and Taylor (Chapter 1) each rejected the development theories that underlay studies of modernisation in the 1950s and 1960s. Berman did so because he perceived them as an American imperial strategy for the Third World. However, Berman's modernity was almost aspatial. He stated that 'Modern environments and experiences cut across all boundaries of geography . . . in this sense, modernity can be said to unite all mankind' (Berman 1983: 15).

Taylor rejected development modernisation theories because they were based upon a 'geo-historical trick' that substituted history for geography. The theories depended on the erroneous idea that all countries went through the same processes of development but at different times. Contemporary contrasts therefore depended on where a country was along a common pathway. Taylor aimed to redress the balance that the trick had pulled by emphasising the 'geo'. In this he was completely correct. But his insight was incomplete. His conceptualisation of modernity, though it cast off the close association between modernity and industrialisation, was too narrow. It relied too much on the idea that capitalism and the capitalist world economy were modernity. His exploration of space and place was too muted.

This book has used the idea of spatial modernity to trace the trajectory of modernity as an experience of time and space through the production of multi-scalar place in Central and Eastern Europe from 1920 to the turn of the third millennium. This has given rise to the core idea of Central and Eastern Europe as Marchlands of Modernities, lands of competition between competing projects of modernity. Three distinct projects that produced modernity through multi-scalar place development – the Nationalist, the Communist and the Neo-liberalist Projects – have been described and the salient discourses, different at different scales, expressed and discussed as contested spatialities. They produced the Marchlands in three geo-historic periods, time/space conjunctions. In each period a new project of modern spatiality strove to remove the earlier project in practices and discourses and establish the naturalness of its own spatial order, within a global spatial modernity. In each geo-historic period the particularities of localities, regions and states within Central and Eastern Europe were produced as new places, by development and resituated in place relations as the overarching discourse of spatial modernity. The particularities of Central and Eastern Europe were thus developed and resituated in European and global place relations. In each period, place particularities and place relations, articulated in multi-scalar meaning, were a key dynamic of the modern experience. Recollections and inventions of the past and aspirations for the future were conjoined in the material and discursive spatiality of the present, but it was this spatiality that was modernity.

In the first and third of these geo-historical periods the production of localities, regions and states within the Marchlands and the production of the Marchlands in European and global space were directly related to western global modernity. However, this western modernity was very much shaped by the European experience and much less by the American. In the second geo-historical period the western influences were displaced by an ideology that had matured in the east. A distinctive place, Eastern Europe was produced in European and global space as a complex layering of contrasting perspectives. Thus the development of Central and Eastern Europe, normally perceived as a discontinuity, is shown to be a sequence of three time/space conjunctions of different modernities and so put into a unifying process and structure of spatial modernity.

In each geo-historical period contemporary spatiality formed the meaning of modernity. In particular the reciprocity of power between individual devel-opment and collective development was reworked in multi-scalar place par-ticularities and place relations. In parallel, the dynamic of development as the interplay of imagined, perceived and experienced place was a prominent arbiter of modernity for locational, societal and cultural practices and discourses.

These dynamics of spatiality reproduced in each distinct geo-historical period the tension between the ideas of Central Europe and Eastern Europe as places in contested spatialities. In each of the geo-historical periods élites, intellectuals and the general public who regarded themselves as Central European emphasised the difference between themselves and those people and places that they perceived as Eastern European. In the communist modernity the idea of Central Europe disappeared until its resurgence as part of the nexus of intellectual and popular opposition to the enforcement of an eastern modernity, particularly among people who identified them-selves with Western European civilisation. Ideas were paralleled by material and spatial practices, and the western areas of the Marchlands always enjoyed lifestyles and material prosperity more abundant than easterly areas.

Established most emphatically after the First World War the idea and practice of the nation-state was a powerful dynamic of spatiality. It was a key place experience of modernity. When transformed into the Party-state it remained in thought and practice central to individual, organisational and institutional spatiality. When released from the constrictions of the Party-state, though challenged by multi-scalar, global and local dualities, the association of nation and state, in all its ambiguities, retains a strong hold in modern spatiality.

Despite advances in transport and communications technology, and the effect they have had on the spatiality of socio-cultural formations, locality and region have remained powerful ideas, conditioning the experience of modernity, as individual and environmental development. A deep sense of locality and region still permeates spatiality and modernity in the Marchlands.

From the 1920s to the present the peoples of the Marchlands have looked for a brighter future, whilst always keeping an eye on the past, but have had to endure an uncertain, threatening, anxious present. Each geo-historical period was a transformation in which the antecedent spatiality was cleared out and new places produced. In each geo-historical period, six key features have been refashioned by the new dominant ideology as it reworked antecedent spatialities. Localities, regions and states were produced as development in a lattice of experience, perception and imagination from insider and outsider perspectives. The identity and meaning of Central and Eastern Europe as Marchlands were reproduced in six key ideas. First, they were Europe's 'other', and the 'Other' Europe, a demi-Orient. Second, they were the battlegrounds fought over by powerful neighbours from west and east. Third, they were economically backward, peripheral to outside cores of economic innovation and power, constantly trying to catch up. Fourth, their ethnic and cultural diversity reproduced an aura and tension of place identity underlain by exclusion and violence. Fifth, they were a zone of received influences, patterned by testing and adapting ideas and practices from west and east to local conditions. Sixth, as a maelstrom of ideas, peoples and events they were produced, in the strongest sense, as 'lands between', as borderlands, formed from contested spatialities as Marchlands of Modernities.

References

Abler, R., Adams, J. and Gould, P. (1972) *Spatial Organisation: the geographer's view of the world*, Prentice Hall, London

Abruden, I. and Turnock, D. (1998) 'A Rural Development Strategy for the Apuseni Mountains, Romania', in D. Turnock (ed.) Rural Diversification in Eastern Europe, *GeoJournal*, Vol. 46(3), 319–36

AEBR (1996) *The Association of European Border Regions: 25 Years of Working Together*, AEBR, Gronau

Agh, A. (1999) 'The Early Consolidation in East Central Europe: parliamentarization as a region-specific way of democratization', *Society and Economy in Central and Eastern Europe*, 3: 83–110

Allen, J., Massey, D. and Cochraine, A. (1998) *Rethinking the Region*, Routledge, London and New York

Altvater, E. (1998) 'Theoretical Deliberations on Time and Space in Post-Socialist Transition', *Regional Studies*, Vol. 32(7), 591–605

Alvarez, A. (1965) *Under Pressure. The writer in society: Eastern Europe and the USA*, Penguin, Harmondsworth

Amin, A. and Thrift, N. (1994) *Globalisation, institutions and regional development in Europe*, Oxford University Press, Oxford

Anderson, B. (1991) *Imagined Communities*, Verso, London

Anderson, P. (1984) 'Modernity and Revolution', *New Left Review*, No. 114, 96–113

Ashworth, G. (1994) 'The Transition to Market Economies and Marketing Cities', in Z. Hajdú and Gy. Horváth (eds) *European Challenges and Hungarian Responses in Regional Policy*, Hungarian Academy of Sciences, Pécs

Ashworth, G. and Tunbridge, J. (1990) *The Tourist Historic City*, Belhaven, London

Ashworth, G. and Voogd, H. (1998) 'Marketing the city: concepts, processes and Dutch applications, *Town Planning Review*, 59, 65–79

Aspaturian, V. (1984) 'Eastern Europe in World Perspective', in Rakowska-Harmstone (ed.) *Communism in Eastern Europe*, 8–49, Manchester University Press, Manchester

Banac, I. (1990) 'Political Change and National Diversity' *Eastern Europe . . . Central Europe . . . Europe DÆDALUS* 119(1), 141–59

Barraclough, G. (ed.) (1986) *The Times Atlas of World History*, Times Books, London

Batchler, J. and Downes, R. (1999) 'Regional Policy in the Transition Countries: a comparative assessment', *European Planning Studies*, 7(6), 793–808

Barisitz, S. (1996) 'Changes in Trade Patterns – the case of Austria', in F.W. Carter, P. Jordan and V. Rey (eds) *Central Europe After the Fall of the Iron Curtain*, 237–44, Lang, Frankfurt

Barta, Gy. (1984) 'The Spatial Impact of Organisational Changes in Industrial Companies', in Gy. Enyedi and M. Pécsi (eds) *Geographical Essays in Hungary*, Hungarian Academy of Sciences, Pécs

Barta, Gy. and Dingsdale, A. (1988) 'Impacts of Changes of Company Organisation on Peripheral Regions: a comparison of Hungary and the United Kingdom', in Linge, G. (ed.) *Peripheralisation and Industrial Change*, 165–83, Croom Helm, London

Bassin, M. (1991) 'Russia between Europe and Asia: the ideological construction of geographical space', *Slavonic Review*, 50, 1, 1–17

BCE (1999) 'Shifting gear' *Business Central Europe*, May 14–19

BCE (2000) 'A Survey of E-Businesses', *Business Central Europe*, March, 50

Beaver, S. (1937) 'The Railways of Great Cities', *Geography*, No. 116, Vol. XXII, Pt 2, 116–20

Beaver, S. (1940) 'Bulgaria: a summary', *Geography*, No. 130, Vol. XXV, Pt 4, 159–69

Beaverstock, J.V., Smith, R.G. and Taylor, P.J. (1999) 'A Roster of World Cities', *Cities*, 16, 445–58

Begg, R. and Pickles, J. (1998) 'Institutions, Social Networks and Ethnicity in the Cultures of Transition: industrial change, mass employment and regional transformation in Bulgaria', in J. Pickles and A. Smith (eds) *Theorising Transition*, 115–46, Routledge, London

Bell, P. (1984) *Peasants in Socialist Transition*, University of California Press, Berkeley

Bender, T. and Schorske, C. (eds) (1994) *Budapest and New York: Studies in Metropolitan Transformation 1870–1930*, Russell Sage Foundation, New York

Bennett, R. (1998) 'Local Government in Postsocialist Cities', in G. Enyedi (ed.) *Social Change and Urban Restructuring in Central Europe*, 35–54, Akadémiai Kiadó, Budapest

Berend, T. and Ránki, G. (1974) *Economic Development in East-Central Europe in the Nineteenth and Twentieth Centuries*, Columbia University Press, New York and London

Berend, T. and Tomaszewski, J. (1986) 'The Role of the State in Industry, Banking and Trade' in M.C. Kaser and E.A. Radice (eds) *The Economic History of Eastern Europe 1919–1975*, II, 3–48, Oxford University Press, London

Berentsen, W. (ed.) (1997) *Contemporary Europe* (Seventh Edition), Wiley, New York

Berman, M. (1983) *All that is Solid Melts into Air: the experience of modernity*, Penguin, New York

Bicik, I. and Dana, F. (1997) 'Second Homes: Case Study of the Kocába Region', *Acta Universitatis Carolinæ Geographia XXXII Supplementum*, 247–53

Bidaleux, R. (1999) '"Europeanisation" versus "Democratization" in East-Central Europe, Ten Years On', *Society and Economy in Central and Eastern Europe*, 3, 24–61

Birzulis, P. (1999) 'Tatu', *Business Central Europe*, June, 62

Bozyk, P. (1988) *Global Challenges and East European Responses*, Polish Scientific Publishers, Warsaw

Bradshaw, M. (ed.) (1997) *Geography and Transition in the Post-Soviet Republics*, Wiley, Chichester

—— (2001) 'The Post-Socialist States in the World Economy: transformation trajectories', *Geopolitics* 6(1), 27–46

Brubaker, R. (1995) 'National Minorities, Nationalising States and External National Homelands in the New Europe', *DÆDALUS*, 124(2), 114–32

Brun, E. and Hersh, J. (1990) *Soviet–Third World Relations in a Capitalist World: the political economy of broken promises*, Macmillan, New York

Bugge, P. (1993) 'The Nation Supreme: the idea of Europe 1914–1954', in K. Wilson and J. van der Dussen (eds) *The History of the Idea of Europe*, 83–149, OU/Routledge, London

Burdack, J., Grimm, F-D. and Paul, L. (1998) 'Introduction', in J. Burdack, F-D. Grimm and L. Paul (eds) *The Political Geography of Current East–West Relations*, 45–52, Leipzig Institut für Länderkunde, Leipzig

Carter, F.W. (1987a) 'Czechoslovakai', in A. Dawson (ed.) *Planning in Eastern Europe*, 103–38, St Martin's Press, New York

—— (1987b) 'Bulgaria', in A. Dawson (ed.) *Planning in Eastern Europe*, 67–102, St Martin's Press, New York

Carter, F.W. and Kaneff, D. (1998) 'Rural Diversification in Bulgaria', in D. Turnock (ed.) Rural Diversification in Eastern Europe, *GeoJournal*, 46(3), 183–91

Carter, F.W., Jordan, P. and Rey, V. (eds) (1996) *Central Europe After the Fall of the Iron Curtain*, Lang, Frankfurt

Chapman, M. (1996) 'Slovenian National Identity Exemplified', in D. Hall and D. Danta (eds) *Reconstructing the Balkans*, 109–14, John Wiley, Chichester

Chossudovsky, M. (1997) *The Globalisation of Poverty: Impacts of IMF and World Bank Reforms*, Zed Books, London

Christaller, W. (1933) *Central Places in Southern Germany*, trans. C.W. Baskin, 1966, Prentice-Hall, London

Compton, P. (1987) 'Hungary', in A. Dawson (ed.) *Planning in Eastern Europe*, 167–94, St Martin's Press, New York

Cosgrove, D. (1994) 'Contested Global Visions: one-world, whole-earth and the Apollo space photographs', *Annals of the Association of American Geographers*, 84(2), 270–94

Crang, M. (1998) *Cultural Geography*, London, Routledge

Csapó, T. (1999) 'Határ menti együttmüködések a munkaerőpiac területén, különös tekintettel Vas és Zala megyére', in M. Nárai and J. Rechnitzer (eds) *Elválast és Össekö-a Határ*, 269–96, MTA RKK, Pécs-Győr

Danish Ministry of Foreign Affairs (1996) *Danish Official Aid to Central and Eastern Europe 1996: an overview of projects and programmes*, DMFA, Copenhagen

Daviddi, R. (1992) 'From the CMEA to the "European Agreements": trade and aid in the relations between the European Community and Eastern Europe', *Economic Systems*, Vol. 16, No. 2, 269–94

Dawson, A. (1979) 'Factories and Cities in Poland', in A. French and F.E.I. Hamilton (eds) *The Socialist City*, 349–86, John Wiley, Chichester

—— (1987) 'Yugoslavia' in A. Dawson (ed.) *Planning in Eastern Europe*, 275–92, St Martin's Press, New York

Delanty, G. (1996) 'The Resonance of Mitteleurope: Habsburg myth or antipolitics', *Theory, Culture and Society*, 13(4), 93–108

Demko, G. (ed.) (1984) *Regional Development Problems in Eastern and Western Europe*, Croom Helm, London

De Rougement, D. (1966) *The Idea of Europe*, Macmillan, New York

Desai, A. (1992) 'East–South Trade: consequences of economic changes in the Soviet Union and Eastern Europe', *Journal of Development Planning*, 23, 73–107

Dicken, P. (1992) *Global Shift: The Internationalisation of Economic Activity* (Second Edition), Paul Chapman Publishing, London

Dingsdale, A. (1986) 'Ideology and Leisure under Socialism: the geography of second homes in Hungary', *Leisure Studies*, 5, 35–55

—— (1991a) 'Socialist Industrialisation in Hungary', in F. Slater (ed.) *Societies, Choices and Environments*, 440–75, Collins, London

—— (1991b) 'Regional Development Policy in Hungary 1948–1990', Paper presented to the Political Association, Annual Conference, Newcastle-upon-Tyne

—— (1996) 'Hungary as a Place of Refuge', in D. Hall and D. Danta (eds) *Reconstructing the Balkans: A New Geography of Southeast Europe*, 197–208, John Wiley, Chichester

—— (1997) 'Hungary in the New Europe: transport policy and development imperatives', in A. Dingsdale (ed.) Transport in Transition: issues in the New Europe, *Trent Geographical Papers*, 1, 64–82

—— (1999a) 'New Geographies of Post-Socialist Europe', *The Geographical Journal*, Vol. 165(2), 145–53

—— (1999b) 'Redefining Eastern Europe: a new regional geography?', *Geography*, vol. 84(3), 205

—— (1999c) 'Budapest's built environment in transition' in Z. Kovács (ed.) Post-socialist urban transition in Eastern and Central Europe, *GeoJournal*, 49(1), 63–78

—— (1999d) 'The Problem is The Map' *Lock Haven International Review*, 13, 10–25, Lock Haven University of Pennsylvania, Lock Haven

Dingsdale, A. and Lóczy, D. (2001) 'The environment challenge of societal transition in East Central Europe' in D. Turnock (ed.) *East Central Europe and the Former Soviet Union: Environment and Society*, 187–99, Arnold, London

Dogan, M. (1997) 'Nationalism in Europe: decline in the west, revival in the east', *Nationalism and Ethnic Politics*, 3(3), 66–85

Donert, K. (2000) 'Virtually Geography: aspects of the changing geography of information and communications', *Geography*, 85(1), 37–45

Doomernik, J. (1997) 'Current Migration in Europe', *Tijdschrift voor Economische en Sociale Geografie*, 88(3), 284–90

Döry, T. (1999) 'Vallalkozások jövőképe osztrák-magyar határ menti térségben', in M. Nárai and J. Rechnitzer (eds) *Elválast és Össekö-a Határ*, 209–34, MTA RKK, Pécs-Győr

Dostál, P. (1997) 'Modernisation, Democratisation and Economic Liberalisation in Post-Communist Countries', *Acta Universitatis Carolinæ Supplementum*, 32, 307–23

Drabek, Z. (1985) 'Foreign Trade Performance and Policy', in M. Kaser and E. Redice (eds) *The Economic History of Eastern Europe 1919–1975*, 1, 379–466

Dragan, A. (1991) 'The Czech–German–Jewish Symbiosis of Prague: The Langer Brothers, *Cross Currents*, 10, 181–93

Drake, G. (1995) *Issues in the New Europe*, Hodder and Stoughton, London

Dunford, M. (1998) 'Economies in Space and Time: economic geographies of development and underdevelopment and historical geographies of modernisation', in B. Graham (ed.) *Modern Europe: place, culture, identity*, 53–88, Arnold, London

Dunford, M. and Kafkalas, G. (eds) (1992) *Cities and Regions in the New Europe*, Belhaven, London

Duró, A. (1992) 'Tanyaközségek a szegedi határban', *Alföldi Tanulmányok*, 1990–1991, XIV, 125–38

EEA (1993) *Environment Action Programme for Central and Eastern Europe: executive summary*, European Environment Agency, Copenhagen (EU Office of Official Publications)

Ehrlich, E. (1985) 'Infrastructure', in M. Kaser and E. Radice (eds) *The Economic History of Eastern Europe 1919–1975*, 1, 323–78, Oxford University Press, Oxford

Elander, I. and Gustafsson, M. (1993) 'The Re-emergence of Local Self-government in Central Europe', *European Journal of Political Research*, 23, 295–322

Englefield, G. (1992) 'Yugoslavia, Croatia, Slovenia: re-emerging boundaries', *Boundary and Territory Briefing*, 3, IBRU, Durham

Enyedi, Gy. (1979) 'Economic Policy and Regional Development in Hungary', *Acta Oeconimica*, 22(1–2), 113–26

—— (1990a) 'Specific Urbanisation in East-Central Europe', *Geoforum*, 21, 2, 163–72

—— (1990b) 'East-Central Europe: a European region', *Geoforum*, 21, 2, 141–3

—— (1994) 'Regional and urban development in Hungary until 2005' in Z. Hajdú and Gy. Horváth (eds) *European Challenges and Hungarian Responses in Regional Policy*, 329–53, Hungarian Academy of Sciences, Centre for Regional Research, Pécs

—— (ed.) (1998a) *Social Change and Urban Restructuring in Central Europe*, Akadémiai Kiadó, Budapest

—— (1998b) 'Central European Capitals in the European Metropolitan System', in J. Burdack, F-D. Grimm and L. Paul (eds) *The Political Geography of Current East–West Relatons*, 45–52, Leipzig Institut für Länderkunde, Leipzig

Enyedi, Gy. and Szirmai, V. (1992) *Budapest: A Central European Capital*, Belhaven, London

European Parliament (1998) *Migration and Asylum in Central and Eastern Europe*, Directorate General for Research Working Paper, Brussels

Faludi, A. (2000) 'The European Spatial Development Perspective – What Next?, *European Planning Studies*, 8(2), 237–50

Fehér, F. (1989) 'On Making Central Europe', *Eastern European Politics and Societies*, 3(5), 412–47

Field, M. (ed.) (1976) *Social Consequences of Modernisation in Communist Countries*, Johns Hopkins University Press, Baltimore

Fisher, J.C. (ed.) (1966) *City and Regional Planning in Poland*, Cornell University Press, Ithaca, New York

Fleischer, T., Pomazi, I. and Tombacz, E. (1991) 'Is There Any Remedy for the Environmental Crisis in the Former Socialist Countries, *Review of Environment, Nature Conservancy, Building and Regional Policy*, 16–19, Hungarian Ministry of the Environment and Regional Policy, Budapest

Fleure, H.J. and Pelham, R.A. (1936) *East Carpathian Studies: Roumania I*, Le Play Society, London

Fleure, H.J. and Evans, E.E. (1939) *South Carpathian Studies: Roumania II*, Le Play Society, London

Forgács, E. (1994) 'Avant-Garde and Conservatism in the Budapest Art World 1910–1932', in *Budapest and New York: studies in metropolitan transformation 1870–1930*, 309–32, Russell Sage Foundation, New York

Frankel, H. (1946) *Poland: the struggle for power 1772–1939*, Lindsey Drummond, London

Franklin, S. (1969) *The European Peasantry: the final phase*, Methuen, London

—— (2001) 'Bialowieza Forest: myth, reality and the politics of dispossession', Paper presented at the Institute of British Geographers Annual Conference, Plymouth

French, R.A. and Hamilton, F.E.I. (eds) (1979) *The Socialist City*, John Wiley, Chichester

Fukuyama, F. (1992) *The End of History and the Last Man*, Penguin, London

Gaffney, C. (1979) 'Kisker: the economic success of a peasant village in Yugoslavia', *Ethnology*, XVIII, 2, 135–51

Garton-Ash, T. (1986) 'Does Central Europe Exist?, *New York Review of Books*, 9 October

Geddes, A. (1940) 'Economic Prospects for the Soviet–Polish Borderlands: a review', *Scottish Geographical Magazine*, 56, 28–33

Gerschenkron, A. (1966) *Economic Backwardness in Historical Perspective*, Harvard University Press, Cambridge, Mass.

Giurescu, D. (1990) *The Razing of Romania's Past*, US/ICOMOS, London

Gorzelak, G. (1996) *The Regional Dimension of Transformation in Central Europe*, Jessica Kingsley, London

—— (1998) *Regional and Local Potential for Transformation in Poland*, Institute for Regional and Local Development, Warsaw

Gowland, D., O'Neill, B. and Reid, A. (eds) (1995) *The European Mosaic: contemporaary politics, economy and culture*, Longman, Harlow

Graham, B. (1998) 'Modern Europe: fractures and faults', in B. Graham (ed.) *Modern Europe: Place, Culture, Identity*, 1–15, Arnold, London

Graham, B. and Hart, M. (1999) 'Cohesion and Diversity in the European Union: irreconcilable forces?', *Regional Studies*, 33, 3, 259–68

Grime, K. and Kovács, Z. (2001) 'Changing Urban Landscapes in East Central Europe' in Turnock, D. (ed.) *East Central Europe and the Former Soviet Union: Environment and Society*, 130–9, Arnold, London

Grozev, V. (1981) 'Major Transformations in Agriculture', in G. Bokov (ed.) *Modern Bulgaria*, 283–297, Sofia Press, Sofia

Grykien, S. (1998) 'Tourist Farms in Lower Silesia', in D. Turnock (ed.) Rural Diversification in Eastern Europe, *GeoJournal*, Vol. 46(3), 279–81

Gutman, F. and Arkwright, P. (1981) 'Tripartite Industrial Co-operation Between East, West, and South' in F.E.I. Hamilton and G. Linge (eds) *Spatial Analysis, Industry and the Industrial Environment*, Vol. 2, 185–214, John Wiley, Chichester

Hajdú, Z. (1984) 'Geography and reforms of administrative areas in Hungary' in Gy. Enyedi and M. Pecs *Geographical Essays in Hungary*, Hungarian Academy of Sciences, Budapest

—— (1992) 'A Tanyai Tanács története: a "szocialista tanypolitika alapvetése és atanyakérdés mrgoldásának radikális, voluntarisztikus kísérlete 1949–1954"', *Alföldi Tanulmáayok*, 1990–1991, XIV, 105–24

Hálasz, Gy. (1928) *Atlas of Central Europe*

Hall, D. (1987) 'Albania', in A. Dawson (ed.) *Planning in Eastern Europe*, 35–66, St Martin's Press, New York

—— (1998) 'Rural Diversification in Albania', in D. Turnock (ed.) Rural Diversification in Eastern Europe, *GeoJournal*, Vol. 46(3), 283–7

—— (1999) 'Representations of Albania', *The Geographical Journal*, 165(2), 161–72

Hall, D. and Danta, D. (eds) (1996) *'Reconstructing the Balkans: A Geography of the New Southeast Europe*, John Wiley, Chichester

Hamilton, F.E.I. (1975) *Poland's Northern and Western Territories*, Oxford University Press, London

—— (1979) 'Urbanisation in Socialist Eastern Europe: The Macro-Environment of Internal City Structure' in French, R.A. and Hamilton, F.E.I. (eds) *The Socialist City*, 167–94, John Wiley, Chichester

—— (1982) 'Regional Policy for Poland: a search for equity', *Geoforum*, 13, 121–32

—— (1995) 'Re-evaluating Space: locational change and adjustment in Central and Eastern Europe', *Geographische Zeitscrift*, 83(2), 67–86

—— (1999) 'Transformation and Space in Central and Eastern Europe', *The Geographical Journal*, Vol. 165(2), 135–44

Hamilton, F.E.I. and Linge, G. (1981) *Spatial Analysis, Industry and the Industrial Environment*, John Wiley, Chichester

Hardi, T. (1999) 'A határ és az ember-Az osztrák-magyar határ mentén élök képe a határról és a "másik oldalról"' in M. Nárai and J. Rechnitzer (eds) *Elválast és Ōssekö-a Határ*, 159–90, MTA RKK, Pécs-Győr

Hardy, J. and Rainnie, A. (1996) *Restructuring Krakow: desperately seeking capitalism*, Cassell Mansell, London

Harloe, M. (1996) 'Cities in Transition', in G. Andruzs, M. Harloe and I. Szelenyi (eds) *Cities after Socialism*, 1–29, Blackwell, Oxford

Harris, F. (1998) 'Capital Offence', *Business Central Europe*, February, 34–5

Hartshorne, R. (1934) 'The Upper Silesian Industrial District', *Geographical Review*, 24, 423–38

Harvey, D. (1989) 'From Managerialism to Entrepreneurialism: the transformation in urban governance in late capitalism', *Geografiska Annaler*, 71B, 3–17

Harvie, C. (1994) *The Rise of Regional Europe*, Routledge, London

Hauner, M. (1985) 'Human Resources', in M.C. Kaser and E.A. Radice (eds) *The Economic History of Eastern Europe 1919–1975*, 1, 68–147, Oxford University Press, London

Haynes, M. and Husan, R. (1998) 'The State and the Market in the Transition Economies: critical remarks in the light of past history and the current experience', *Journal of European Economic History*, Winter, 609–44

Heffernan, M. (1998) *The Meaning of Europe: geography and geopolitics*, Arnold, London

Helecki, O. (1952) *Borderlands of Western Civilisation*, The Roland Press Company, New York

Heller, A. (1992) 'Europe: an epilogue?', in B. Nelson, D. Roberts and W. Veit (eds) *The Idea of Europe: problems of national and transnational identity*, 12–25, Berg, Oxford

Herrschel, T. and Light, D. (2001) 'The Meaning of Post-Socialism Across Global Regions: concepts and interpretations', Paper presented to the Institute of British Geographers Annual Conference, Plymouth

Hobson, H.A. (1938) *Report on the Economic and Commercial Conditions in Latvia*, HMSO, London

Holmes, L. (1997) *Post-Communism: an introduction*, Polity Press, Cambridge

Hörcher, N. (1998) *Regional Development in Hungary*, Ministry of Agriculture and Regional Development, Budapest

Horváth, Gy. (1995) 'Economic Reforms in East-Central Europe', in S. Hardy, M. Hart, L. Albrechts and A. Katos (eds) *An Enlarged Europe: regions in competition?*, 35–52, Regional Science Association, J. Kingsley, London

Horváth, M. (ed.) (1980) *Budapest története a forradalmak korától a felszadabulásig*, Akadémiai Kiadó, Budapest

Hunge, G. (1996) 'Contribution of FDI to Economic Recovery and Restructuring', in G. Csak, G. Foti and D. Mayes (eds) 'Foreign Direct Investment and Transition: the case of the Visegrád countries', *Trends in the World Economy*, 78, 73–9, Hungarian Academy of Sciences, Institute for World Economy, Budapest

Hupchick, D. (1994) *Culture and History in Eastern Europe*, St Martin's Press, New York

Hupchick, D and Cox, H. (1996) *A Concise Historical Atlas of Eastern Europe*, St Martin's Press, New York

Iglicka, K. (1998) 'Are They Fellowcountrymen or Not? The Migration of Ethnic Poles from Kazakhstan to Poland', *International Migration Review*, 32(4), 995–1014

Illés, Gy. (1971) *People of the Puszta*, Chatto and Windus, London

Inkeles, A. (1976) 'The Nature of "Modern" Man', in M. Field (ed.) *Social Consequences of Modernisation in Communist Countries*, 50–9, Johns Hopkins University Press, Baltimore

IOM (1994) *Ukriane: A Comfortable Waiting Room to the West*, Press Release, No. 722

Izsák, E. (1999) 'A határmentiség és a határ menti regionál együttmüködések a sajtó tükrében 1989-től napjainig' in M. Nárai and J. Rechnitzer (ed.) *Elválast és Össekö-a Határ*, 191–208, MTA RKK, Pécs-Győr

Jackson, L. (1998) 'Identity, Language and Transformation in Eastern Ukraine: a case study of Zaporizhzhia', in T. Kuzio (ed.) *Contemporary Ukraine*, 99–113, M.E. Sharpe, London

Jeszenszky, G. (1998) 'The New European Frontier', *Society and Economy*, Vol. XX(1), 45–57

Jezernik, B. (1998) 'Western Perceptions of Turkish Towns in the Balkans', *Urban History*, 25(2), 211–30

Johnston, J. (1994) 'One World, Millions of Places: the end of history and the ascendency of geography', *Political Geography*, 13(2), 111–21

Jordan, P. (1998) Central and Nation State Concepts in Eastern Europe as Obstacles for European Integration', in J. Burdack, F-D. Grimm and L. Paul (eds) *The Political Geography of Current East–West Relations*, Beiträge zur Regionalen Geographie, 47, 266–72, Institut für Länderkunde, Leipzig

Jordan, T. (1973) *The European Culture Area: A systematic geography*, Harper Collins, New York

—— (1988) *The European Culture Area: A systematic geography*, (2nd Edition), Harper Collins, New York

—— (1996) *The European Culture Area: a systematic geography* (3rd Edition), Harper Collins, New York

Kaczmarek, J. (1995) 'The daily routine of inhabitants as an element of the spatial differentiation of living conditions in Łódź' in D. Smith (ed.) *Łódź: geographical studies of a Polish city*, 30–40, Research Paper 8, Queen Mary and Westfield College, London

Kaczmarek, S. (1995) 'Urban Space, Living Conditions and Residential Preference in Łódź', in D. Smith (ed.) *Łódź: geographical studies of a Polish city*, 19–29, Research Paper 8, Queen Mary and Westfield College, London

Keegen, J. (1971) *Opening Moves August 1914*, Pan/Ballantine, London

Khmelko, V. and Wilson, A. (1998) 'Regionalism and Ethnic and Linguistic Cleavage in Ukraine', in T. Kuzio (ed.) *Contemporary Ukraine*, 60–79, M.E. Sharpe, London

Kirk, D. (1946) *Europe's Population in the Interwar Years*, League of Nations, Geneva

Klosowski, S. (1995) 'City Identity to City Marketing, Planning for Economic Development: the case of Łódź', in T. Markowski and T. Marszal (eds) *Redevelopment and Regeneration of Urban Areas: international experiences*, 36–44, University Press Łódź, Łódź

Kocsis, K. (1993) *Jugoszlávia: Egy felrobbant etnikai mozaik esete*, Teleki László Alapítvány, Budapest

Kocsis, K. and Kocsis-Hodosi, E. (1998) *Ethnic Geography of the Hungarian Minorities in the Carpathian Basin*, Hungarian Academy of Sciences, Geographical Research Institute and Minority Studies Programme, Budapest

Kolk, A. and van der Weij, E. (1998) 'Financing Environmental Policy in East Central Europe', in S. Barber and P. Jehlicka (eds) Dilemmas of Transition: The Environment, Democracy and Economic Reform in East Central Europe, *Environmental Politics*, 7(1), 53–68

Köszegi, L. (1969) 'Development Problems in Backward Areas of Hungary', in E. Robinson (ed.) *Backward Areas in Advanced Countries*, 296–314, Macmillan, London

Kovács, Z. (1994) 'City at the Crossroads' *Urban Studies*, 3, 1081–96

Kovács, Z. and Dingsdale, A. (1998) 'Whither Eastern European Democracies?', *Political Geography*, 17(4), 437–58

Kowalczyk, A. (1997) 'The Influence of Socio-Political and Economic Transformations on the Development of Border Towns in Poland', *Acta Universitatis Carolinae Geographia, Supplementum*, 369–75

Kramer, H. (1993) 'The European Community's Response to the "New Eastern Europe"', *Journal of Common Market Studies*, 31(2), 213–43

Kramer, J. (1983) 'The Environmental Crisis in Eastern Europe', *Slavonic Review*, 42(2), 204–20

Kristof, L. (1994) 'The Image and Vision of the Fatherland: the case of Poland in comparative perspective', in D. Hooson (ed.) *Geography and National Identity*, 221–32, Blackwell, Oxford

Kuus, M. and Raagmaa, G. (1998) 'The Changing Role of Tallinn, Estonia in the Baltic Sea Region', in J. Burdack, F-D. Grimm and L. Paul (eds) *The Political Geography of Current East–West Relations*, Beiträge zur Regionalen Geographie, 47, 52–64, Institut für Länderkunde, Leipzig

Lackó, L. (1985) 'Regional Policies in Eastern Europe and Hungary', Paper presented to the Regional Science Association, XXVth European Congress, Budapest

Lambert, A. (1989) 'Return of the Vampire', *Geographical Magazine*, February, 16–20

Lampe, J. (1984) 'Interwar Sophia versus the Nazi-style Garden City: the struggle over the Muesmann Plan', *Journal of Urban History*, 11(1), 39–62

Levigne (1992) 'Implications of Reform in Eastern Europe and the Soviet Union for Developing Country Members of CMEA', *Journal of Development Planning*, 23, 155–80

Lingeman, E.R. (1938) *Report on the Economic and Commercial Conditions in Finland*, HMSO, London

Linz, J. and Stepan, A. (1996) *Problems of democratic transition and consolidation: southern Europe, South America and post-communist Europe*, Johns Hopkins University Press, Baltimore and London

Liszewski, S. (1995) 'The Łódź Urban Region: genesis, present state and restructuring', in D. Smith (ed.) *Łódź: Geographical Studies of a Polish City*, 8–18, Research Paper 8, Queen Mary and Westfield College, London

Little, A. (1994) 'Disputed Borderlands' *Assignment*, BBC 2, 2 May

Maas, W. (1951) 'The "Dutch" Villages in Poland', *Geography*, No. 174, Vol. XXXVI, Pt 4, 263–8

Macartney, C.A. and Palmer, A.W. (1962) *Independent Eastern Europe*, Macmillan, New York

Maciulyte, J. and Maurel, M-C. (1998) 'Décollectivisation et Indépendance. La Trajectoire Agraire Lithuanienne', *Bulletin de l'Association de Géographes Francais*, 4, 455–69

Mackinder, H. (1919) *Democractic Ideals and Reality: a study in the politics of reconstruction*, Constable, London and Holt, New York

Marer, P. (1984) 'Eastern European Economies: achievements, problems, prospects', in T. Rakowska-Harmstone (ed.) *Communism in Eastern Europe* (2nd Edition) 283–328, Manchester University Press, Manchester

Marples, D. (1994) 'The Ukraine Economy in the Autumn of 1994: status report', *Post-Soviet Geography*, 35(8), 484–91

Masser I., Sviden, O. and Wegener, M. (1992) *The Geography of Europe's Futures*, Belhaven, London

Matczak, A. (1997) 'The Recreational Travels of Łódź Residents to the Suburban Zone as an Example of Links Connecting the City and its Hinterland, *Acta Universitatis Carolinae Geographia, Supplementum*, 209–13

Matykowski, R. and Schaefer, K. (1998) 'Socio-Economic Aspects of the Operation of the Split Town of Guben-Gubin', in J. Burdack *et al.* (eds) *The Political Geography of Current East–West Relations*, 205–12, Beiträge zur Regionalen Geographie 47, Institut für Länderkunde, Leipzig

Matykowski, R. and Tobolska, A. (1998) 'Ethnic Minorities and Regional Organisations in the Polish Elections of 1991 and 1993', in J. Burdack *et al.* (eds) *The Political Geography of Current East–West Relations*, 242–51 Beiträge zur Regionalen Geographie 47, Institut für Länderkunde, Leipzig

Mellor, R.E.H. (1975) *Eastern Europe: a geography of the Comecon countries*, Macmillan, London

—— (1987) 'The German Democratic Republic', in A. Dawson (ed.) *Planning in Eastern Europe*, 139–66, St Martin's Press, New York

Meurs, M. (1998) 'Imagined and Imagining Equality in East Central Europe: gender and ethnic differences in the economic transformation of Bulgaria', in J. Pickles and A. Smith (eds) *Theorising Transition*, 330–46, Routledge, London

Miall, H. (ed.) (1994) *Redefining Europe*, Royal Institute of International Affairs, London

Michalak, M. (1995) 'Foreign Aid and Eastern Europe in the "New World Order"', *Tijdschrift voor Economische en Sociale Geografie*, 86(3), 260–77

Michalak, W.Z. and Gibb, R.A. (1992) 'Political geography and eastern Europe' *Area*, 24, 341–9

Mihailovic, K. (1972) *Regional Development: experience and prospects in Eastern Europe*, Mouton, The Hague

Mikkeli, H. (1998) *Europe as an Idea and an Identity*, Macmillan, London

Milosz, C. (1991) 'The Budapset Roundtable', *Cross Currents*, 10, 17–30

Ministry of Building (1977) *Regional Planning in Hungary*, Ministry of Building and Urban Development, Budapest

Misiunas, R. and Taagepera, R. (1993) *The Baltic States: years of dependence 1940–1990*, C. Hurst, London

Mitchell, B.R. (1975) *European Historical Statistics 1750–1970*, Macmillan, London

Moore, P. (1984) 'Bulgaria', in T. Rakowska-Harmstone (ed.) *Communism in Eastern Europe* (2nd Edition), 186–212 Manchester University Press, Manchester

Motyl, A. (1998) 'State, Nation and Elites in Independent Ukraine', in T. Kuzio (ed.) *Contemporary Ukraine*, 3–16, M.E. Sharpe, London

Nagy, G. and Turnock, D. (1998) 'The Future of Eastern Europe's Small Towns', *Regions*, No. 213/214, 18–22, 16–21

Nárai, M. (1999) 'A hartár menti, mint élettér – A határmentiséjelentosége az emberek életben', in M. Nárai and J. Rechnitzer (eds) *Elválast és Őssekö-a Határ*, 129–58, MTA RKK, Pécs-Győr

Nárai, M. and Rechnitzer, J. (eds) (1999) *Elválast és Őssekö-a Határ*, 209–34, MTA RKK, Pécs-Győr

National Geographic (2000) Map supplement to Vol. 197(2)

Nemes-Nagy, J. and Ruttkay, E. (1985) 'Geographical Distribution of Innovation in Hungary', Paper presented to the Regional Science Association, XXV Congress, Budapest

Neumann, I. (1996) *Russia and the Idea of Europe*, Routledge, London

North, D. (1990) *Institutions, Institutional Change and Economic Performance*, Cambridge University Press, Cambridge

O'Dell, A. (1935) 'European Air Services, June 1934', *Geography*, XX, 109(3), 196–200

Okey, R (1982) *Eastern Europe 1740–1980: feudalism to communism*, Hutchinson, London

Örn, T. (1994) *The Baltic Sea: past, present and future area*, David Davies Memorial Institute, London

Orton-Jones, C. (ed.) (2001) 'Baltic States', Special Report, *Eurobusiness*, Vol. 2(8), 69–98

Osborne, R.H. (1967) *East-Central Europe*, Chatto and Windus, London

Palmer, A.W. (1970) *The Lands Between: a history of East Central Europe since the Congress of Vienna*, Macmillan, New York

Pándi, J. (1997) *Kösztes-Európa*, Osiris, Budapest

Pasvolsky, L. (1928) *Economic Nationalism in the Danubian States*, Macmillan, New York

Pávlinek, P. (1998) 'Foreign Direct Investment in the Czech Republic', *Professional Geographer*, 50, 71–85

Pávlinek, P. and Pickles, J. (2000) *Environmental Transitions*, Routledge, London

Pearson, R. (1983) *National Minorities in Eastern Europe 1848–1945*, Macmillan, London

Perényi, I. (1978) *Town and Environs*, Akadémiai Kiadó, Budapest

Petrakos, G. (1997a) 'The Regional Structures of Albania, Bulgaria and Greece: implications for cross-border co-operation and development', *European Urban and Regional Studies*, 4(3), 195–210

—— (1997b) 'A European Macro-Region in the Making? The Balkan trade relations of Greece', *European Planning Studies*, 5(4), 515–33

Pickles, J. and Smith, A. (eds) 1998 *Theorising Transition: the political economy of post-communist transition*, Routledge, London

Pickvance, C. (1992) 'The Transition from State Socialism: Towards New Local Power Structures?', in Gy. Enyedi (ed.) *Social transition and urban restructuring in Central Europe*, 9–13, European Science Foundation, Budapest

—— (1996) 'Environmental and Housing Movements in Cities after Socialism: the case of Budapest and Moscow', in G. Andruzs, M. Harloe and I. Szelenyi (eds) *Cities after Socialism*, 232–67, Blackwell, Oxford

Pinder, D. (ed.) (1998) *The New Europe*, John Wiley, Chichester

Potrykowski, M. (1998) 'Border Regions and Trans-border Co-operation: the case of Poland', in G. Enyedi (ed.) *Social Change and Urban Restructuring in Central Europe*, 233–254, Akadémiai Kiadó, Budapest

Potter, R. and Unwin, T. (1995) 'Urban-Rural Interaction: physical form and political processes', *Third World Cities*, 12(1), 67–73

Pounds, N.G.J. (1969) *Eastern Europe*, Longman, London

Pundeff, M. (1969) 'Bulgarian Nationalism', in P. Sugar and I. Lederer (eds) *Nationalism in Eastern Europe*, 93–165, University of Washington Press, Seattle

Pye, R. (1997) *Foreign Direct Investment in Central Europe: the experience of major western investors*, City University Business School, London

Rechnitzer, J. (1999) 'Az ostrák-magyar határ menti együttmüködések a kilencvenes években', in M. Nárai and J. Rechnitzer (eds) *Elválast és Ôssekö-a Határ*, 73–128, MTA RKK, Pécs-Győr

Rechnitzer, J. and Döményová, M. (1998) 'Vienna, Bratislava, Győr – a potential Euroregion', in Gy. Enyedi (ed.) *Social Change and Urban Restructuring in Central Europe*, 255–74, Akadémiai Kiadó, Budapest

Rédei, M. (1995) 'Internal Brain Drain', in M. Fullerton, E. Sik and J. Tóth (eds) *Refugees and Migrants: Hungary at a Crossroads*, Hungarian Academy of Sciences, Institute for Political Science, Budapest

Regulska, J. (1998) 'The Political and its Meaning for Women: transition politics in Poland', in J. Pickles and A. Smith (eds) *Theorising Transition*, 309–329, Routledge, London

Rey, V. (1996) 'The New Central Europe: waiting for convergence?', in F. Carter, P. Jordan and V. Rey (eds) (1996) *Central Europe After the Fall of the Iron Curtain*, 45–62, Lang, Frankfurt

Riley, R. (1997) 'Central Area Activities in a Post-Communist City: Łódź, Poland', *Urban Studies*, 34(3), 453–70

—— (1999) 'A Top-Down, Bottom-Up Approach to Manufacturing Change. Some Evidence from Łódź, Poland', *Geographia Polonica*, 72(1), 7–28

Robertson, G. (1999) 'Go East', *Business Central Europe*, Annual 2000, 14

Robinson, E. (ed.) (1969) *Backward Areas in Advanced Countries*, Macmillan, London

Rónai, A. (1993) *Atlas of Central Europe 1945*, Digital Facsimile Edition, Púski Publishing House, Budapest

Rootes, C. (1999) 'Acting Globally, Thinking Locally? Prospects for a Global Environmental Movement', *Environmental Politics*, 8(1), 290–310

Rostow, W. (1961) *The Stages of Economic Development: a non-communist manifesto*, Cambridge University Press, Cambridge

Roucek, J. (1935) 'Economic Geography of Bulgaria', *Economic Geography*, 11, 307–23

Royal Geographical Society (1997) 'Interpreting the Balkans', *Geographical Intelligence Paper* 2, Royal Geographical Society, London

Rugg, D. (1985) *Eastern Europe*, Longman, London

Rupnik, J. (1988) *The Other Europe*, Weidenfeld and Nicholson, London

—— (1990) 'Central Europe or Mitteleuropa?', *DÆDALUS*, 119, 1, Winter, 249–78

Said, E.W. (1978) *Orientalism*, Pantheon Books, New York

Sailer-Fliege, U. (1999) 'Characteristics of post-socialist urban transformation in East Central Europe' in Z. Kovács (ed.) Post-socialist urban transition in Eastern Central Europe, *GeoJournal*, 49(1), 7–16

Sandner, G. (1989) 'Historical Studies of German Political Geography', *Political Geography Quarterly*, Special Issue, 8, 4

Sármány-Parsons, I. (1998) 'Aesthetic Aspects of Change in Urban Space in Prague and Budapest During Transition', in Gy. Enyedi (ed.) *Social Change and Urban Restructuring in Central Europe*, 35–54, Akadémiai Kiadó, Budapest

Schöpflin, G. (1993) *Politics in Eastern Europe 1945–1992*, Blackwell, London
Schöpflin, G. and Wood, N. (eds) (1989) *In Search of Central Europe*, Polity Press, London
Seton-Watson, H. (1945) *Eastern Europe Between the Wars, 1918–1941*, Cambridge University Press, Cambridge
—— (1985) 'What is Europe, Where is Europe?', *Encounter*, 65(2), 9–17
Shackleton, M.R. (1925) 'Economic Resources and Problems of Yugoslavia', *Scottish Geographical Magazine*, 41, 346–65
Sharp, S.L. (1979) 'The Peasantry of Eastern Europe under Communism', in I. Volges (ed.) *The Peasantry of Eastern Europe: 20th Century Developments*, 213–18, Pergamon Press, New York
Short, J.R. (1996) *The Urban Order*, Blackwell, London
Sidaway, J. and Pryke, M. (2000) 'The Strange Geographies of "Emerging Markets"', *Transactions of the Institute of British Geographers*, New Series, Vol. 25(2), 187–202
Simpson, F. (1999) 'Tourist Impact in the Historic Centre of Prague: resident and visitor perceptions of the historic built environmnet', *The Geographical Journal*, Vol. 165(2), 173–82
Sinnhuber, K. (1954) Central Europe–Mitteleuropa–Europe Centrale: an analysis of a geographical term', *Transactions and Papers of the Institute of British Geographers*, 20, 15–39
Sjöberg, O. and Josefson, M. (1998) 'The Geography of Foreign Direct Investment in Central and Eastern Europe: inputs to an extended analysis', in J. Burdack *et al.* (eds) *The Political Geography of Current East–West Relations*, 121–34, Beiträge zur Regionalen Geographie 47, Institut für Länderkunde, Leipzig
Smith, A. (1997) 'Breaking the Old and Constructing the New? Geographical uneven development in Central and Eastern Europe', in R. Lee and J. Wills (eds) *Geographies of Economies*, Arnold, London
Smith, G. (1989) *Planning in the Socialist World*, Cambridge University Press, Cambridge
—— (1993) 'Ends, Geopolitics and Transition', in R. Johnson (ed.) *The Challenge for Geography*, 76–109, Blackwell, Oxford
Staddon, C. (1998) 'Democratisation and the Politics of Water in Bulgaria: local protest and the 1994–95 Sofia water crisis', in J. Pickles and A. Smith (eds) *Theorising Transition: the political economy of post-communist transition*, 347–72, Routledge, London
—— (1999) 'Localities, Natural Resources and Transition in Eastern Europe', *The Geographical Journal*, Vol. 165(2), 200–8
Stambrook, F. (1963) 'A British Proposal for the Danubian States: the Customs Union Project of 1932', *Slavonic and East European Review*, 42, 64–88
Sugar, P. and Lederer, I. (eds) (1969) *Nationalism in Eastern Europe*, University of Washington Press, Seattle
Sugar, P. and Treadgold, D. (eds) (1974) *A History of East-Central Europe* (eleven volumes), University of Washington, Seattle
Swain, A. and Hardy, J. (1998) 'Globalisation, Institutions, Foreign Investment and the Reintegration of East and Central Europe and the Former Soviet Union with the World Economy', *Regional Studies*, Vol. 32(7), 587–90
Swedish Ministry of Foreign Affairs (1996) *Swedish Co-operation with Central and Eastern Europe*, Information Sheet
Sýkora, L. (1994) 'Local Urban Restructuring as a Mirror of Globalisation Processes: Prague in the 1990s', *Urban Studies*, 31, 1149–66
—— (1998) 'Commercial Property Development in Budapest, Prague and Warsaw', in Gy. Enyedi (ed.) *Social Change and Urban Restructuring in Central Europe*, 35–54, Akadémiai Kiadó, Budapest
Szelényi, I. (1996) 'Cities under Socialism – and After', in G. Andrusz, M. Harloe and I. Szelenyi (eds) *Cities After Socialism*, 286–317, Blackwell, Oxford

Szörényi-Kukorelli, I. (1999) 'A női egyéni vállalkozások néhány jellemzője a határ mentén', in M. Nárai and J. Rechnitzer (eds) *Elválast és Összekö-a Határ*, 235–68, MTA RKK, Pécs-Győr

Szücs, J. (1988) 'Three Historical Regions of Europe: an outline', in J. Keane (ed.) *Civil Society and the State: new European perspectives*, 291–332, Verso, London and New York

Tatai, Z. (1978) 'Selective Industrialisation and Dispersal of Factories from Budapest: their influence on the growth of the agglomeration', in Gy. Enyedi (ed.) *Urban Development in the USA and Hungary*, 97–108, Akadémiai Kiadó, Budapest

Taylor, P. (1991) 'A Theory and Practice of Regions: the case of Europe', *Environment and Planning D: Space and Society*, 9, 183–95

—— (1989) *Political Geography: world economy, nation-state and locality*, Longman, London

—— (1999) *Modernities: a geohistorical interpretation*, Polity Press, Cambridge

—— (2001) 'Urban Hinterworlds: geographies of corporate service provision under conditions of contemporary Globalisation', *Geography*, Vol. 86, Pt 1, 51–60

Taylor, P. and Johnston, R. (1993) '1989 and All That: a response to Michalak and Gibb', *Area*, 25, 300–5

Taylor, P. and Hoyler, M. (2000) 'The Spatial Order of European Cities under Conditions of Contemporary Globalisation', *Tijdscrift voor Economische en Sociale Geografie*, Vol. 91(2), 176–89

Tchoubarian, A. (1994) *The European Idea in History in the Nineteenth and Twentieth Centuries: a view from Moscow*, Frank Cass, London

Teichova, A. (1985) 'Industry', in M.C. Kaser and E.A. Radice (eds) *The Economic History of Eastern Europe 1919–1975*, 1, 222–321, Oxford University Press, London

Teplán, I. (1994) 'St. Imre Garden City: an urban community', in T. Bender and C. Schorske (eds) (1994) *Budapest and New York: studies in metropolitan transformation 1870–1930*, 161–80, Russell Sage Foundation, New York

Therborn, G. (1995) *European Modernity and Beyond: the trajectory of European societies 1945–2000*, Sage, London

Thomas, A. (1999) 'Modernisation Versus the Environment? Shifting Objectives of Progress', in T. Skelton and T. Allen (eds) *Culture and Global Change*, 45–57, Routledge, London

Tiltman, H.H. (1934) *Peasant Europe*, Jarrods Publishers, London

Tinar, T. (1992) 'The Changing Role of Telecommunications in the Restructuring of the City of Budapest', in Z. Kovács (ed.) *New Perspectives in Hungarian Geography*, 97–111, Akadémiai Kiadó, Budapest

Tivenius, M. (1996) *Sweden and the Baltic Countries*, Swedish Ministry of Foreign Affairs, Stockholm

Todorova, M. (1994) 'The Balkans: from discovery to invention', *Slavonic Review*, 53, 453–82

Tomaszewski, J. (1993) 'The National Question in Poland in the Twentieth Century', in M. Teich and R. Porter (eds) *The National Question in Europe in Historical Context*, 293–316, Cambridge University Press, Cambridge

Tomlinson, J. (1991) *Cultural Imperialism*, Pinter, London

—— (1999) *Globalisation and Culture*, Polity Press, Cambridge

Turnock, D. (1978) *Eastern Europe: An Industrial Geography*, Wm Dawson and Sons, Folkestone

—— (1987) 'Romania', in A. Dawson (ed.) *Planning in Eastern Europe*, 229–74 St Martin's Press, New York

—— (1989) *Eastern Europe: An Historical Geography 1815–1945*, Routledge, London

—— (1991) 'The Planning of Rural Settlement in Romania', *The Geographical Journal*, 157(3), 251–64

—— (1996a) 'The Rural Transition in Eastern Europe', *GeoJournal*, 36(4), 420–6

—— (1996b) 'Frameworks for Understanding Post-Socialist Processes in Eastern Europe and the Former Soviet Union', *GeoJournal*, 39(4), 409–12

—— (1997a) *The Eastern European Economy in Context: communism and transition*, Routledge, London

—— (1997b) 'Cross-Border Co-operation as a Factor in the Development of Transport in Eastern Europe', in A. Dingsdale (ed.) Transport in Transition: issues in the new Central and Eastern Europe, *Trent Geographical Papers*, 1, 5–29

—— (1998) 'Introduction', in D. Turnock (ed.) Rural Diversification in Eastern Europe, *GeoJournal*, Vol. 46(3), 171–81

—— (1999) 'Sustainable Rural Tourism in the Romanian Carpathians', *The Geographical Journal*, Vol. 165(2), 192–99

—— (ed.) (2001) *East Central Europe and the Former Soviet Union*, Arnold, London

Ulram, P. and Plasser, F. (1999) 'The Political Climate in East-Central Europe: mainly sunny with scattered clouds', *Society and Economy in Central and Eastern Europe*, 3, 111–40

UNCTAD (1999) *World Investment Report: foreign direct investment and the challenge of development*, United Nations, New York

UNDP (1998) *Poverty in Transition?*, United Nations, New York

Unwin, T. (1998) 'Rurality and the Construction of Nation in Estonia', in J. Pickles and A. Smith (eds) *Theorising Transition*, 284–308, Routledge, London

Van der Heijden, H-A. (1999) 'Environmental Movements, Ecological Modernisation and Political Opportunity Structures', *Environmental Politics*, 8(3), 199–221

Van Ham, P. (1993) *The EC, Eastern Europe and European Unity: discord, collaboration and integration since 1947*, Pinter Publishers, London

Van Zon, H. (1998) 'The Mismanaged Integration of Zaporizhzhya with the World Economy: implications for regional development in peripheral regions', *Regional Studies*, Vol. 32(7), 607–18

Varfolomeyer, O. (1997) 'Rival "clans" mix business, politics and murder', *Transition* 3(6), 31–4

Velikonja, J. (1994) 'The Quest for Slovene National Identity', in D. Hooson (ed.) *Geography and National Identity*, 249–56, Blackwell, London

Von der Porten, E. (1994) 'The Hanseatic League: Europe's first common market', *National Geographic*, 186(4), 56–79

Vujakovic, P. (1992) 'Mapping Europe's Myths', *Geographical Magazine*, Vol. LXIV(9), 15–17

Wanklyn, H. (1941) *The Eastern Marchlands of Europe*, G. Philip and Son, London

White, G. (1996) 'Place and its Role in Serbian National Identity', in D. Hall and D. Danta (eds) *Reconstructing the Balkans*, 39–52, John Wiley, Chichester

White, R. (1989) 'The Europeanism of Coudenhove-Kalergi', in P. Stirk (ed.) *European Unity in Context: the Inter-War Period*, Pinter Publishers, London

Willett, J. (1991) 'Is there a Central European Culture?', *Cross Currents*, 10, 1–16

Wolff, L. (1994) *Inventing Eastern Europe*, Stanford University Press, Stanford, California

World Bank (2000) *Entering the 21st Century: World Development Report 1999/2000*, Oxford University Press, Oxford

Young, C. (1997) 'Urban Governance and Local Economic Development in Łódź, Poland: the emergence of an "entrepreneurial" local state', in A. Dingsdale (ed.) *Urban Regeneration and Development in Post-Socialist Towns and Cities*, Trent Geographical Papers, No. 2, 32–51

Young, C. and Kaczmarek, S. (1999) 'Changing Perception of the Post-Socialist City: place promotion and imagery in Łódź, Poland', *The Geographical Journal*, Vol. 165(2), 183–91

Young, C. and Kaczmarek, S. (2001) 'Decontextualising History and Heritage in the Post-Socialist City: history as cultural capital in the promotion of Łódź,

Poland', Paper presented to the Institute of British Geographers Annual Conference, Plymouth

Z (1990) 'To the Stalin Mausoleum Eastern Europe . . . Central Europe . . . Europe', *DÆDALUS*, 119(1), 295–344

Zacek, J.F. (1969) 'Nationalism in Czechoslovakia', in P. Sugar and I. Lederer (eds) *Nationalism in Eastern Europe*, 166–206, University of Washington, Seattle

Zaprudnik, J. (1993) *Belarus: at a crossroads in history*, Westview Press, Boulder

Zaremba, J. (1966) 'Regional Planning in Poland: theory, methods and results', in J.C. Fisher (ed.) *City and Regional Planning in Poland*, 271–97, Cornell University Press, Ithaca, New York

Index

Note: Page numbers in *italics* refer to maps, figures and tables.